CW00922795

Pattern Recognition with Neural Networks in C++

Pattern Recognition with Neural Networks in C++

Abhijit S. Pandya
Robert B. Macy
Florida Atlantic University
Boca Raton, Florida

A CRC Book Published in Cooperation with IEEE Press

LIMITED WARRANTY

CRC Press warrants the physical diskette(s) enclosed herein to be free of defects in materials and workmanship for a period of thirty days from the date of purchase. If within the warranty period CRC Press receives written notification of defects in materials or workmanship, and such notification is determined by CRC Press to be correct, CRC Press will replace the defective diskette(s).

The entire and exclusive liability and remedy for breach of this Limited Warranty shall be limited to replacement of defective diskette(s) and shall not include or extend to any claim for or right to cover any other damages, including but not limited to, loss of profit, data, or use of the software, or special, incidental, or consequential damages or other similar claims, even if CRC Press has been specifically advised of the possibility of such damages. In no event will the liability of CRC Press for any damages to you or any other person ever exceed the lower suggested list price or actual price paid for the software, regardless of any form of the claim.

CRC Press SPECIFICALLY DISCLAIMS ALL OTHER WARRANTIES, EXPRESS OR IMPLIED, INCLUDING BUT NOT LIMITED TO, ANY IMPLIED WARRANTY OF MERCHANTABILITY OR FITNESS FOR A PARTICULAR PURPOSE. Specifically, CRC Press makes no representation or warranty that the software is fit for any particular purpose and any implied warranty of merchantability is limited to the thirty-day duration of the Limited Warranty covering the physical diskette(s) only (and not the software) and is otherwise expressly and specifically disclaimed.

Since some states do not allow the exclusion of incidental or consequential damages, or the limitation on how long an implied warranty lasts, some of the above may not apply to you.

Library of Congress Cataloging-in-Publication Data

Pandya, Abhijit S.
Pattern recognition with neural network in C++ / Abhijit S. Pandya, Robert B. Macy.
p. cm.
Includes bibliographical references and index.
ISBN 0-8493-9462-7
1. Neural networks (Computer science) 2. C++ (Computer program language) 3. Pattern recognition systems. I. Macy, Robert B. II. Title.
QA76.87.P36 1995
006.4—dc20 95-4558
CIP

Direct all inquiries to CRC Press, Inc., 2000 Corporate Blvd., N.W., Boca Raton, Florida 33431.

International Standard Book Number 0-8493-9462-7
Library of Congress Card Number 95-4558
Printed in the United States of America 1 2 3 4 5 6 7 8 9 0
Printed on acid-free paper

Preface

Why do we feel a need to write a book about pattern recognition when many excellent books are already available on this classical topic? The answer lies in the depth of our coverage of neural networks as natural pattern classifiers and clusterers. Artificial neural network computing has emerged as an extremely active research area with a central focus on manipulation of pattern-formatted information, information containing an underlying pattern. This has given rise to a new coherent approach to pattern recognition which builds upon both the contributions of the past and the rapid progress in neural network research. Pattern recognition has grown to encompass a wider scope of methodology than is available in the traditional domain of statistical pattern recognition. The addition of artificial neural network computing to traditional pattern recognition has given rise to a new, different, and more powerful methodology which we intended to present in this book for the practitioner.

Pattern recognition systems are systems which automatically classify or cluster complex patterns or objects based on their measured properties or on features derived from these properties. With this viewpoint a neural network can be seen as a system that recognizes patterns. The discovery of underlying regularities is, however, precisely the task at which neural networks excel. In this very real sense the study of neural networks and the study of pattern recognition converge. Neither subject is truly complete in the absence of the other. We suggest that many of the effective applications of neural networks in domains not generally thought of explicitly as pattern recognition (e.g., control) can be viewed as pattern recognition in the sense that they still depend on the network ability to detect and identify subtle underlying regularities in the input space.

The extent to which neural networks are or should be reflective of biological systems has been a contentious subject. Primarily, we take an engineering approach, foregoing any extensive treatment of biological plausibility. Nevertheless, we recognize that when designing a system it can be useful to observe other systems which perform the desired function well. Biological systems are superb pattern recognizers. By the way of analogy, in designing the first airplane one might do well to observe birds in an attempt to isolate characteristics enabling flight (e.g., wing shape, mass/volume ratio, etc.). However, at some point in the process, one must break free of slavishly following the biological metaphor. Otherwise, there could be no aircraft capable of supersonic flight. We therefore believe that while we may look to biological systems for inspiration, artificial neural systems must ultimately take on their own identity to be truly effective.

Our objective is to produce a practical guide to the application of artificial neural networks. Many neural network texts do exist and serve as excellent references, but in our opinion they do not completely satisfy the needs of our "applied" view of the field. Here are some of the reasons:

- Books on artificial neural systems are mostly written by academics who cover the formal theory very well, but often omit practical information on the application of various network architectures and learning paradigms. While they frequently contain code segments and results of computer implementations of individual paradigms, they rarely provide the complete code that could assist experimental design.
- The largest category of recent neural network books consists of article compilations focusing on research of one or few individuals, which presupposes experience in that field of specialization. For some of the neural network algorithms, even though they may be clearly stated, the details of the computing carried out by the network are not easily accessible, and their effects are not easily traced.
- Several cookbooks with neural network recipes have been recently published. They contain simplified formulas and C++ complete programs that implement methods but they are written for laypersons.
- There are many books which deal with the subject of statistical pattern recognition but do not cover neural network methodologies. By contrast, texts on neural networks seldom cover the pattern recognition problem in any depth.

Our book is an attempt to cover pattern classification and neural network approaches within the same framework geared toward the practitioner. Neural networks should not be considered a black box, governed by complicated mathematics, with answers that may surprise or disappoint us. Armed with an understanding of underlying theory and practical examples the practitioner will be better able to make judicious design choices which will render neural application predictable and effective.

For each network paradigm we have attempted to give an intuitive explanation of each method. This discussion is augmented and amplified by a rigorous mathematical approach where appropriate. C++ has emerged as a rich and descriptive means by which concepts, models, or algorithms can be unambiguously described. For most of the neural network models we discuss, we present the C++ programs for actual implementation. We provide pictorial diagrams to explain the topics and a verbal description of the method. We provide necessary derivative steps for the mathematical models so that astute readers can incorporate new ideas into their programs as the field advances with new developments.

For each approach we have tried to state simply the known theoretical results, the known tendencies of the approach, and our recommendations for how to get the best results from the method. Some may disagree with one or another of these recommendations, but at least the practitioner will have advice with which to proceed and which can be tested empirically. We have attempted to make the methods presented here accessible to working engineers with little or no explicit background in neural networks. However, the material is presented in sufficient depth that those with prior background should find this book beneficial.

The topics covered are suitable for courses in neural networks at advanced undergraduate and graduate levels in computer science and most engineering specialties. In this connection the pattern recognition coverage provides a tangible vehicle through which the operation and application of neural networks can be easily observed and understood. Additionally this book would be suitable for a course in pattern recognition, individually if neural methods are to be emphasized or in combination with another text covering statistical methods.

Anyone who has practical experience with any particular neural network paradigm, perhaps having used a programmed application of one of the network architectures, should be able to apply the material of this book. Our hope is that this text will be valuable toward strengthening the fundamentals for academic as well as practical research.

Acknowledgment

We would like to thank the many people who helped us in the preparation of this book. Thanks are due to the more than 300 students who patiently sat through and have been the guinea pigs for the course material of CAP5615 in the Department of Computer Science and Engineering (CSE) at Florida Atlantic University (FAU) from 1989 through 1994. They have been most helpful assistants in improving the material of the text.

The greatest thanks are owed to Lynn and Robert Lloyd. They did the majority of the figures, layout, and initial proofreading. Without their contributions this book would not exist. We further thank them for their friendship, sharp eyes, and support. We would like to thank Okechukwu Uwechue for helping us in the writing of the section on higher-order neural networks.

We are deeply indebted to the many reviewers who have given freely of their time to read through the manuscript. In particular, we are most grateful to Professors Erich Harth, Yo-Han Pao, and James Anderson for their generous help. Their many comments and criticisms were of enormous value.

Along the way we have received important understanding and encouragement from several people. These supporters include Dr. Mitsuo Kawato, Professors Hiroshi Shimizu, Ryoji Suzuki, Sun-Ichi Amari, Lotfi Zadeh, and Dr. Richard Sutton. We are grateful for those enjoyable and strengthening interactions. My sincere appreciation goes to my professor, Erich Harth, for many years of inspiration and guidance and for the review of the book (A.P.).

This book is a product of a love for neural networks that can only be obtained from top educators and researchers. We would like to include in our thanks Professors J. A. Scott Kelso, Yohsuke Kinouchi, Hirofumi Nagashino, Peramber Neelakanta, Evangelia Tzanakou, and Dr. Unnikrishanan for their valuable guidance.

Craig Hartley, Dean of Engineering at FAU, has been extremely generous in supporting the development of this book during the past two years. Special thanks are also due to Professors Neal Coulter and Mohammad Ilyas, the successive chairs of the FAU CSE Department over the past three years for their unwavering support. Additionally, the lively interdisciplinary interaction in neural networks at FAU, across various engineering departments and the Center for Complex Systems, has been of great value to this project. Special thanks go to our friend and colleague, William McLean. He has been a constant source of new insights and ideas, as well as a superb critic and evaluator.

The love, patience, understanding, and encouragement of my wife, Bhairavi, are greatly appreciated (A.P.).

We would like to thank the professionals at CRC Press, namely Ms. Andrea Demby, for the design and production, and Ms. Paula Clodfelter, for the editing of this version. Finally, we want to thank Robert Stern, also of CRC Press, and Dudley Kay, of IEEE Press, for their encouragement, patience, and wise counsel.

We would appreciate reports of any errors you may find. Our E-Mail address is: pandya@acc.fau.edu.

We dedicate this book to our parents,
whose lifelong commitments to research have inspired us.

Table of Contents

Chapter 7

Chapter 8

Chapter 9

1 Introduction

What is a pattern? A pattern is essentially an arrangement or an ordering in which some organization of underlying structure can be said to exist. We can view the world as made up of patterns. Watanabe [1985] defines a pattern as an entity, vaguely defined, that could be given a name.

A pattern can be referred to as a quantitative or structural description of an object or some other item of interest. A set of patterns that share some common properties can be regarded as a pattern class. The subject matter of pattern recognition by machine deals with techniques for assigning patterns to their respective classes, automatically and with as little human intervention as possible. For example, the machine for automatically sorting mail based on 5-digit zip code at the post office is required to recognize numerals. In this case there are ten pattern classes, one for each of the 10 digits. The function of the zip code recognition machine is to identify geometric patterns (each representing an input digit) as being a member of one of the available pattern classes.

A pattern can be represented by a vector composed of measured stimuli or attributes derived from measured stimuli and their interrelationships. Often a pattern is characterized by the order of elements of which it is made, rather than the intrinsic nature of these elements. Broadly speaking, pattern recognition involves the partitioning or assignment of measurements, stimuli, or input patterns into meaningful categories. It naturally involves extraction of significant attributes of the data from the background of irrelevant details. Speech recognition maps a waveform into words. In character recognition a matrix of pixels (or strokes) is mapped into characters and words. Other examples of pattern recognition include: signature verification, recognition of faces from a pixel map, and friend-or-foe identification. Likewise, a system that would accept sonar data to determine whether the input was a submarine or a fish would be a pattern recognition system.

1.1 PATTERN RECOGNITION SYSTEMS

For a typical pattern recognition system the determination of the class is only one of the aspects of the overall task. In general, pattern recognition systems receive data in the form of "raw" measurements which collectively form a stimuli vector. Uncovering relevant attributes in features present within the stimuli vector is typically an essential part of such systems (in some cases this may be all that is required). An ordered collection of such relevant attributes which more faithfully or more clearly represent the underlying structure of the pattern is assembled into a feature vector.

Class is only one of the attributes that may or may not have to be determined depending on the nature of the problem. The attributes may be discrete values, Boolean entities, syntactic labels, or analog values. Learning in this context amounts to the determination of rules of associations between features and attributes of patterns.

Practical image recognition systems generally contain several stages in addition to the recognition engine itself. Before moving on to focus on neural network recognition engines we will briefly describe a somewhat typical recognition system [Chen, 1973].

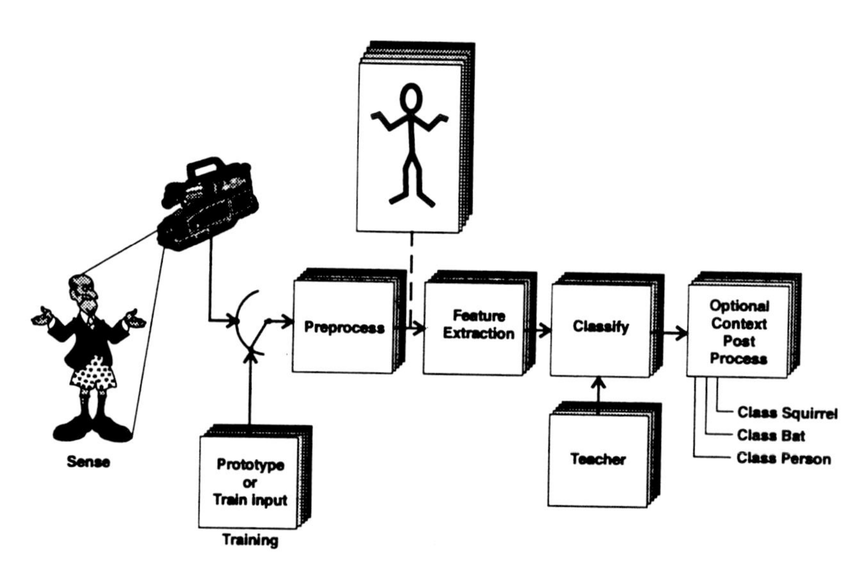

FIGURE 1.1 Components of a pattern recognition system.

Figure 1.1 shows all the aspects of a typical pattern recognition task:

Preprocessing partitions the image into isolated objects (i.e., characters, etc.). In addition it may scale the image to allow a focus on the object.

Feature extraction abstracts high level information about individual patterns to facilitate recognition.

The *classifier* identifies the category to which the pattern belongs or, in general, the attributes associated with the given pattern.

The *context processor* increases recognition accuracy by providing relevant information regarding the environment surrounding the object. For example, in the case of character recognition it could be the dictionary and/or language model support.

Figure 1.2 shows the steps involved in the design of a typical pattern recognition system. The choice of adequate sensors, preprocessing techniques, and decision-making algorithm is dictated by the characteristics of the problem domain. Unlike the expert systems, the domain-specific knowledge is implicit in the design and is not represented by a separate module.

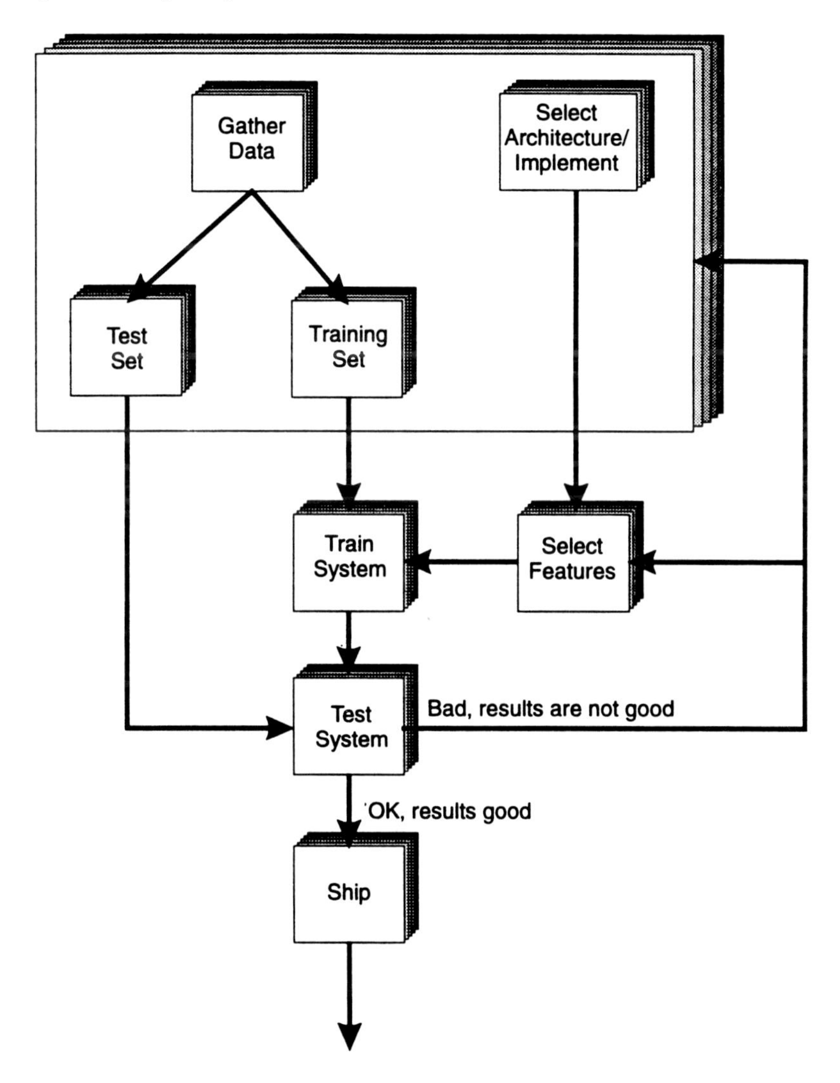

FIGURE 1.2 A flow chart of the process of designing a learning machine for pattern recognition.

A pattern classification system is expected to perform (1) supervised classification, where a given pattern has to be identified as a member of already known or defined classes; or (2) unsupervised classification or clustering, where a pattern needs to be assigned to a so far unknown class of patterns.

Pattern recognition may be static or dynamic. In the case of asynchronous systems, the notion of time or sequential order does not play any role. Such a paradigm can be addressed using static pattern recognition. Image labeling/understanding falls into this category. In cases of dynamic pattern recognition, where relative timing is of importance, the temporal correlations between inputs and outputs may a major role. The learning process has to determine the rules governing these temporal correlations. This category includes such applications as control using artificial neural networks or forecasting using neural nets. In the case of recognizing handwritten characters, for example, the order in which strokes emerge from a digitizing tablet provides much information that is useful in the recognition process.

The task of pattern recognition may be complicated when classes overlap (see Figure 1.3). In this case the recognition system must attempt to minimize the error due to misclassification. The classification error is significantly influenced by the number of samples in the training set. Several researchers (for example, see Jain and Chandrasekaran [1982], Fukunaga and Hayes [1989], Foley [1972]) have addressed this issue.

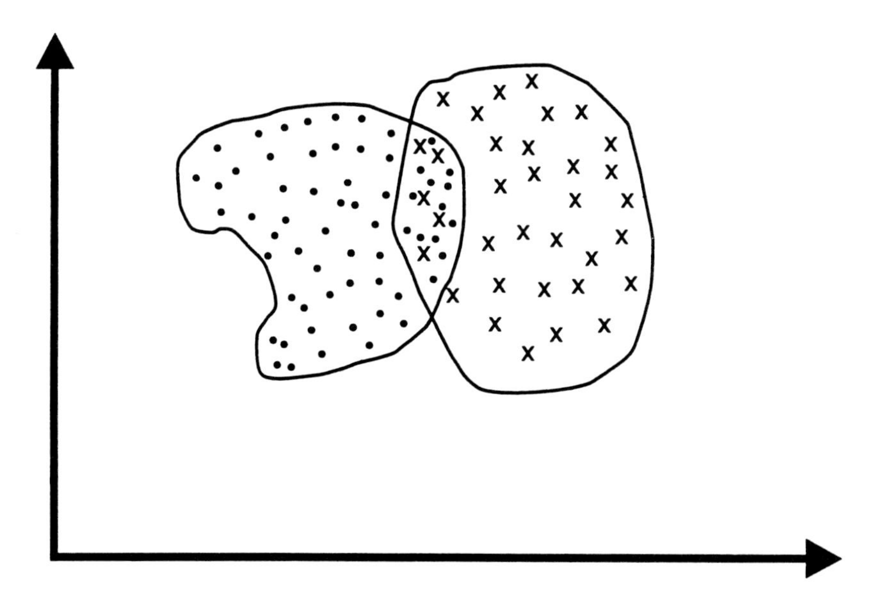

FIGURE 1.3 Two categories of patterns plotted in the pattern space. Patterns belonging to both classes can be observed in the overlapping region.

The three major approaches for designing a pattern recognition are (1) statistical, (2) syntactic or structural, and (3) artificial neural networks. Statistical pattern recognition techniques use the results of statistical communication and estimation theory to obtain a mapping from the representation space to the interpretation space. They rely on the determination of an appropriate combination of feature values that provides measures for discriminating between classes. However, in some cases, the features are not important in themselves. Rather the critical information regarding pattern class, or patterns attributes, is contained in the structural relationships among the features. Applications involving recognition of pictorial patterns (which are characterized by recognizable shapes) such as character recognition, chromosome identification, elementary particle collision photographs, etc. fall into this category. The subject of syntactic pattern recognition deals with this aspect, since it possesses the structure-handling capability lacked by the statistical pattern recognition approach. Many of the techniques in this field draw from the earlier work in mathematical linguistics and results of research in computer languages. A large body of literature exists in this field which includes Watanabe [1972], Fu [1974, 1977], Gonzalez and Thomason [1978].

Despite the existence of a number of good statistical, syntactic (grammar-based), and graphical approaches to pattern recognition, we limit the scope of this book to the discussion of the various artificial neural network based modules. However, where statistical methods are strongly related to corresponding neural network techniques, the applicable statistical methods are discussed. Additionally, it should not be overlooked that neural recognizers can and have been used in combination with other types of recognition engines such as elastic pattern matchers.

1.2 MOTIVATION FOR ARTIFICIAL NEURAL NETWORK APPROACH

The development of a computer as something more than a calculating machine marked the birth of the field of pattern recognition. We have witnessed increased interest in research involving use of machines for performing intelligent tasks normally associated with human behavior. Pattern recognition techniques are among the most important tools used in the field of machine intelligence. Recognition after all can be regarded as a basic attribute of living organisms. The study of pattern recognition capabilities of biological systems (including human beings) falls in the domain of such disciplines as psychology, physiology, biology, and neuroscience. The development of practical techniques for machine implementation of a given recognition task and the necessary mathematical framework for designing such systems lies within the domain of engineering, computer science, and applied mathematics. With the advent of neural network technology a common ground between engineers and students of living systems (psychologists, physiologists, linguists, etc.) was established. We would like to point out that mathematical operations used in theories on pattern recognition and neural networks are often formally similar and identical. Thus, there is good mathematical justification for teaching the two areas together.

Recognizing patterns (and taking action on the basis of the recognition) is the principal activity that all living systems share. Living systems, in general, and human beings, in particular, are the most flexible, efficient, and versatile pattern recognizers known; and their behavior provides ample data for studying the pattern recognition problem. For example, we are able to recognize handwritten characters in a robust manner, despite distortions, omissions, and major variations. The same capabilities can be observed in the context of speech recognition. Humans also have the ability to retrieve information, when only a part of the pattern is presented, based on associated cues. Take, for example, the cocktail party phenomena. At a party you can pick up your name being mentioned in a conversation all the way across the hall even when most of the conversation is inaudible due to a clutter of noise. Similarly, you can recognize a friend in the crowd at a distance even when most of the image is occluded.

Decision-making processes of a human being are often related to the recognition of regularity (patterns). Humans are good at looking for correlations and extracting regularities based on them. Such observations allow humans to act based on anticipation which cuts down the response time and gives an edge over reactionary behavior. Machines are often designed to perform based on reaction to the occurrence of certain events which slows them down in applications such as control.

The nature of patterns to be recognized could be either *sensory recognition* or *conceptual recognition.* The first type involves recognition of concrete entities using sensory information, for example, visual or auditory stimulus. Recognition of physical objects, characters, music, speech, signature, etc. can be regarded as examples of this type of act. On the other hand, conceptual recognition involves acts such as recognition of a solution to a problem or an old argument. It involves recognition of abstract entities and there is no need to resort to an external stimulus in this case. In this book, we shall be concerned with recognition of concrete items only.

The real problem of pattern recognition, however, is to generate a theory that specifies the nature of objects in such a way that a machine will be able to robustly identify them. A study of the way living systems operate provides great insight into addressing this problem. The image in Figure 1.4 indicates the complexity of the type of problem we have been discussing. The image in Figure 1.4(a) shows the face with distinct boundaries between pixels. Thus an image understanding/pattern recognition algorithm, which labels areas with different intensities as parts of different surfaces, would have difficulties in recognizing this pattern of a face. On the other hand, for a human observer it is easier to see that blurring of the boundaries between pixels, as shown in Figure 1.4(b), would result in a easily recognizable face. The ability may be attributed to the existence of interacting high and low spatial frequency channels in the human visual system.

One strong objective of the engineering and the artificial intelligence community has been the creation of "intelligent" systems which can exhibit human-like behavior. Such intelligent behavior would enable humans/machine interactions to occur in some fashion that is more natural for the human being. That is, we would like to provide perceptual and cognitive capabilities enabling computers to communicate with us in a fashion that is natural and intuitive to us. One of the goals is to design machines with decision-making capabilities. To accomplish this it is essential that such machines achieve the same pattern information processing capabilities that human beings possess.

FIGURE 1.4(a) A facial image with low resolution seen with pixel grid.

Some of the early work in building pattern recognition systems was indeed biologically motivated. The most common historical references are to the devices called perceptron and adaptive linear combiner (ADALINE), respectively. The objective of these studies was to develop a recognition system whose structure and strategy followed the one utilized by humans. The pessimism concerning the limited capability of the perceptron, caused by the book by Minskey and Pappert [1969], effectively curtailed most research in this field. Subsequently, with the advent of other, more powerful neural techniques, the field of neural network research is again

FIGURE 1.4(b) Same image blurred to deemphasize the boundaries between pixels.

vigorous. The current serious activity in the area of artificial neural networks and connectionist paradigms is reminiscent of the early period when neurocomputing research flourished.

Some of the early disappointments with the perceptron approach, led some of the researchers to concentrate on the mathematical or computer science aspects of pattern-formatted information processing. For example, emphasis shifted to statistical pattern recognition and classification of patterns with syntactic structures. The neural network, or connectionist paradigm, provides a promising path toward computer systems possessing truly intelligent capabilities. The recent advances in the field of artificial neural networks over the last decade has therefore brought us that much closer to the goal of creating systems exhibiting human-like behavior.

Jain and Mao [1994] provide a good discussion on common links between artificial neural network approaches and the statistical pattern recognition approach.

In summary, neural networks are natural classifiers with significant and desirable characteristics. Among these are

1. Resistance to noise
2. Tolerance to distorted images/patterns (ability to generalize)
3. Superior ability to recognize partially occluded or degraded images
4. Potential for parallel processing

1.3 A PRELUDE TO PATTERN RECOGNITION

A pattern can be represented by a set of n numerical measurements or an n-dimensional pattern or measurement vector, $\mathbf{Z}$:

$$\mathbf{Z} = z_1, z_2, \ldots, z_{N_m} \tag{1}$$

Subsequently a feature vector, $\mathbf{X}$, may be derived from the pattern vector:

$$\mathbf{X} = x_1, x_2, \ldots, x_N \tag{2}$$

Thus a pattern can be viewed as a point in either N_m-dimensional measurement hyperspace or the N-dimensional feature hyperspace. Typically, feature spaces are chosen to be of lower dimensionality than the corresponding measurement space. Pattern classification involves mapping a pattern correctly from the feature/measurement space into a class-membership space. Thus the decision-making process in pattern classification can be summarized as follows. Consider a pattern represented by an n-dimensional feature vector:

$$\mathbf{X} = \left(x_1, x_2, \ldots, x_n\right)^T \tag{3}$$

where T indicates a transpose.

The task is to assign it to one of the K categories, C_1, C_2, ..., C_K. Note that the measurement vector represents the sensed data, where N_m is the number of measurements. If, for example, an image is represented by an m × m array of pixels with 16 gray levels, we have the dimensionality of the pattern vector, $n = m^2$. Each component $\mathbf{z}_i$ of the vector $\mathbf{Z}$ assumes the appropriate gray level, from the 16 possible values.

Consider the problem of recognizing speech patterns. In this case, the acoustic signals are a function of time. The entities of interest are continuous functions of a variable t, unlike the discrete gray-scale values in the previous example. In order to perform this type of classification we must first measure the observable characteristics of the samples, which involves observing the speech waveform over a period of time in this case. A pattern vector can be formed by sampling these functions at discrete time intervals, t_1, t_2, ..., t_n, etc. Figure 1.5 shows measurements of time sampled values for a waveform given by $\mathbf{z}(t_1)$, $\mathbf{z}(t_2)$, ..., $\mathbf{z}(t_n)$.

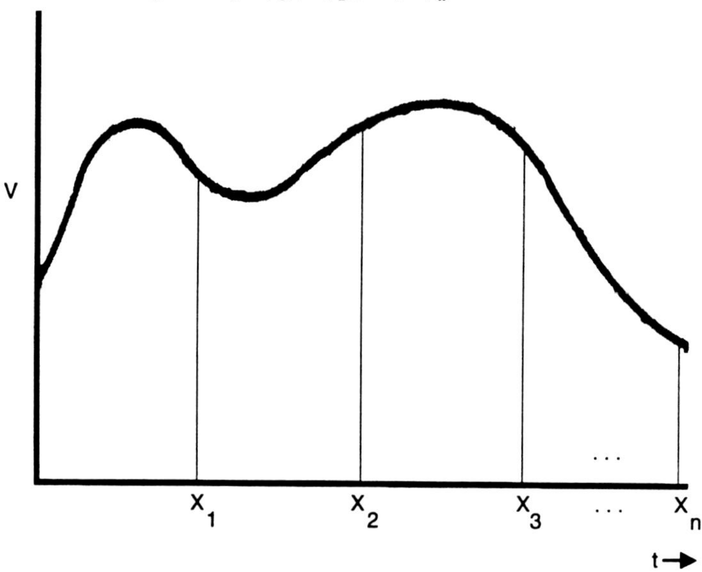

FIGURE 1.5 Sampling of a waveform at discrete time intervals.

A feature vector for speech recognition might, for example, consist of the first *N* Fourier coefficients of the captured waveform.

Design of a pattern recognition system also involves choosing an appropriate approach into the description of patterns in a form acceptable to the machine in consideration. This decision is also influenced by the nature of the problem domain to which the recognition system will be applied. For example, in the face recognition problem, the image may be converted to an array of pixels with gray-scale representation by means of a photosensitive matrix device (or a camera with a frame grabber). In an application involving color codes, it may be more appropriate to use intensity levels of each of the red, blue, and green (RBG) signals.

Thus, first the feature extractor is designed to find the appropriate features for representing the input patterns, such that the difference between patterns from different classes is enhanced in this feature space. After the feature set is defined and the feature extraction algorithm is in place, a typical recognition process involves two phases: training and prediction. Once the mapping into the feature space has been established, the training phase may begin. Training data that are representative of the problem domain must be obtained. The recognition engine is adjusted such that it maps feature vectors (derived from the training data) into categories with a minimum number of misclassifications. In the second phase (prediction phase), the trained classifier assigns the unknown input pattern to one of the categories/clusters based on the extracted feature vector. The process could be iterative where if prediction results are not acceptable, the choice of features can be revisited or the training can be performed again with different parameters (see Figure 1.2).

Neither raw data representation (bit map or stroke in the case of character recognition) is particularly good for direct input to a neural recognizer. As will be seen, the degree of "badness" will, to a varying extent, differ with the characteristics of the recognizer in question. Some of the problems inherent in using the raw data input formats above as direct inputs to a neural recognizer are

1. They are nonorthogonal.
2. They are unlikely to represent salient features of the patterns to be recognized.
3. They are verbose. Unnecessarily large input vectors lead to a larger than necessary network in which performance during both training and recognition are degraded.
4. They are sensitive to slight variations in the image, i.e., font/stroke variations in characters.
5. They are likely to contain a good deal of extraneous or nonrelevant information thus providing an invitation to overfitting/overtraining in the recognizer.
6. They will not be invariant with respect to translation rotation, etc.

Given these problems a good deal of attention will be paid to the subject of uncovering other and better representation of the data to present to the recognizer.

1.4 STATISTICAL PATTERN RECOGNITION

In this field the problem of pattern classification is formulated as a statistical decision problem. Statistical pattern recognition is a relatively mature discipline and a number of commercial recognition systems have been designed based on this approach. Several classic books are available on this subject [Tou and Gonzalez, 1974; Duda and Hart, 1973; Fu, 1977; Fukunaga, 1990]. Pao [1989] is an excellent source of the most relevant techniques from the perspective of practical engineering applications. These books present pattern recognition as a problem of estimating density functions in a high-dimensional space and dividing this hyperspace into regions of categories or classes. Decision making in this case is performed using appropriate discriminant functions. Thus mathematical statistics forms the foundation of this subject.

This discipline is also referred to as the decision-theoretical approach since it utilizes decision functions to partition the pattern space. These functions, which are also called discriminant functions, are scalar functions of the pattern $\mathbf{x}$. Regions in the pattern space enclosed by these boundaries provided by the decision functions are labeled as individual classes. A decision function, for n-dimensional pattern space can be expressed as:

$$d_k(\mathbf{x}) = w_k l(\mathbf{x}) \qquad k = 1, 2, \ldots, M \tag{4}$$

where w's are coefficients of the decision function corresponding to class C_k and the $l(\mathbf{x})$ are real, single-valued functions of the pattern, **x**.

The approach is to establish M decision functions $d_1(\mathbf{x}), d_2(\mathbf{x}), \ldots, d_k(\mathbf{x})$, one for each class, such that if a pattern **x** belongs to class C_i, then:

$$d_i(\mathbf{x}) > d_j(\mathbf{x}) \qquad j = 1, 2, \ldots, M,\ j \neq i \tag{5}$$

Thus we have a relationship that specifies a decision rule. In order to classify a given pattern it is first substituted into all decision functions. Then the pattern is assigned to the class which yields the largest numerical value. Then we have the equation of the decision boundary:

$$d_i(\mathbf{x}) - d_j(\mathbf{x}) = 0 \tag{6}$$

which separates classes C_i and C_j. Figure 1.6 shows the block diagram of an automatic classification scheme using discriminant function generators (DFGs).

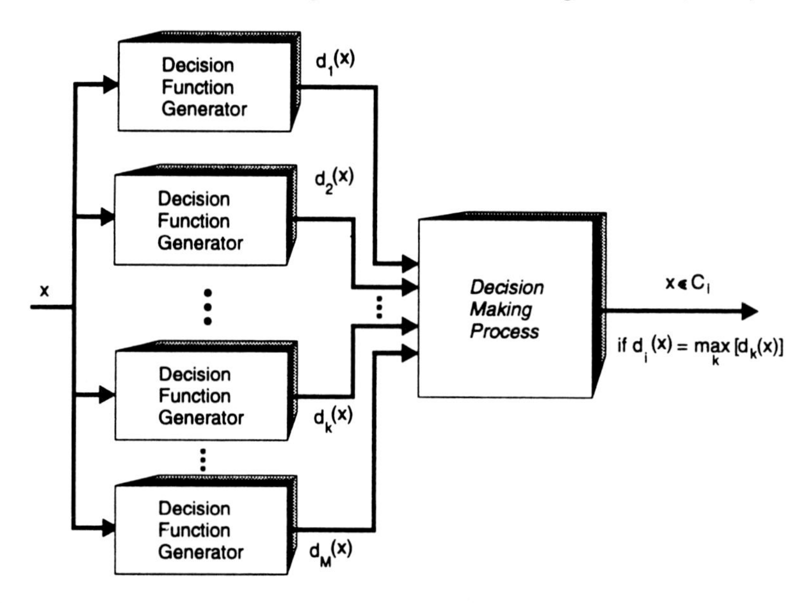

FIGURE 1.6 Block diagram of a pattern classifier which uses discriminant function generators (DFGs). (Adapted from Tou and Gonzalez [1974]).

Let us consider a simple example where two measurements are performed on each entity yielding a two-dimensional pattern space which is easy to visualize, for example, the class consisting of professional football players and the class of professional jockeys. Each pattern in this case can be characterized by two measurements: height and weight. Figure 1.7 shows two pattern classes C_1 and C_2 in this two-dimensional pattern space.

Thus, $M = 2$, and for all patterns of class C_1:

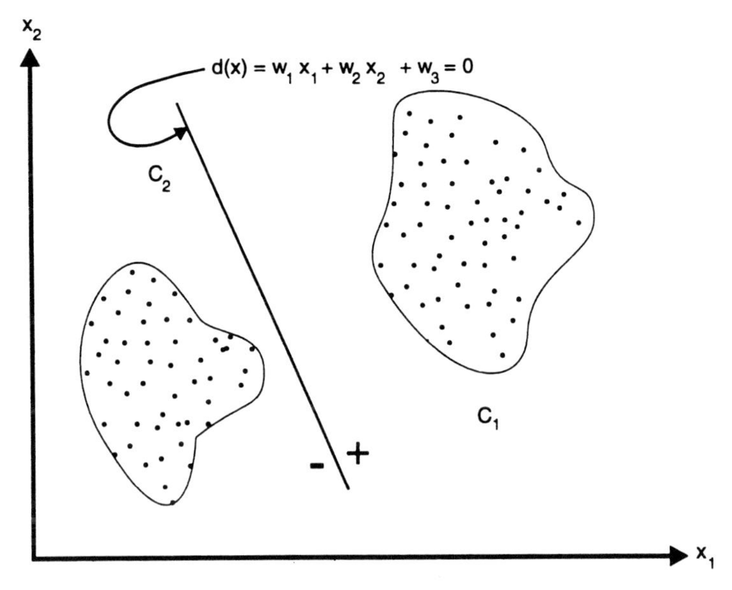

FIGURE 1.7 Scatter diagram for the feature vectors of two disjoint pattern classes. A simple linear decision function can be used to separate them. (Adapted from Tou and Gonzalez [1974]).

$$d_1(\mathbf{x}) > d_2(\mathbf{x}) \tag{7}$$

and, conversely, for all patterns in class C_2:

$$d_2(\mathbf{x}) > d_1(\mathbf{x}) \tag{8}$$

We can now define:

$$d(\mathbf{x}) = d_1(\mathbf{x}) - d_2(\mathbf{x}) \tag{9}$$

such that it leads to the condition:

$$d(\mathbf{x}) > 0 \quad \text{for} \quad \mathbf{x} \in C_1 \tag{10}$$

and

$$d(\mathbf{x}) < 0 \quad \text{for} \quad \mathbf{x} \in C_2 \tag{11}$$

In the case of two classes in Figure 1.7 it can be seen that a straight line can separate them. Then we have:

$$d(\mathbf{x}) = w_1 x_1 + w_2 x_2 + w_3 = 0 \tag{12}$$

which is a special case of the decision rule stated in equation 4. Note that (x_1, x_2) represents a pattern in this case, and the w's are parameters. The patterns of class C_2 lie on the negative side of this boundary; conversely, all patterns in class C_1 lie on the positive side.

Note that the decision function in equation 4 is quite general in the sense it can represent a variety of complex (including nonlinear) boundaries in n-dimensional pattern space.

There are various classification methods that can be used to design a recognition engine for the system. The choice depends on the kind of information that is available about the class-conditional densities. Class-conditional density is the probability function (which estimates the distribution) of pattern **x**, when **x**, is from class C_i, and can be given as follows:

$$p(\mathbf{x}/C_i), \quad i = 1, 2, \ldots, K \tag{13}$$

If all the class-conditional densities are completely known *a priori,* the decision boundaries between pattern classes can be established using the optimal Bayes decision rule (see Figure 1.8). Since the problem in this case is statistical hypothesis testing, the Bayes classifier gives the smallest error we can achieve from the given distributions. Thus the Bayes classifier is optimal. However, in practical applications the pattern vectors are often of very high dimensionality. This is due to the fact that the number of measurements, *n,* becomes high in order to ensure that the measurements carry all of the information contained in the original data. In such cases, implementation of the Bayes classifier turns out to be quite difficult due to its complexity.

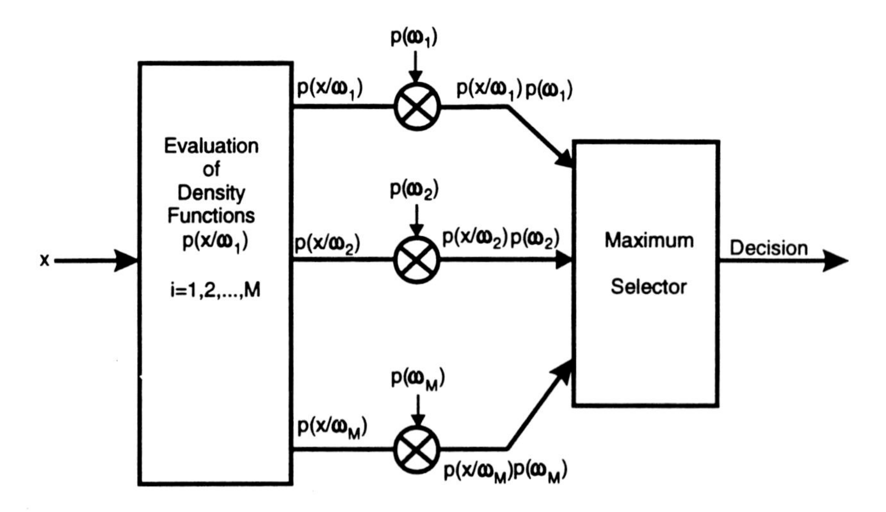

FIGURE 1.8 Bayes classifier. (Adapted from Tou and Gonzalez [1974]).

Also, in practice, the class-conditional densities are rarely known beforehand and a set of training patterns is needed to determine them. In some cases the functional form of the class-conditional densities is known and the task is to determine the exact values of some of the parameters that are not known. Such a problem is referred to as the parametric decision-making problem. In such cases simpler parametric classifiers are considered. Classifiers with linear, quadratic, and piecewise discriminant functions are the most common choices.

In cases where the precise form of the density function is not known, either it must be estimated or nonparametric methods must be used to obtain a decision rule. While our focus in this book is not on statistical pattern recognition as such, statistical techniques do play a role in some neural approaches. The K-nearest-neighbor algorithm, which is of nonparametric category, is discussed in chapter 8. The Karhunen-Loeve technique is often useful in determining which particular features set is accurate within the degree of tolerance, and is described in chapter 7. The radial basis function networks also rely on statistical clustering methods for training the hidden layer neurons. This reliance is evident in chapter 5. Figure 1.9 shows a tree diagram of various dichotomies which appear in the design of statistical pattern recognition (see Jain and Mao, [1994] for details). Interested readers can consult, for example, the book by Tou and Gonzalez [1974] for further study of these algorithms.

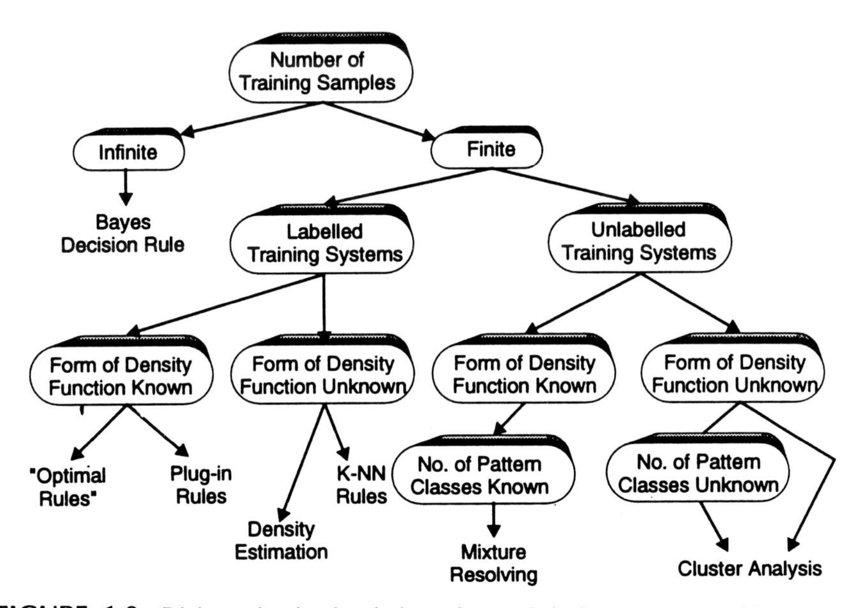

FIGURE 1.9 Dichotomies in the design of a statistical pattern recognition system. (Adapted from Jain and Mao [1994]).

1.5 SYNTACTIC PATTERN RECOGNITION

In applications involving patterns that can be represented meaningfully, using vector notations the statistical pattern recognition approach is ideal. However, this approach lacks a suitable formalism for handling pattern structures and their relationships. For example, in applications like scene analysis, the structure of a pattern plays an important role in the classification process. In this case, a meaningful recognition scheme can be established only if the various components of fundamental importance are identified and their structure, as well as relationships among them, are adequately represented.

In the 1950s several researchers (for example, see Chomsky, [1956]) in the field of formal language theory developed mathematical models of grammar. The linguists attempted to apply these mathematical models for describing natural languages, such as English. Once the model is successfully developed it would be possible to provide the computers with the ability to interpret natural languages for the purpose of translation and problem solving. So far these expectations have been unrealized, but such mathematical models of grammar have significantly impacted research in the areas of compiler design, computer languages, and automata theory. Syntactic pattern recognition is influenced primarily by concepts from formal language theory. Thus, the terms linguistic, grammatical, and structural pattern recognition are also often used in the literature to denote the syntactic approach.

In the syntactic approach the patterns are represented in a hierarchical fashion. That is, patterns are viewed as being composed of subpatterns. These subpatterns may be composed of other subpatterns or they can be primitives. Figures 1.10 (a) and (b) show the different chromosome structures. Figure 1.10 (a) shows a prototype pattern for the class named submedian chromosomes, while Figure 1.10 (b) shows the prototype for the second class, called telocentric chromosomes. These patterns can be decomposed in terms of primitives which define various curved shapes (see Figure 1.10 [c]). Each chromosome shown in Figure 1.10 can now be encoded as a string of qualifiers by tracking each structure boundary in a clockwise direction. For the submedian chromosome we detect these primitives which can be represented in the form of a string *abcbabdbabcbabdb*. The telocentric chromosome can be represented by the string *ebabcbab*.

We can view the underlying similarities within various structures belonging to the class, submedian chromosomes, as a set of rules of syntax for generation of strings from primitives. A set of rules governing the syntax can be viewed as a *grammar* for the generation of sentences (strings) from the given symbols. Each primitive can be interpreted as a symbol permissible in some grammar. Thus, we can envision two grammars G1 and G2 whose rules allow the generation of strings that correspond to submedian and telocentric chromosomes, respectively. In other words the language L(G1), consisting of sentences (strings) representing submedian chromosomes, can be generated by G1. Similarly, the language L(G2) generated by G2 would consist of sentences representing telocentric chromosomes.

Thus, for the determination of the class of the chromosome using the syntactic pattern recognition approach first the two grammars G1 and G2 have to be established. In order to establish a given input pattern (i.e., determine which type of chromosome it is) it is decomposed into a string of primitives. This sentence represents the input pattern and now the problem is to determine the language in which this input pattern represents a valid sentence. In the chromosome identification application if the sentence corresponding to the input pattern belongs to the language L(G1), it is classified as a submedian chromosome. On the other hand, if it belongs

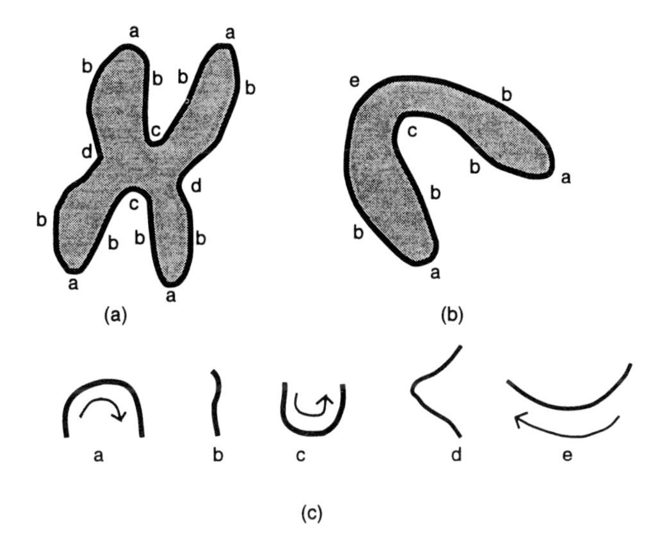

FIGURE 1.10(a) A submedian chromosome; **(b)** a telocentric chromosome; **(c)** five primitives that can be used to code the two types of chromosomes. (Adapted from Ledely, R. S., *Science,* vol. 146, no. 3461, pp. 216–223, 1964.)

to language L(G2), it is classified as telocentric chromosome. If it belongs to both L(G1) and L(G2), it is declared ambiguous. If the sentence representing the input pattern is found to be invalid over both the languages, the input pattern is assigned to a rejection class consisting of all invalid patterns. Techniques for establishing the class membership of syntactic structures, as well as issues involved in forming grammars, are discussed in Gonzalez and Thomason [1978].

For multiclass pattern recognition problems more grammars (at least one for each class) have to be determined. The pattern is assigned to class i if it is a sentence of only L(Gi) and no other language. Thus the syntactic pattern recognition approach in this case is the same as that described for the two-class problem.

The foregoing concepts are valid even in cases where patterns are represented by other data structures instead of strings (i.e., trees and webs (undirected, labeled graphs)).

Figure 1.11 shows a typical pattern recognition system designed for classifying patterns using a syntactic approach.

1.6 THE CHARACTER RECOGNITION PROBLEM

The problem of designing machines that can recognize patterns is highly diverse. It appears in many different forms in a variety of disciplines. The problems range from practical to the profound. The great variety of pattern recognition problems makes it difficult to say what pattern recognition is. However, a good idea of the scope of the field can be given by considering some typical pattern recognition tasks.

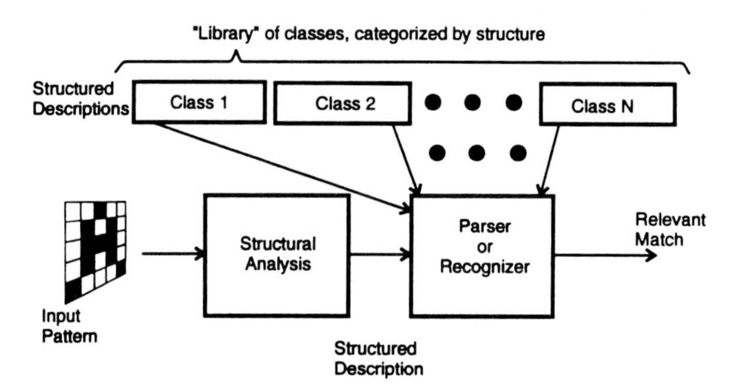

FIGURE 1.11 Block diagram of a syntactic pattern recognition system for classification.

Frequently a great deal of preprocessing may be required before the "act of recognition" can even begin. We therefore will focus primarily on character recognition since there is an abundance of available data and a minimum of preprocessing. After all, character recognition is one of the classic examples of pattern recognition. There is little loss of generality in that the fundamental neural recognition techniques will be similar to those used in other problem domains. Furthermore, our objective of taking a hands-on practitioners approach is facilitated by using character recognition as an exemplar problem.

The character recognition problem is widely studied in the pattern recognition literature, yet is far from being a solved problem. Nevertheless, it is tractable in the sense that a great amount of data can be easily obtained. One of the most common divisions between character recognition systems lies in whether the recognizer is focused on handwritten text or machine printed text.

Handwritten text data presented to the system may be either on-line or off-line. On-line handwritten text input from a tablet is presented as a sequence of coordinates $v(x, y, t)$ where t is time. Stroke order is available in this context as an aid to recognition. The down side is that handwritten text is significantly less constrained than printed text. Another issue that arises in the context of handwritten text is that of both word and character segmentation. Determination of which strokes should be grouped together to form characters and of where word boundaries exist is a nontrivial problem. There are three major categories into which noncursive text may be grouped from a segmentation point of view. These are

1. Box mode — Characters are written in a predefined box.
2. Ruled mode — Characters (and words) are written on a predefined line.
3. Unruled mode — Characters (and words) may be written anywhere on the input surface and may also slope arbitrarily.

In box mode, segmentation is trivial. In ruled and unruled mode, segmentation problems could turn out to be very difficult. The crucial importance of segmentation should not be underestimated. Incorrect segmentation can and will lead to poor recognition by the overall system. These and other issues render the recognition of handwritten characters more formidable than the machine printed character recognition problem even where the goal is omnifont recognition.

In the case of optical character recognition (OCR) (which can also be regarded as off-line), printed or handwritten text will be represented by a bit-mapped image typically from a scanner. Segmentation is less of a problem in this context although a preprocessing stage is still required. The preprocessing of bit-mapped images will be covered in chapter 3. Even though printed text is more constrained than handwritten, the recognition of machine printed text remains challenging. Figure 1.12 below illustrates character confusion in machine printed text recognition.

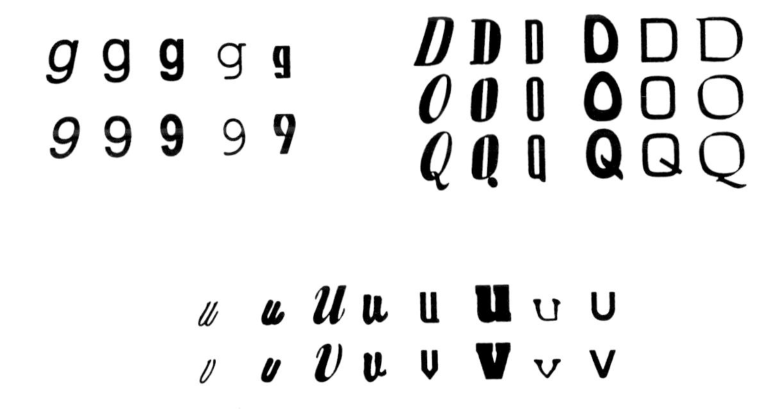

FIGURE 1.12 Some machine printed text with noisy characters.

In what follows we will not make a great deal of distinction between handwritten as opposed to printed characters in that there will be no attempt to deal with using stroke order information. In this sense we deal with handwritten characters (as seems to be the case in most of the literature) as though they were a kind of very highly unconstrained printed text. The discussion of cursive text recognition will be limited to a brief description of selective attention in the neocognitron.

One very nice aspect of research into character recognition is the abundant supply of test data in the form of the National Institute of Standards and Technology (NIST) database. This database contains over 1,000,000 sample characters. This is especially important given the variability of handwritten text and its being crucial to creating recognition systems with sound decision boundaries.

Many results for neural network based recognizers are reported in the literature. Given the ambiguities inherent in varying sample sizes, in varying test data, and in many cases varying objectives, it is not reasonable to draw conclusions about the relative effectiveness of the systems based on the raw accuracy data reported. Results ranging from about 80% to high 90% have been encountered. Some questions that are salient in this connection are

1. What level of recognition accuracy is good enough?
2. What level of reliability is good enough?

These questions are largely rhetorical and certainly depend on context. The reliability measure (discussed in detail in chapter 12), for example, is much more crucial to a banking system than it would be in a personal data assistant (PDA). Conversely, recognition accuracy rates in the low to mid-90% range may sound quite good but will frustrate a PDA user beyond endurance. Likewise, in the PDA context, processing power and performance will become pertinent problems. The potential for parallel processing is ubiquitous in discussions of neural systems but the dedicated hardware to achieve this advantage is frequently unavailable in the context of commercial systems.

1.7 ORGANIZATION OF TOPICS

In what follows we will discuss a wide range of methods for pattern recognition by neural networks. In general, we will begin with a discussion of underlying theory. This more generic discussion will be followed by specific implementations and practical suggestions. Finally, we will provide examples and variations found in the literature along with reported results.

REFERENCES AND BIBLIOGRAPHY

Chen, C. H., *Statistical Pattern Recognition,* Hayden, Washington, D.C., 1973.

Chomsky, N., "Three models for the description of language," *Proc. Group. Inform. Th.,* vol. 2, no. 2, pp. 113–124, 1956.

Devijer, P. and Kittler, J., *Pattern Recognition: A Statistical Approach,* Prentice-Hall, Englewood Cliffs, NJ, 1982.

Duda, R. O. and Hart, P. E., *Pattern Classification and Scene Analysis,* John Wiley & Sons, New York, 1973.

Foley, D. H., Considerations of sample and feature size, *IEEE Trans. Inf. Theory,* IT-18, pp. 618–626, 1992.

Fu, K. S., *Syntactic Methods in Pattern Recognition,* Academic Press, New York, 1974.

Fu, K. S., *Syntactic Methods in Pattern Recognition: Applications,* Springer-Verlag, New York, 1977.

Fukunaga, K. and Hayes, R. R., "Effects of sample size in classifier design," *IEEE Trans. PAMI,* PAMI-11, pp. 873–885, 1989.

Fukunaga, K., *Introduction to Statistical Pattern Recognition,* 2nd ed. Academic Press, New York, 1990.

Gonzalez, R. C. and Thomason, M. G., *Syntactic Methods in Pattern Recognition,* Addison-Wesley, Reading, MA, 1978.

Jain, A. K. and Chandrasekaran, B., "Dimensionality and sample size considerations in pattern recognition practice," In *Handbook of Statistics 2,* P. R. Krishnaiah and Kanal, L. N. (eds.), North-Holland, Amsterdam, 1982.

Jain, A. and Mao, J., Neural Networks and Pattern Recognition, In *Computational Intelligence: Imitating Life,* J. Zurada, R. J. Marks II, and C. J. Robinson (Eds.), IEEE Press, Piscataway, NJ, 1994.

Ledley, R. S., "High speed automatic analysis of biomedical pictures," *Science,* vol. 146, no. 3461, pp. 216–223, 1964.

Miclet, L., *Structural Methods in Pattern Recognition,* Springer-Verlag, New York, 1986.

Minsky, M. and Pappert, S., *Perceptrons: An Introduction to Computational Geometry,* MIT Press, Cambridge, MA, 1969.
Pao, Y., *Adaptive Pattern Recognition and Neural Networks,* Addison-Wesley, Reading, MA, 1989.
Pavilidis, T., *Structural Pattern Recognition,* Springer-Verlag, New York, 1977.
Tou, J. T. and Gonzalez, R. C., *Pattern Recognition Principles,* Addison-Wesley, Reading, MA, 1974.
Ullman, J. R., *Pattern Recognition Techniques,* Butterworths, London, 1973.
Watanabe, S., *Frontiers of Pattern Recognition,* Academic Press, New York, 1972.
Watanabe, S., *Pattern Recognition: Human and Mechanical,* John Wiley & Sons, New York, 1985.

2 Neural Networks: An Overview

2.1 MOTIVATION FOR OVERVIEWING BIOLOGICAL NEURAL NETWORKS

Neurobiological analogy as a source of inspiration is likely to have a major impact on the design and understanding of artificial neural systems. Engineers often look to neurobiology to gather new ideas for neurocomputing architectures, and to solve problems more complex than those easily addressable using conventional techniques.

Human vision, for example, is a highly complex *information-processing* task [Marr, 1982; Levine, 1985; Churchland and Sejnowski, 1992]. The brain can perform a routine perceptual task, where the visual system provides a representation of the environment and supplies necessary information needed to interact with it in a matter of 100–200 msec. However, huge conventional computers often take days for tasks of much lesser complexity [Churchland, 1986]. The retina processes visual information by extracting edge information using lateral inhibition between retinal neurons. For a patch of such images bounded by edges the cortex computes the lightness via a lateral excitation process. Then a comparison of images from two eyes results in formation of depth perception. Such determination of depth often takes numerous trials before the cortical net finds a "solution". Understanding of visual system has resulted in retina and cochlea [Mead, 1989] chips for engineering applications.

Another example is that of an echo-locating bat whose ability to pursue and capture its target would be the envy of a radar or sonar engineer. The sonar of the bat, whose brain is merely the size of a plum, is an active echo system. For a target, such as a flying insect, the bat sonar conveys information about its distance, azimuth, elevation, relative velocity, and size. The brain performs the complex neural computations to extract all this information from the target echo [Suga, 1990a; 1990b].

A bat inspired model sonar receiver has been developed by Simmons et al. [1992]. The model consists of the following three stages:

1. A front end mimics the inner ear of the bat. This stage encodes the incoming waveform.
2. A subsystem of delay lines computes echo delays.
3. A subsystem computes the spectrum of echoes. In the case of multiple target traces, this spectrum is used to estimate time separation of the echoes.

Many such biological processing examples have provided useful clues for the development of artificial neural networks. With this in mind, we shall now discuss the structural organization of the brain and some interesting aspects of neurophysiology.

2.2 BACKGROUND

Our primary interest in this book is confined to a study of artificial neural systems from a computational intelligence perspective. Thus we view neural networks as a class of mathematical algorithms. We discuss how these algorithms provide solutions to a number of specific problems. There exists a scientific community that views neural networks as synthetic structures that emulate biological neural networks found in living organisms. For a perspective on the philosophy of neural networks and computational aspects of the brain see Churchland and Sejnowski [1992].

For scholars working in the field of computing applications of neural models, knowledge of the following two aspects of neurobiology is useful:

1. Basic morphology of neurons including, axons, dendrites, cell bodies or somata, and synapses
2. The chemical transmitters at synapses and how connection of nerve impulses is affected by the actions of various ions in and around the cells

Another level would address the structural and functional overview of the brain regions, especially the cognitive functions and the pathways between the regions.

We give an extremely cursory summary of the biological facts, which provides some insight into the functioning of the brain. For a more detailed coverage of topics related to fundamental electrical and chemical processes at the neuronal level we refer the reader to Katz [1966] or Byrne and Shultz [1988]. Shepherd [1988, 1990a, 1990b] gives a strong insight into the cognitive aspects of neural structures. For a perspective on physiological psychology, especially the psychological aspects of conditioning and memory, the reader should refer to Thompson [1967] or Carlson [1977]. There are several good textbooks on neuroanatomy of different brain regions, including Truex and Carpenter [1969], Nauta and Feirtag [1986], and Kandel and Schwartz [1991]. Anderson [1994], Levine [1991], Koch and Segev [1989], etc. discuss neural networks with an emphasis on neurophysiological considerations, neural modeling, and cognition.

Some of the useful lessons from neuroscience, in the context of engineering applications of neural networks, may lie in the details of the data representation where Mother Nature has no peer and the device characteristics, which may account for the massive parallelism. These issues are really useful to know and are immensely practical because they have a half-billion years of free research and development behind them. Readers may refer to Anderson (1994) for a detailed discussion. A description of neural network systems would not be complete without including at least a brief discussion on biological neural systems.

2.3 BIOLOGICAL NEURAL NETWORKS

The current view of the nervous system owes much to the two pioneers, Ramony Cajal [1934] and Sherrington [1933] who introduced the notion that the brain is composed of distinct cells (neurons). The brain has approximately 100 billion (10^{11}) nerve cells (neurons) and it is estimated that there are about 100 trillion (10^{14}) connections (synapses) having a density of 1000 connections per neuron. As a result of this truly staggering number of neurons and synapses, the brain is an enormously efficient structure, even though neurons are much slower computing elements compared with the silicon logic gates. In a silicon chip, events happen at the rate of few nanoseconds (10^{-9} seconds), while neural events occur at the rate of milliseconds (10^{-3} seconds).

The nervous system of humans and other primates consists of three stages [Arbib, 1987] as shown in Figure 2.1. The sensory stimuli from the environment or human body is converted into electrical impulses by the receptors, such as eyes, ears, nose, skin, etc. These *information-bearing* signals are then passed on through forward links to the brain which is central to the nervous system. The brain in Figure 2.1 is represented by a neural network.

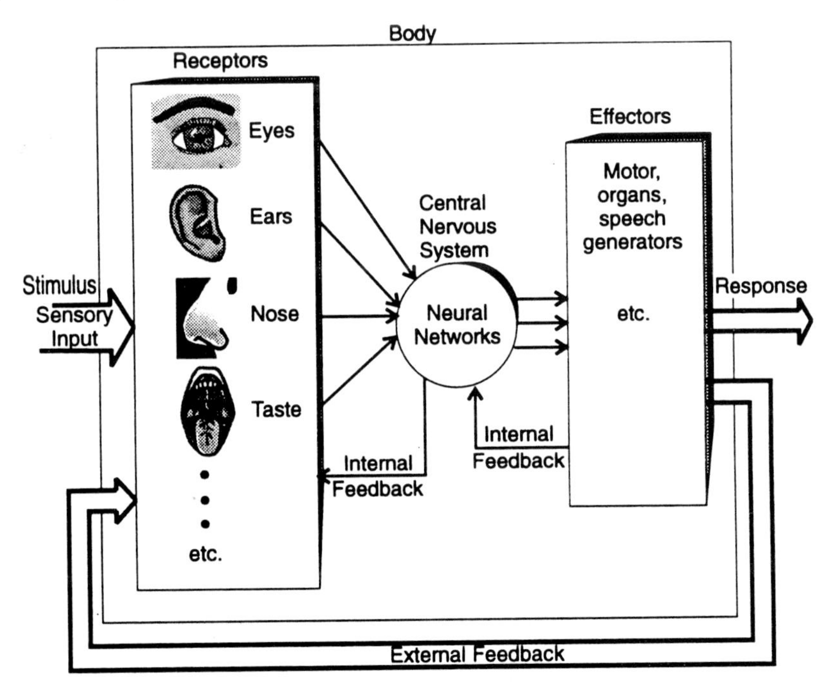

FIGURE 2.1 Block diagram of the nervous system showing the information flow through the links.

The brain continually receives information which it processes, evaluates, and compares to the stored information and makes appropriate decisions. The necessary commands are then generated and transmitted to the effectors (motor organs like tongue, vocal cords, etc. for speech) through forward links. The effectors convert these electrical impulses into discernible responses as system outputs. At the same time motor organs are monitored in the central nervous system by feedback links that verify the action. The implementation of these commands function through both external and internal feedback for acts, such as hand-eye coordination. Thus, the overall system bears some resemblance to a closed-loop control system.

Figure 2.2 shows the schematic diagram of a "generic neuron" with its main components labeled as axon, cell body (soma), dendrites, and synapses. Figure 2.2 also shows the characteristic ions (Na^+ K^+, and Cl^-) where they are prevalent inside and outside the cell membrane.

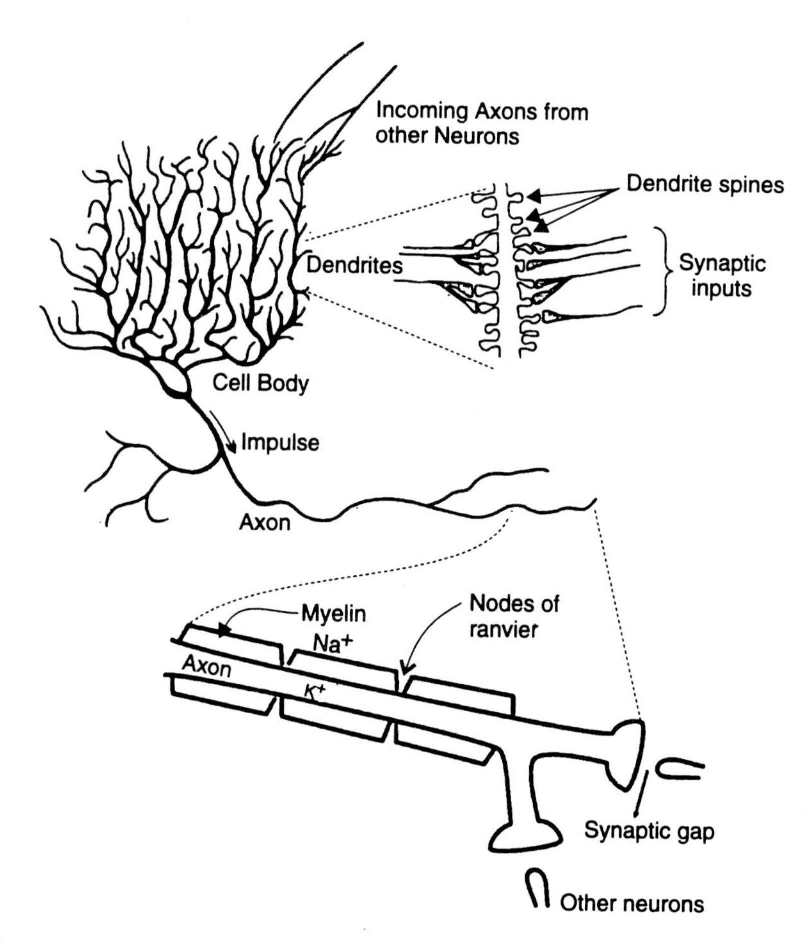

FIGURE 2.2 A neuron with its components labeled. (Adapted from Dayhoff [1990].)

Dendrites (with many small branches resembling a tree) are the receptors of electrical signals from other cells. Axons, the transmission lines, carry the signals away from the neuron. They have a smoother surface, fewer branches, and greater length compared with dendrites which have an irregular surface. The soma (cell body) contains the cell nucleus (the carrier of the genetic material) and is responsible for providing the necessary support functions to the entire neuron. These support functions include energy generation, protein synthesis, etc. The soma acts as an *information processor* by summing the electrical potentials from many dendrites.

The interactions between the neurons are mediated through elementary structural and functional units, called synapses. The synapses could be electrical, where the action potential (shown in Figure 2.2) travels between cells by direct electrical condition. However, chemical synapses, where conduction (information transfer) is mediated by a chemical transmitter, are more common. The synapse can impose excitation or inhibition on the receptive neuron. For a good discussion on synapses, the reader is referred to Dayhoff [1990, chapter 8].

The chemical synapse operates as follows [Shepherd and Koch, 1990]:

1. The transmitting neuron, called *presynaptic cell,* liberates a transmitter substance that diffuses across the synaptic junction. Thus an electrical signal is converted to a chemical signal.
2. The chemical neurotransmitter causes a positive increase (for an excitatory connection) and a decrease (for an inhibitory connection) in the *postsynaptic* membrane potential. The receiving neuron is called the *postsynaptic cell.*
3. Thus, at the postsynaptic cell, the chemical signal is converted back into an electrical potential which now propagates through to the other components of the neural network.

In various parts of the brain there are a wide variety of neurons, each with a different shape and size. Also, the number of different types of synaptic junctions between the cells is quite large [Shepherd, 1983]. The cell membrane, shown in Figure 2.1, consists of myelin sheaths (electrically insulating layer) with nodes of Ranvier acting as channels for ion transfer. It plays a very important role in the activities of the nerve cell, such as impulse propagation (for details of action potential generation and transmission see Shepherd [1983, p. 107].

Figure 2.3 (a) shows a trace of the nerve impulse waveform (action potential) as it would appear on an oscilloscope. Such a nerve impulse train can be recorded by placing a microelectrode near an axon. Figure 2.3 (b) shows the corresponding nerve impulse train.

Long-term memories are thought to be defined in the nervous system in terms of variations in synaptic strengths. The changes in synaptic efficiency are mediated through biochemical changes associated with learning and memory. Experimental evidences supporting this universal assumption are as follows:

1. Changes in strength in specific synapses in hippocampal neurons depend on combined activity of multiple inputs;
2. Changes in the morphology of dendritic spines contribute substantially to learning and memory in central neurons;

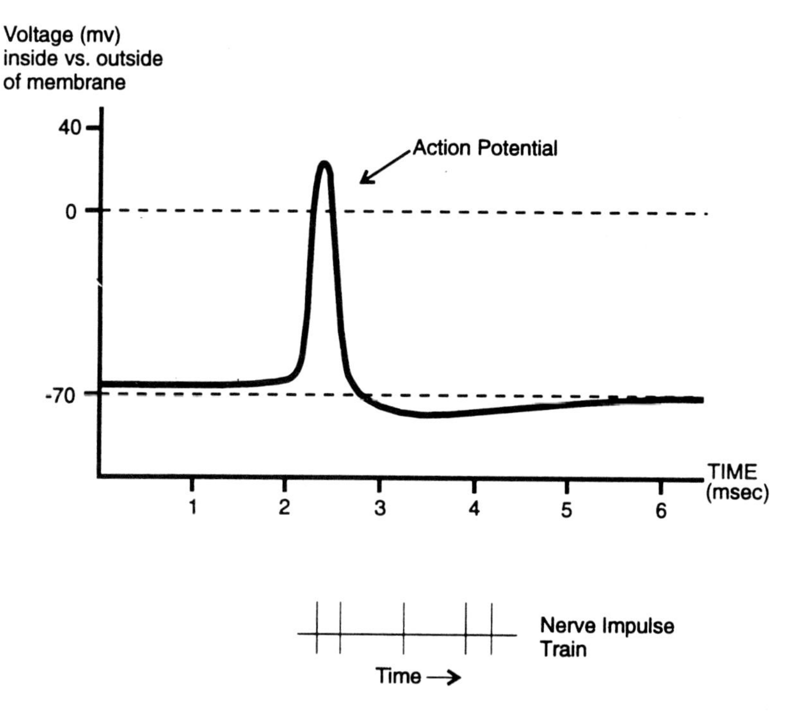

FIGURE 2.3 (a) Trace of a nerve impulse waveform; (b) corresponding nerve impulse train.

3. Calcium ions mediate changes in synaptic efficiency contributing to increases post-synaptic receptors, proteins synthesis involved in spine swelling, transport of dendritic microtubules, and output of pre-synaptic transmitter.

MacGregor [1987] contains a good discussion on this subject.

2.4 HIERARCHICAL ORGANIZATION IN THE BRAIN

Extensive research on the analysis of local regions in the brain [Churchland and Sejnowski, 1992; Shepherd, 1988] has revealed the structural organization of the brain with different functions taking place at higher and lower levels in the brain. Figure 2.4 shows the hierarchy of the levels of organization in the brain. The traditional digital computers can also be viewed as a structured system [Tanenbaum, 1990], but the organization is radically different. Thus, a look into the levels of organization of the nervous system may provide new insights into designing computers with a radically different organization.

At the top the behavior of an individual is determined at the "whole brain" level. The behavior is mediated by *topographic maps,* systems and pathways at the inter-regional circuit level beneath it. Topographic maps involve multiple regions located

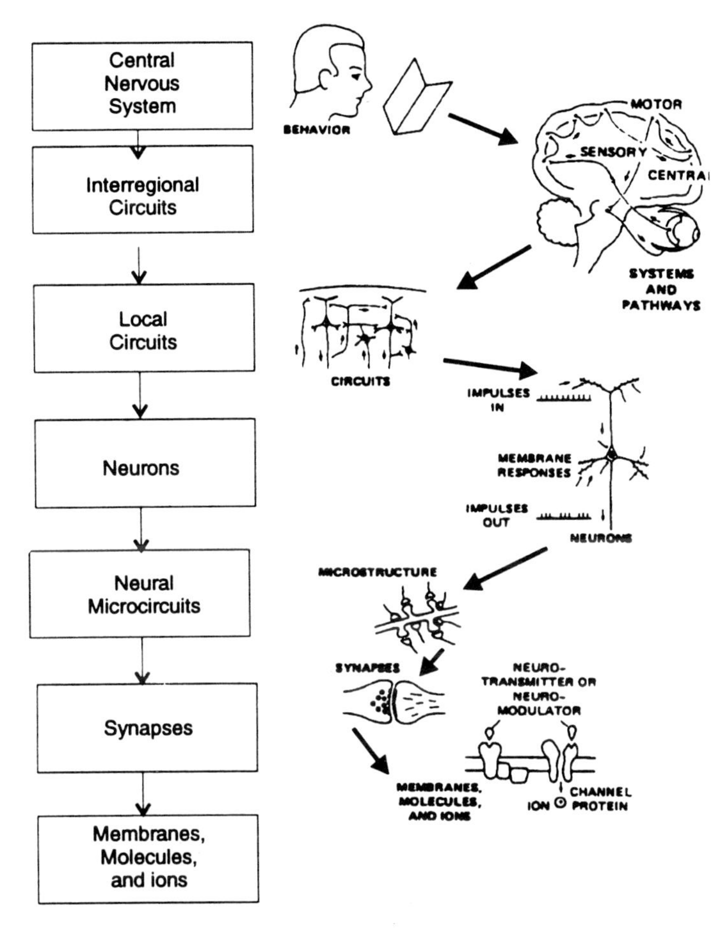

FIGURE 2.4 Structural organization of levels in biological nervous systems. (Adapted from Dayhoff, J., *Neural Network Architectures: An Introduction,* Van Nostrand Reinhold, New York, 1990.)

in the different parts of the brain and they are organized to respond to the sensory information. In fact the visual system, the motor system, and the auditory system taken as a whole fit into this category. The third level of complexity is called *local circuitry* and is made up of neurons with similar or different properties. These neuronal assemblies are responsible for local processing. The next level is the neuron itself, about 100 micrometer in size, containing several *dendric subunits.* Below this level lies the *neural microstructure* (like a silicon chip made up of an assembly of transistors in the case of a computer) which produces various functional operations. These are structures that affect areas around the synapses and are of the size of a few microns, with a speed of operation of a few milliseconds (quite slow compared with a speed of nanoseconds for transistors in traditional computers). The fact that neurons are such slow, millisecond devices may partially account for the massive parallelism required for a large biological computer (i.e., the brain). The next level consists of *synaptic junctions* where cells transmit signals from one to another. Synapses in turn rely on the actions of the molecules and ions at the level below.

One of the critical differences between the human brain and that of the lower animals is that the cortex that covers the human brain is much larger in size and complexity. Figure 2.5 shows the side view of the human brain covered with cerebral cortex.

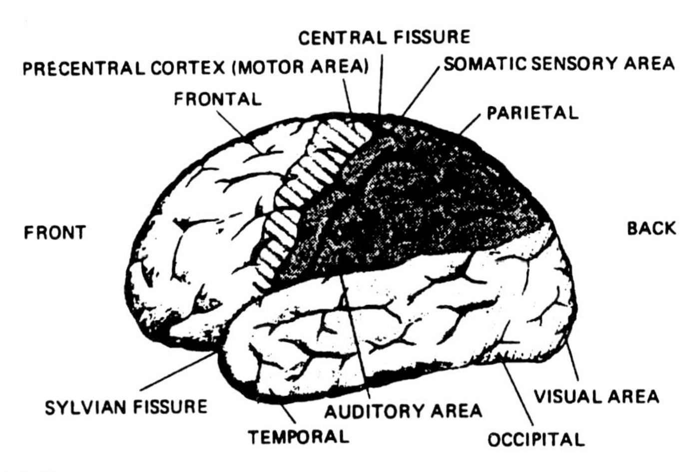

FIGURE 2.5 The human brain and the different regions of the cerebral cortex. (From Thompson, R. F., *The Brain: An Introduction to Neuroscience,* W. H. Freeman, San Francisco, 1985. With permission.)

Neuroscientists have identified different regions of the brain in terms of their specialization for complex tasks, such as vision, speech, hearing, etc. The visual information is analyzed in the back side of the brain (in the occipital lobe). The auditory sensors send information to the auditory cortex (upper part of the temporal lobe) where they are analyzed. The visual cortex in fact contains a map that reflects the layout of the surface of the retina. The cochlea is the part of the inner ear that receives auditory input and the map on the auditory cortex reflects the sheets of receptors in the cochlea. Figure 2.5 also shows the front of the central fissure labeled as precentral cortex which is responsible for organizing motor activity through control of muscular movements. The parietal lobe (middle region in Figure 2.5) participates in processing information from the skin and body. The association cortex carries out higher brain functions like cognition, perception, etc.

The major neural structures within or below the cerebral cortex are shown in Figure 2.6. The cerebellum, at the bottom, is involved in muscular activities such as walking, jumping, playing a musical instrument, etc., which require sensory-motor coordination [Albus, 1981]. Next to it the brainstem is concerned with respiration, heart rhythm, and gastrointestinal functions. The spinal cord, below the brain, transmits signals to and from the brain and generates appropriate reflex actions. The hippocampus in primitive animals participates in finding appropriate responses to various smells in the environment, but in humans it takes on new roles.

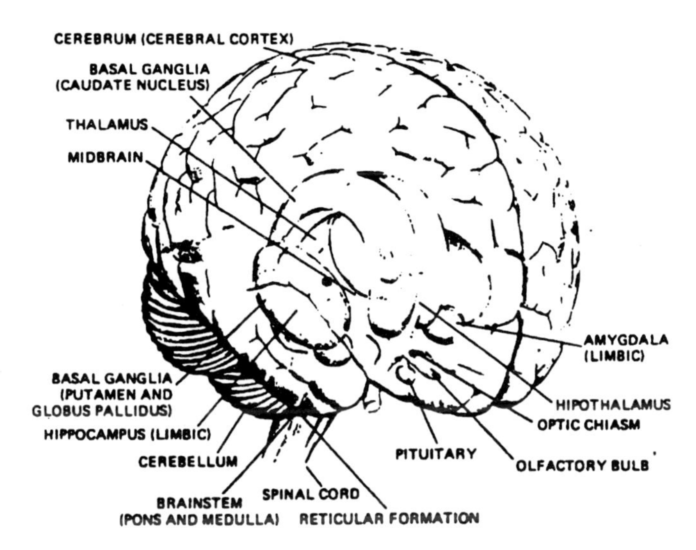

FIGURE 2.6 Major neural structures in the brain. (Adapted from Nauta, W. J. H. and Feirtag, M., *Sci. Am.*, Sept. 1979, p. 102.)

In the case of the visual system, the retina is the sensory organ that acts as a transducer. This transducer converts the photons (stimulus energy) into corresponding neural signals which are subsequently processed in the brain by the visual cortex (see Figure 2.5). The retina which senses the stimulus for the visual system is shown in Figure 2.7 (a). The retina itself has five layers, with receptor cells (consisting of rods and cones) receiving light signals from outside. These signals are then transmitted through different layers of cells where some horizontal preprocessing occurs. Finally the ganglion cells transmit the signals to the primary visual cortex, such that they encode the local areas of the visual stimulus. Figure 2.7 (b) shows the visual pathways, originating from the retina, via the lateral geniculate nucleus (LGN), to the primary visual cortex. The brain makes a map of the visual field, called topographic maps, at the visual cortex. Extensive studies by Hubel and Weisel [1962, 1965] showed that various neurons in the cat's visual system respond selectively to borders, orientation, motion, length of line, etc.

Similarly, in the auditory system the cochlea (the sensory organ) converts the sound waves (stimulus) into neural signals which are subsequently processed by the auditory cortex (see Figure 2.5). Figure 2.8 shows the basic pathways for the auditory system from the ear to the auditory cortex.

In fact, a tremendous amount of data exists regarding the anatomy of the brain. The locations and functions of various major structures within the nervous system are very well understood. However, the precise conclusions about the role of each part of the neural circuitry are lacking. Neural network models are likely to contribute toward gaining a better understanding of mechanisms and circuitry involved in various functions carried out by the brain.

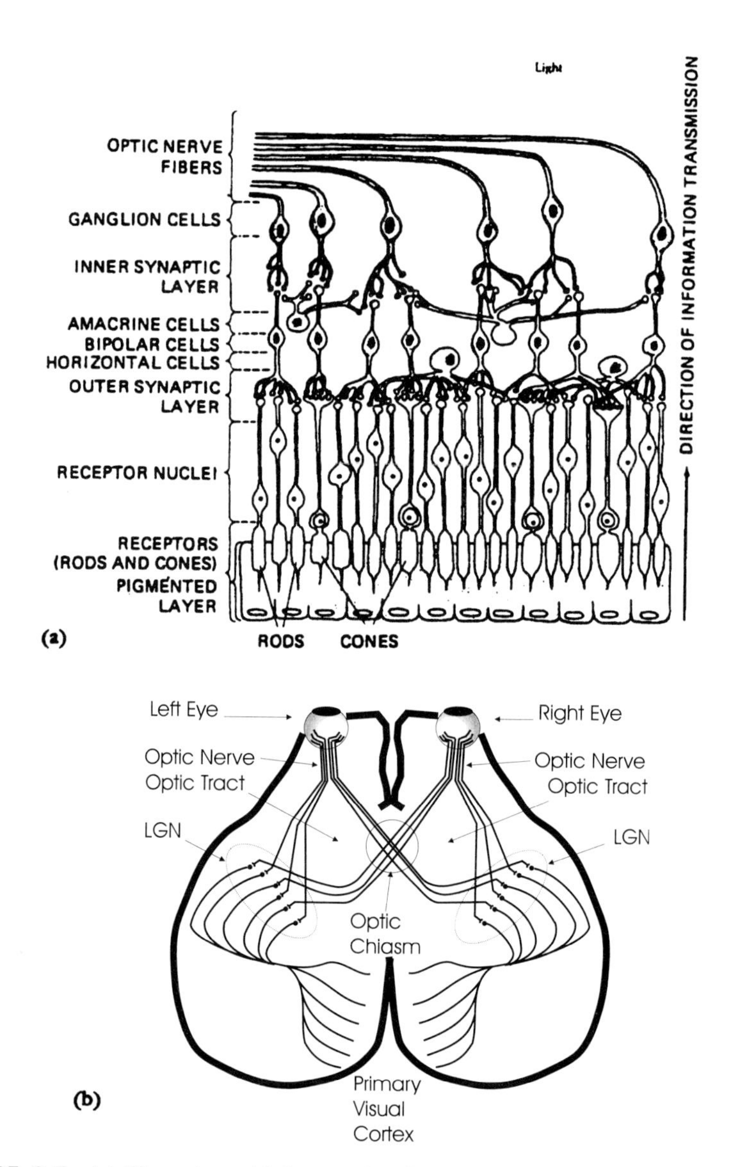

FIGURE 2.7 (a) The retina with layers of cells. (From Cornsweet, *Visual Perception,* Academic Press, New York, 1970.) (b) a schematic of the human visual system. (Adapted from Hubel, P. H. and Wiesel, T. N., *Sci. Am.,* Sept. 1979, p. 154.)

For example, Suzuki, Kawato, and colleagues [Kawato et al., 1987; 1988] have developed a series of models of voluntary movements. Figure 2.9 shows the control circuitry driven by sensory stimulus. Their model is inspired by known anatomy and physiology of several brain areas, including the pathways from the cortex to the muscles, spinal cord, cerebrum, and cerebellum (see Figure 2.6). These networks in

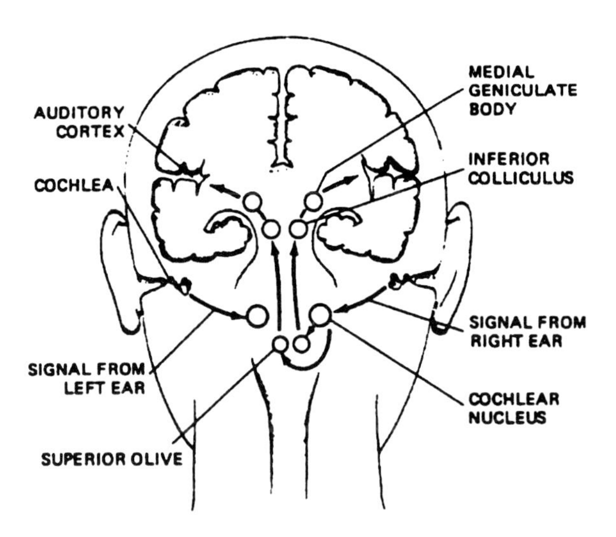

FIGURE 2.8 Schematic view of the auditory pathways. (From Lindsay, P. and Norman, D., *Human Information Processing: An Introduction to Psychology,* 2nd ed., Harcourt, Brace, and Javanovich, New York, 1977.)

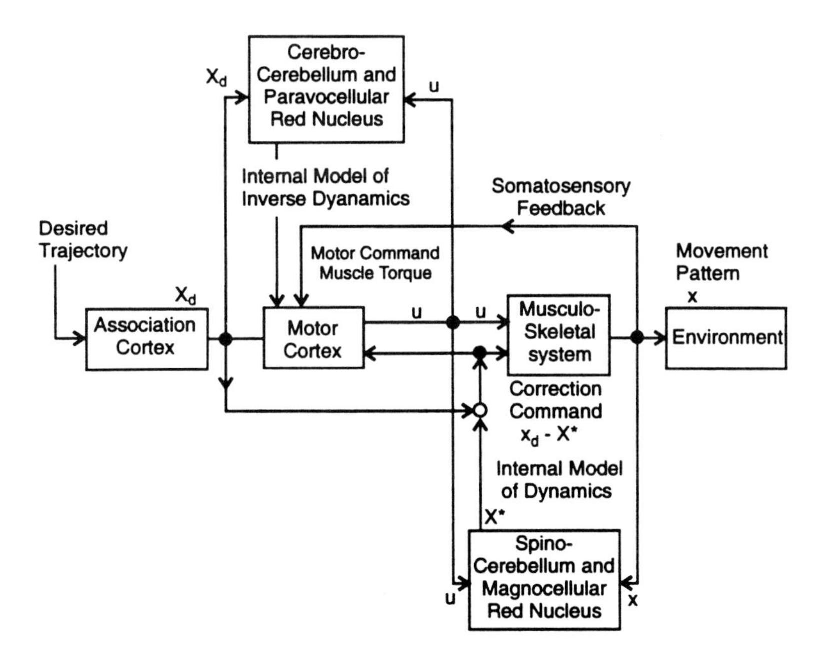

FIGURE 2.9 Neural network model for movement control and learning. (Proposed by Kawato, M., et al., *Biol. Cybern.,* vol. 57, pp. 169–185, 1987.)

fact learn a movement at one speed and then perform the same movement at a different speed.

The question of how topographic maps are formed is quite fascinating. Willshow and van de Malsberg [1979] discuss a theoretical model of this process which bears a close relation to the connectionist learning technique. Hinton et al. [1986] proposed a technique called coarse coding, which allows information to be accurately represented in a population of cells, each of which has a large receptive field. In fact, experimental observations show that there are many cells with large receptive fields in the nervous system, even in areas capable of great precision. This could be explained using coarse coding without contradiction which arises from the common sense view that high accuracy requires small receptive fields. Walters [1987] has analyzed an entire set of possible representational schemes and has shown that the observed receptive fields of neurons correspond to a very efficient representation. The topic of the degree of distribution found in the neural code and details of such code is quite interesting and is discussed in Anderson [1995].

There exists a massive body of literature on the brain and the understanding of nervous system function. MacGregor [1987] contains a detailed account with appropriate references. In our opinion, the following areas may provide useful lessons in the context of computer and information sciences:

1. Information processing paradigm of the brain function; and
2. Details of data representation in relation to learning and memory.

In the context of the information processing paradigm, the primary purpose of the system is to support signals which, in turn, are characterizable in terms of the information they carry (MacGregor [1987]). Another feature is that the system is organized in terms of paths along which information is transferred and the operations which are performed on the signals. The paths through the system may be represented by an operational flowchart. Algorithms represent the operations which are performed on the signals in the component parts of the system.

Several researchers have proposed representative overall flowcharts for the processes of learning and memory. Kessner [1973] suggests that memory can be subdivided into cue access, short-term memory, and long-term memory storage and retrievable systems. These, in turn, are controlled by operations such as match-mismatch, decay, selective attention, expectance, rehearsal, arousal, consolidation, and readout processes. He goes on to propose the primary locations for these operations in the nervous system.

Neelakanta and DeGroff [1994] contains a good discussion on the concepts of mathematical neurobiology and the information-theoretics aspects of neural networks.

2.5 HISTORICAL BACKGROUND

Neural networks as we know today began with the pioneering work of McCulloch and Pitts [1943] and has its roots in a rich interdisciplinary history dating from the early 1940s. McCulloch was trained as a psychiatrist and neuroanatomist, while Pitts was a mathematical prodigy. Their classical study of all-or-none neurons described the logical calculus of neural networks.

Figure 2.10 shows a McCulloch-Pitts model of a neuron with inputs x_i, for i = 1, 2, ..., N.

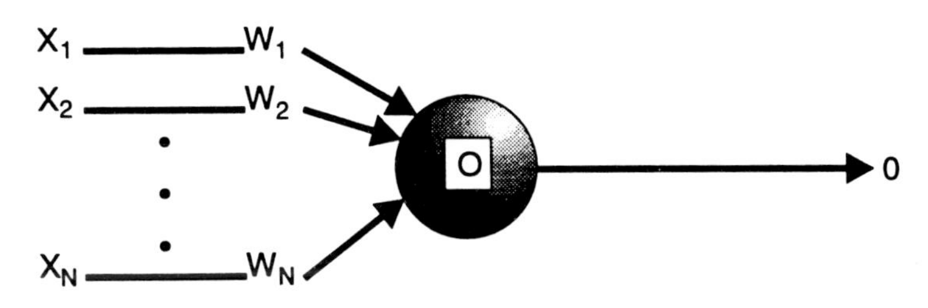

FIGURE 2.10 A McCulloch-Pitts model of a neuron.

W_i denotes the multiplicative weight (synaptic strength) connecting the ith input to the neuron. Theta is the neuron threshold value, which needs to be exceeded by the weighted sum of inputs for the neuron to fire (output = O is the output of the neuron. The weight, W_i, is positive if the connection (synapse) is excitatory and negative if the connection is inhibitory. The inputs, x_i are binary (0 or 1) and can be from sensors directly or from other neurons. The following relationship defines the firing rule for the neuron:

$$O = g\left(\sum_{i=1}^{N} W_i x_i\right) \quad (1)$$

where $g(x)$ is the activation function defined as:

$$g(x) = \begin{cases} 1 \text{ if } x \geq \theta \\ 0 \text{ if } x < \theta \end{cases} \quad (2)$$

This simplistic model could demonstrate substantial computing potential, since by appropriate choice of weights it can perform logic operations such as AND, OR, NOT, etc. Figure 2.11 shows the appropriate weights for performing each of these operations. As we know, any multivariate combinatorial function can be performed using either the NOT and AND gates, or the NOT and OR gates.

If we assume that a unit delay exists between input and output of a McCulloch-Pitts neuron (as shown in Figure 2.10), we can indeed build sequential logic circuitry with it. Figure 2.12 shows an implementation of a single memory cell that can retain the input.

As seen from the figure, an input of 1 at x_1 sets the output (O = 1) while the input of 1 at x_2 resets it (O = 0). Due to the feedback loop the output will be sustained in the absence of inputs.

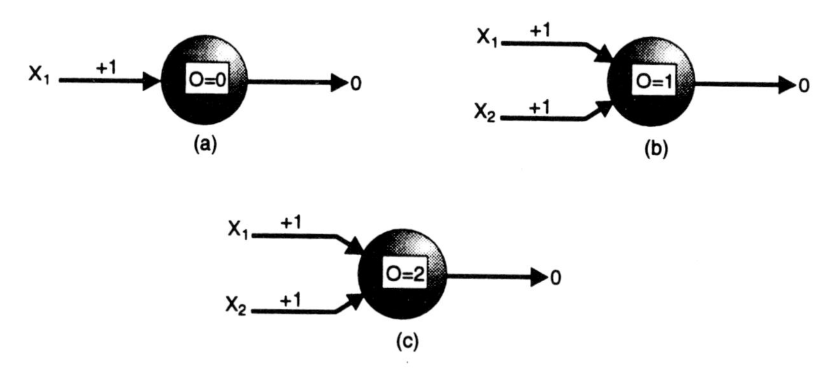

FIGURE 2.11 (a) Implementation of a NOT gate; (b) an OR gate; and (c) an AND gate implementation.

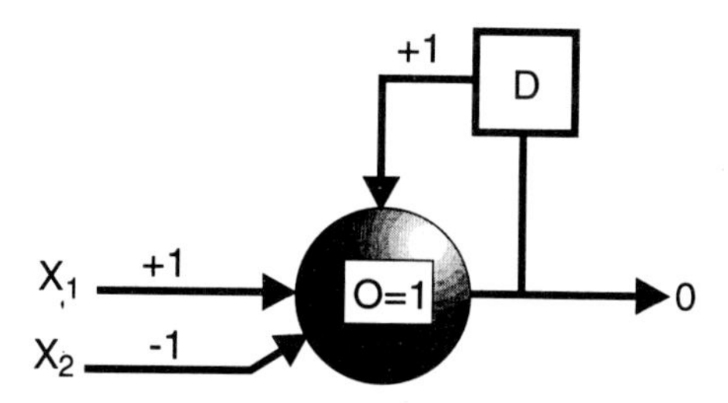

FIGURE 2.12 Implementation of a memory cell by using a feedback and assuming a delay of one unit of time.

This led to the computer-brain analogy, called *cybernetics* [Wiener, 1948], based on the fact that neurons are binary, just like switches in a digital computer. Wiener [1948] described important concepts of control, communications, and signal processing based on his perception of similarities between computers and brains, which spurred interest in developing the science of cybernetics. He discussed the significance of statistical mechanics in the context of learning systems, but it was Hopfield [1982, 1984] who established the real linkage between statistical mechanics and neural assemblies.

Von Neumann used the idealized switch-delay elements derived from the neuronal models of McCulloch and Pitts to construct the EDVAC computer [Aspray and Burks, 1986]. He in fact suggested that research in using "brain language" to design brain-like processing machines might be interesting [von Neumann, 1958].

The next major development came when a psychologist, Hebb [1949] proposed a learning scheme for updating the synaptic strengths between the neurons. He proposed that as the biological organisms learn different functional tasks, the connectivity in the brain continually changes. He was also first to propose that neural assemblies are created by such changes. His famous *postulate of learning,* which we now refer to as the Hebbian learning rule, stated that information can be stored in synaptic connections and the strength of a synapse would increase by the repeated activation of one neuron by the other one across that synapse. Quoting from Hebb [1949]:

> When an axon of cell A is near enough to excite a cell B and repeatedly or persistently takes part in firing it, some growth process or metabolic changes take place in one or both cells such that A's efficiency as one of the cells firing B, is increased.

This learning rule, called the *Hebb rule* or *correlation learning rule,* has had a profound impact on the future developments in the field of computational models of learning and adaptive systems. The original Hebb rule did not contain a provision for selectively weakening (or eliminating) a synapse. Rochester et al. [1956] performed simulations on digital computers to test Hebb's theory of learning in the brain, on an assembly of neurons. They demonstrated that it was essential to add inhibition for the theory to actually work for a neuronal assembly.

In the 1950s Frank Rosenblatt, a psychologist by training, proposed a neuron-like element called *perceptron* [Rosenblatt, 1958]. His intention as he states in his book was as follows [Rosenblatt, 1958, p. 387]:

> ...illustrate some of the fundamental properties of intelligent systems in general, without becoming too deeply enmeshed in the special, and frequently unknown, conditions which hold for particular biological organisms.

He chose to depart from the ideas of symbolic logic and used probability theory to analyze these models. Perceptron was a trainable machine and it learned to classify certain patterns by modifying the synaptic strengths. The perceptron architecture generated a lot of excitement in the early days of pattern recognition. Perceptron is of historical interest, though it is occasionally used. We shall briefly describe this approach as it has been taught in pattern recognition texts for decades [Duda and Hart, 1973; Tou and Gonzalez, 1974].

Figure 2.13 shows a simple perceptron with sensory elements (S elements), association units (A units), and response units (R units). The sensors could be photoreceptive devices for optical patterns in analogy to the retina where the light impinges.

Several S elements, which respond in all-or-none fashion, are connected to each A unit in the association area through fixed excitatory or inhibitory connections. A units in turn are connected to R units in the response area through modifiable connections.

A perceptron with a single R unit can perform classification when only two classes are involved. For classification involving more than two categories, several R units are required in the response layer. Rosenblatt [1958, 1962] provided the learning procedure (algorithm) for adjusting the free parameters in the network shown in Figure 2.13. The proof of convergence of the algorithm, known as the *perceptron convergence algorithm,* states that if the parameters used to train the perceptron are drawn from two linearly separable classes, then the perceptron algorithm converges and positions the decision surface in the form of a hyperplane between the two classes.

Also, during this period (early 1960s) Widrow and Hoff [1960, 1962] proposed a learning mechanism where the summed square error in the network output was

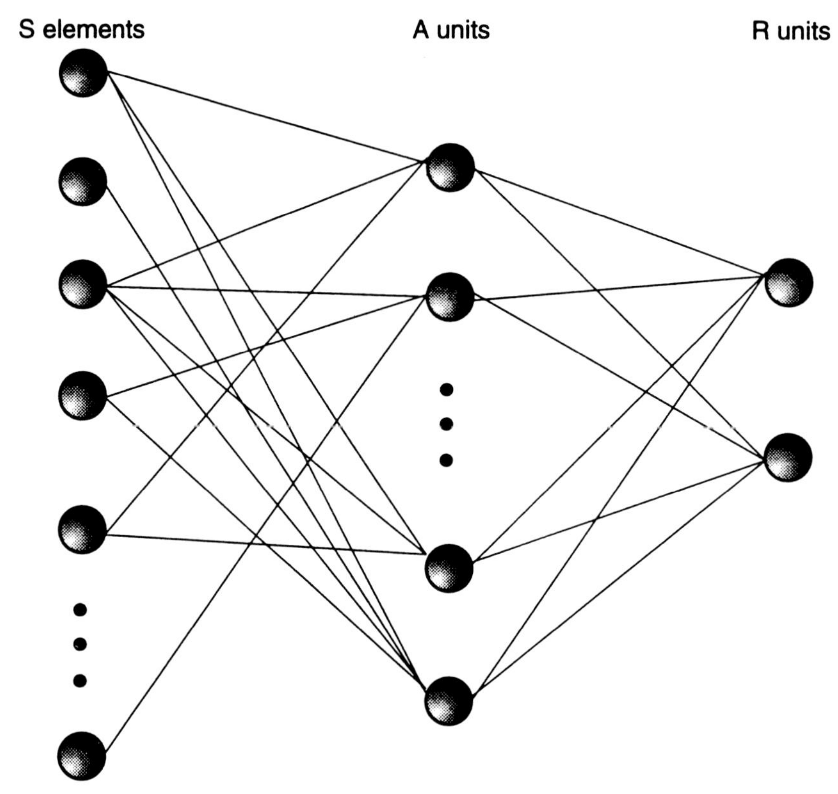

FIGURE 2.13 A simple perceptron structure with connections between units in three different areas.

minimized. They introduced a device called ADALINE (for adaptive linear combiner) based on this powerful learning rule (also called the *Widrow-Hoff learning rule*). During the 1960s ADALINE and its extensions to MADALINE (for many ADALINEs) were used in several pattern recognition and adaptive control applications. In the communications industry they were applied as adaptive filters for echo suppression in long-distance telephone communication.

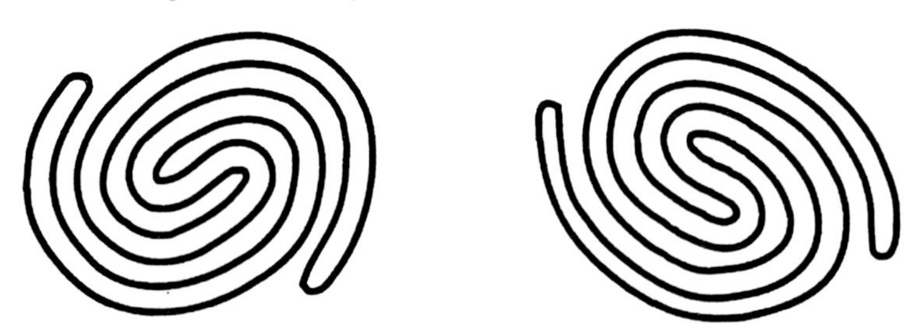

FIGURE 2.14 Two patterns, one with connected arc and the other with disjoint arcs. (Adapted from Minsky, M. L. and Pappert, S. A., *Perceptrons,* MIT Press, Cambridge, MA, 1969.)

During the 1960s it seemed that there were no fundamental limits of perceptrons and they could do anything. Minsky and Pappert [1969] used elegant mathematics to demonstrate the computational limitations of single-layer perceptrons. Their theorem stated that a single-layer perceptron would fail to achieve nonlinear separation patterns in a hyperspace (like the separation required for the EXCLUSIVE OR (XOR) problem, f(0, 0) = 0, f(1, 0) = 1, f(0, 1) = 1 and f(1, 1) = 0). Figure 2.14 shows two patterns, where the one on the left is connected unlike the one on the right which consists of disjoint arcs. Minsky and Pappert proved that a finite-order perceptron (as per their definition) cannot classify these patterns into different classes.

The abstract computational geometry used by Minsky and Pappert to study perceptrons was actually a subclass of Rosenblatt's perceptrons. They chose to depart from the probabilistic approach championed by Rosenblatt, and returned to the ideas of predicate calculus in their analysis of perceptrons, by imposing additional constraints on the structure of a perceptron. They proved the elegant theorems for their abstract form of perceptrons to show that single-layer perceptrons would fail for certain important geometric classifications. The authors in their book go on to make a conjecture (which proved to be totally unjustified later) that the limitations of the kind they had discovered for single-layer perceptrons would also hold true for its variants, more specifically multilayer neural networks. In section 13.2 of their book [Minsky and Pappert, 1969], they state:

> The perceptron has shown itself worthy of study despite (and even because of!) its severe limitations. It has many features to attract attention: its linearity; its intriguing learning theorem; its clear paradigmatic simplicity as a kind of parallel computation. There is no reason to suppose that any of these virtues carry over to many-layered version. Nevertheless, we consider it to be an important research problem to elucidate (or reject) our intuitive judgement that the extension to multi-layer systems is sterile.

As a result, their theorems were widely interpreted as discrediting the utility of all perceptron-like devices as learning machines. Thus, a campaign led by Minsky and Pappert, who were then involved in establishing at MIT an artificial intelligence (AI) laboratory, to discredit neural network research efforts and divert research funding to the field of AI was a great success. As Cowan [1990] said later, the lack of computational power in those days coupled with major psychological as well as financial (lack of support from funding agencies for such research) reasons, resulted in a major discouragement to work on perceptron-like architectures. However, Minsky later said that, in retrospect, the discrediting of perceptrons was an overreaction [Rumelhart and McClelland, 1986, pp. 158–159].

Nilsson [1965], in his book called *Learning Machines,* had shown that multilayer perceptrons can be used to separate patterns nonlinearly (as in the case of the XOR problem) in a hyperspace, but the *perceptron convergence theorem* applied only to learning in single-layer perceptrons. Thus, a mechanism for learning in multilayer perceptrons was not clear at that time. Rosenblatt [1962, p. 262] in fact did state:

> The procedure to be described here is called the "backpropagating error correction procedure" since it takes its cue from the error of the R-units, propagating corrections back towards the sensory end of the network if it fails to make a satisfactory correction quickly at the response end.

The difficulty was how to determine the error in the hidden neurons in the network in order to update weights connected to them. During the 1970s many of the ideas and concepts necessary to address this problem were formulated. For example, a mathematical framework for a new training scheme for multilayer perceptrons was proposed by Paul Werbos [1974] in his Ph.D. thesis at Harvard University. Similarly, Harth [1976] and colleagues proposed a learning paradigm based on cross-correlation which could update weights connected to hidden neurons without explicitly computing errors in their neuronal outputs [Pandya and Szabo, 1991; Pandya and Venugopal, 1994]. However, we had to wait until 1980s for the solutions of these basic problems to emerge.

During the 1970s a handful of researchers pursued research in neural networks and accomplished pioneering work. Several neural network models were developed during this period which contributed towards gaining a better understanding of mechanisms and circuitry involved in various functions carried out by the brain. One of the bewildering paradoxes in brain science at that time was that Lashley [1929] had shown that memory of events in the brain is distributed throughout rather than localized. However, neuroscientists like Mountcastle [1957] and Hubel and Wiesel [1962, 1965] had found localized (well-organized topographic) encoding of visual and somatosensory (touch-sensitive) information in regions of the brain. It was Erich Harth and colleagues [Harth et al., 1970; Anninos et al., 1970] who resolved this paradigm by proposing a model of a neural network with random connections. Through computer simulations of random net models they demonstrated the principle of "randomness in small but structure in large". Evidence of their proposed netlets was indeed found in the visual and somatosensory areas of the brain. They characterized the activity levels of random nets as oscillatory, monostable, bistable, etc., and investigated stability criteria. The Harth-Anninos model provides an elegant theory that captures the essential features of internally sustained activity in recurrently connected neuronal populations. Later, Harth [1976] developed a model for visual perception based on an "alopex" principle in which positive feedback is emphasized. This feedback causes feature-specific enhancement of input. In this model, "percepts" are produced by feature extraction using spatial derivatives and positive feedback.

Amari [1972, 1977] developed an adaptive model of threshold neuron, and used it to study the dynamic behavior of random nets. Nakano, also at Tokyo University, proposed a model of associative memory, called *associatron* [Nakano, 1972] and demonstrated its functioning through actual robots. Kohonen [1972, 1977, 1980] in Finland and Anderson [1972, 1994] and Anderson et al. [1973] pursued research on associative memories. Fukushima, in Japan, proposed a model, called *cognitron* [Fukushima, 1975], and its variations, called *neocognitron* [Fukushima, 1980] for visual pattern recognition based on knowledge about visual pathways in the brain.

Grossberg, introduced a number of architectures and theories including an adaptive model of a neuron [Grossberg, 1967, 1968] and showed its use as a short-term memory. He then built on his earlier work by introducing top-down template matching and bottom-up adaptive filtering. This allowed pattern recognition by learned feedback matching and adaptive resonance. They [Carpenter and Grossberg, 1987a, 1987b] coined the networks based on this phenomenon as *adaptive resonance theory* (ART).

In the 1980s the era of renaissance started with several exceptional publications that significantly furthered the potential of artificial neural networks. Hopfield's papers introduced fully connected network of neurons and addressed their potential as associative memories [Hopfield, 1982, 1984].

In 1986, with the publication of two volumes on parallel distributed processing, edited by Rumelhart and McClelland [1986], another revitalization was experienced. The concepts and learning paradigms introduced in this book addressed the criticisms of Minsky and Pappert [1969] which had resulted in underestimating the potential of multilayer perceptrons. This work successfully removed the training barriers that had essentially grounded the mainstream efforts of the mid-1960s.

2.6 ARTIFICIAL NEURAL NETWORKS

In its most general form a *network of artificial neurons,* as information processing units, is inspired by the way in which the brain performs a particular task or function of interest. Aleksander and Morton [1990] define a neural network in a broader sense such that the neural nets of the actual brain are included in the field of study and provide room for a consideration of biological findings. Their definition is as follows:

> Neural computing is the study of networks of adaptable nodes which, through a process of learning from task examples, store experiential knowledge and make it available for use.

Learning algorithms are procedures used for modifying synaptic weights in an orderly fashion. Linear adaptive filter theory, which is widely applied in various fields [Haykin, 1991; Widrow and Sterns, 1985], uses a similar approach. However, neural networks which are inspired by the brain (where cells die and regenerate all the time) can also incorporate *plasticity* (ability to modify its own topology).

The following statement from Hecht-Nielsen [1990, p. 2] defines neural networks as follows:

> A *neural network* is a parallel, distributed information processing structure consisting of *processing elements* (which can possess a local memory and can carry out localized information processing operations) interconnected via unidirectional signal channels called *connections.* Each processing element has a single output connection that branches ("fans out") into as many collateral connections as desired; each carries the same signal — the *processing element output signal.* The processing element output signal can be of any mathematical type desired. The information processing that goes

on within each processing element can be defined arbitrarily with the restriction that it must be completely local; that is it must depend only on the current values of the input signals arriving at the processing element via impinging connections and on values stored in the processing element's local memory.

Figure 2.15 shows a general model of a neuron with synaptic connections and the simple processing unit which is capable of performing nonlinear transformations.

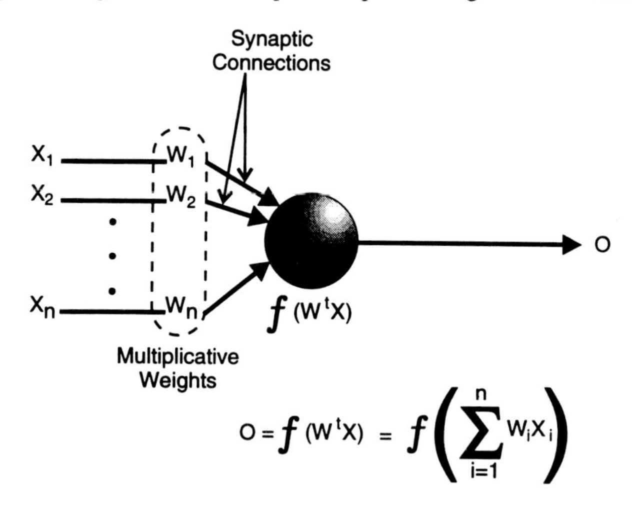

$$O = f(W^t X) = f\left(\sum_{i=1}^{n} W_i X_i\right)$$

FIGURE 2.15 A general neuron capable of providing continuous-valued outputs.

A typical computer, in its most basic form, consists of a central processing unit (CPU) that can execute a wide variety of instructions. The CPU also addresses an array of memory locations in order to load and store information. The serial computer also consists of a memory unit where data and instructions are stored in various locations. In a typical computation cycle, the CPU fetches an instruction and any data required by that instruction. Then it executes the instruction and stores the results, if any, back into appropriate memory cells.

In contrast, neural networks do not consist of a separate memory array for the storage of information (i.e., data and instruction). They also do not have a single general purpose CPU capable of executing a wide variety of instructions. Instead, a neural net is composed of many simple processing elements that can typically perform only the weighted summation of inputs.

Unlike a traditional computer, neural nets do not execute a series of instructions but, rather, respond to the variety of inputs presented to them. Neural nets do not store the results in specific memory locations, but represent information through the overall state of the network after it has reached some equilibrium condition.

In the case of a traditional computer we can easily access information from, for instance, memory location addressed by 3541 and retrieve the current value of the variable X. Since neural nets store information through various interconnections between processors and the importance of each input to the processing element, information is more a function of the architecture or structure of the network rather than the contents of a particular memory location in the network.

On the other hand, this very approach of accessing items in memory using their addresses (location s in memory) makes it hard for traditional computers to discover the address of an item from an arbitrary subset of its contents. In case of Artificial Neural Network Systems (ANNS), since the idea that the basic method for retrieving items is via their address is abandoned in favor of parallel computation using interconnected simple elements it is easy to achieve content-addressable memory, i.e., partial contents of an item can be used to retrieve the remaining contents.

Neural networks are also referred to in literature as artificial neural systems, connectionist networks, connectionism, neurocomputers, parallel distributed processors (systems), layered adaptive systems, self-organizing networks, network computation, neuromorphic systems, etc.

In other words, *artificial neural systems* are physical cellular systems which can acquire, store, and utilize experiential knowledge. The following characteristics of neural networks have played an important role in a wide variety of applications:

1. **Adaptiveness** — Powerful learning algorithms and self-organizing rules allow it to self-adapt as per the requirements in a continually changing environment.
2. **Nonlinear processing** — Ability to perform tasks involving nonlinear relationships and noise-immunity make it a good candidate for classification and prediction.
3. **Parallel processing** — Architectures with a large number of processing units enhanced by extensive interconnectivity provide for concurrent processing as well as parallel distributed information storage.

REFERENCES AND BIBLIOGRAPHY

Albus, J., *Brains, Behavior and Robotics,* McGraw-Hill, Peterborough, NH, 1981.

Aleksander, I. and Morton, H., *An Introduction to Neural Computing,* Chapman & Hall, London, 1990.

Amari, S. I., "Learning patterns and pattern sequences by self-organizing nets," *IEEE Trans. Comput.,* vol. 21, pp. 1197–1206, 1972.

Amari, S. I., "Neural theory of association and concept formation," *Biol. Cybern.,* vol. 26, pp. 175–185, 1977.

Anderson, J. A., "A simple neural network generating an interactive memory," *Math. Biosci.,* vol. 14, pp. 197–220, 1972.

Anderson, J. A. and Bower, G. H., *Human Associative Memory,* V. H. Vincent, Washington, D.C., 1973.

Anderson, J. A., Introduction to Practical Neural Modeling, MIT Press, Cambridge, MA, 1994.

Anninos, P. A., Beek, B., Csermely, T. J., Harth, E. M., and Pertile, G., "Dynamics of neural structures," *J. Theor. Biol.,* vol. 26, pp. 121–148, 1970.

Arbib, M. A., *Brains, Machines, and Mathematics,* 2nd ed., Springer-Verlag, Berlin, 1987.

Aspray, W. and Burks, A., Papers on John von Neumann on Computing and Computer Theory, Charles Babbage Institute Reprint Series for *History of Computing,* vol. 12, MIT Press, Cambridge, MA, 1986.

Byrne, J. H. and Schultz, S. G., *An Introduction to Membrane Transport and Bioelectricity,* Raven Press, New York, 1988.

Carlson, N. R., *Physiology of Behavior,* Allyn & Bacon, Boston, 1977.

Carpenter, G. A. and Grossberg, S. "A massively parallel architecture for a self-organizing neural pattern recognition machine," *Comput. Vision, Graphics Image Process.,* vol. 37, pp. 54–115, 1987a.

Carpenter, G. A. and Grossberg, S. "ART 2: Self-organization of stable category recognition codes for analog input patterns," *Appl. Optics,* vol. 26, pp. 4919–4930, 1987b.

Churchland, P. S. and Sejnowski, T. J., *The Computational Brain*, MIT Press, Cambridge, MA, 1992.

Churchland, P. S., *Neurophilosophy: Toward a Unified Science of the Mind/Brain*, MIT Press, Cambridge, MA, 1986.

Cowan, J. D., "Neural Networks: The early days," in *Advances in Neural Information Processing Systems 2*, D. S. Touretzky (Ed.), Morgan Kaufmann, San Mateo, CA, pp. 828–848, 1990.

Dayhoff, J., *Neural Network Architectures: An Introduction*, Van Nostrand Reinhold, New York, 1990.

Duda, R. O. and Hart, P. E., *Pattern Classification and Scene Analysis*, John Wiley & Sons, New York, 1973.

Fukushima, K., "Cognition: a self-organizing multilayered neural network," *Biol. Cybern.*, vol. 20, pp. 121–136, 1975.

Fukushima, K., "Neocognition: a self-organizing neural network model for a mechanism of pattern recognition unaffected by shift in position," *Biol. Cybern.*, vol. 36, pp. 193–202, 1980.

Grossberg, S., "Nonlinear difference-differential equations in prediction and learning theory," *Proc. Natl. Acad. Sci.*, vol. 58, pp. 1329–1334, 1967.

Grossberg, S., "A prediction theory for some nonlinear functional-difference equations," *J. Math. Anal. Appl.*, vol. 22, pp. 643–694, 1968.

Harth, E., "Visual perception: a dynamic theory," *Biol. Cybern.*, vol. 22, pp. 169–180, 1976.

Harth, E. M., Csermely, T. J., Beek, B., and Lindsay, R. D., "Brain functions and neural dynamics," *J. Theor. Biol.*, vol. 26, pp. 93–120, 1970.

Haykin, S., *Adaptive Filter Theory*, 2nd ed., Prentice-Hall, Englewood Cliffs, NJ, 1991.

Haykin, S., *Neural Networks: A Comprehensive Foundation*, Macmillan College Publishing, New York, 1994.

Hebb, D. O., *The Organization of Behavior: A Neuropsychological Theory*, John Wiley & Sons, New York, 1949.

Hetch-Nielsen, R., *Neurocomputing*, Addison-Wesley, Reading, MA, 1990.

Hopfield, J. J., "Neural networks and physical systems with emergent collective computational abilities," *Proc. Natl. Acad. of Sci.*, vol. 79, pp. 2554–2558, 1982.

Hopfield, J. J., "Neurons with graded response have collective computational properties like those of two-state neurons," *Proc. Natl. Acad. of Sci.*, vol. 81, pp. 3088–3092, 1984.

Hubel, D. H. and Wiesel, T. N., "Receptive fields, binocular interaction and functional architecture in cat's visual cortex," *J. Physiol.*, vol. 160, pp. 106–154, 1962.

Hubel, D. H. and Wiesel, T. N., "Receptive fields and functional architecture in two non-striate visual areas (18 and 19) of the cat," *J. Neurophysiol.*, vol. 28, pp. 229–298, 1965.

Kandel, E. R. and Schwartz, J. H., *Principles of Neural Science*, 3rd ed., Elsevier, New York, 1991.

Katz, B., *Nerve, Muscle and Synapse*, McGraw-Hill, New York, 1966.

Kawato, M., Furukawa, K., and Suzuki, R., "A hierarchical neural-network model for control and learning of voluntary movement," *Biol. Cybern.*, vol. 57, pp. 169–185, 1987.

Kawato, M., Isobe, M., Maeda, Y., and Suzuki, R., "Coordinates transformation and learning control for visually-guided voluntary movement with iteration: a Newton-like method in function space," *Biol. Cybern.*, vol. 59, pp. 161–177, 1988.

Kessner, R., "A neural system analysis of memory storage and retrieval," *Psychol. Bull.*, vol. 80, pp. 177–203, 1973.

Koch, C. and Segev, I., *Methods in Neuronal Modeling: From Synapses to Networks*, MIT Press, Cambridge, MA, 1989.

Kohonen, T., "Correlation matrix memories," *IEEE Trans. Comput.*, vol. C-21, no. 4, pp. 353–359, 1972.

Kohonen, T., *Associative Memory: A System-Theoretical Approach*, Springer-Verlag, Berlin, 1977.

Kohonen, T., *Content-Addressable Memories*, Springer-Verlag, Berlin, 1980.

Kung, S. Y., *Digital Neural Networks*, Prentice-Hall, Englewood Cliffs, NJ, 1993.

Lashley, K., *Brain Mechanisms and Intelligence*, University of Chicago Press, Chicago, 1929.

Levine, D. S., *Introduction to Neural and Cognitive Modelling*, Lawrence Erlbaum, Hillside, NJ, 1991.

Levine, M., *Man and Machine Vision*, McGraw-Hill, New York, 1985.

MacGregor, R. J., *Neural and Brain Modeling*, Academic Press, London, 1987.

Marr, D., *Vision*, W. H. Freeman, New York, 1982.

McCulloch, W. S. and Pitts, W., "A logical calculus of the ideas immanent in nervous activity," *Bull. Math. Biophys.*, vol. 5, pp. 115–133, 1943.

Mead, C. A., *Analog VLSI and Neural Systems,* Addison-Wesley, Reading, MA, 1989.
Minsky, M. L. and Pappert, S. A., *Perceptrons,* MIT Press, Cambridge, MA, 1969.
Mountcastle, V. B., "Modality and topographic properties of single neurons of cat's somatic sensory cortex," *J. Neurophysiol.,* vol. 20, pp. 408–434, 1957.
Nakano, K., "Associatron — A model of associative memory," *IEEE Trans. Syst. Man and Cybern.,* Vol. SMC-2, pp. 380–388, 1972.
Nauta, W. J. H. and Feirtag, M., *Fundamental Neuroanatomy,* W. H. Freeman, New York, 1986.
Neelakanta, P. S. and DeGroff, D. F., *Neural Network Modeling: Statistical Mechanics and Cybernetic Perspectives,* CRC Press, Boca Raton, FL, 1994.
Nilsson, N. J., *Learning Machines: Foundations of Trainable Pattern-Classifying Systems,* McGraw-Hill, New York, 1965.
Pandya, A. S. and Szabo, R., "A fast learning algorithm for neural network applications," *Proc. of IEEE Conf. Syst. Man Cybern.,* pp. 1569–1573, 1991.
Pandya, A. S. and Venugopal, K. P., "A stochastic parallel algorithm for supervised learning in neural networks," *IEICE Trans. Inf. Syst.,* vol. E77-D, no 4, pp. 376–384, 1994.
Rámon Y. Cajál, "Les preuves objectives de l'unit'e anatomique des cellules nurveuses," *Trob. Lab. Inest. Biol. Univ. Madrid,* vol. 29, pp. 1–37, 1934. (translation: Purkiss, M. V. and Fox, C. A., Madrid: Instituto "Ramon y Cajal", 1954).
Rochester, N., Holland, J. H., Haibt, L. H., and Duda, W. L., "Tests on a cell assembly theory of the action of the brain, using a large digital computer," *IRE Trans. Inf. Theory,* vol. IT-2, pp. 80–93, 1956.
Rosenblatt, F., "The Perceptron: a probabilistic model for information storage and organization in the brain," *Psychol. Rev.,* vol. 65, pp. 386–408, 1958.
Rosenblatt, F., *Principles of Neurodynamics,* Spartan Books, Washington, D.C., 1962.
Rumelhart, D. E. and McClelland, J. L. (Eds.), *Parallel Distributed Processing: Explorations in the Microstructure of Cognition,* Vol. 1, MIT Press, Cambridge, MA, 1986.
Shepherd, G. M., *Neurobiology,* Oxford University Press, New York, 1983; 2nd ed., 1988.
Shepherd, G. M., *The Synaptic Organization of the Brain,* Oxford University Press, New York, 1990a.
Shepherd, G. M., The significance of real neuron architectures for neural network simulations, in *Computational Neuroscience,* E. L. Schwartz (Ed.), MIT Press, Cambridge, MA, 1990b.
Shepherd, G. M. and Koch, C., Introduction to Synaptic Circuits, in *The Synaptic Organization of the Brain,* G. M. Shepherd (Ed.), Oxford University Press, New York, 1990.
Sherrington, C. S. *The Brain and Its Mechanism,* Cambridge University Press, New London, 1933.
Suga, N., "Cortical computational maps for auditory imaging," *Neural Networks,* vol. 3, pp. 3–21, 1990a.
Suga, N., Computations of Velocity and Range in the Bat Auditory System for Echo Location, in *Computational Neuroscience,* E. L. Schwartz (Ed.), MIT Press, Cambridge, MA, 1990b.
Simmons, J. A., Saillant, P. A., and Dear, S. P., "Through a bat's ear," *IEEE Spectrum,* vol. 29, no. 3, pp. 46–48, 1992.
Tanenbaum, A. S., *Structured Computer Architecture,* 3rd ed., Prentice-Hall, Englewood Cliffs, NJ, 1990.
Tou, J. T. and Gonzalez, R. C., *Pattern Recognition Principles,* Addison-Wesley, Reading, MA, 1974.
Thompson, R. F., *Foundations of Physiological Psychology,* Harper & Row, New York, 1967.
Truex, R. C. and Carpenter, M. B., *Human Neuroanatomy,* Williams & Wilkins, Baltimore, 1969.
von Neumann, J., *The Computer and the Brain,* Yale University Press, New Haven, CT, 1958.
Werbos, P. J., Beyond Regression: New Tools for Prediction and Analysis in the Behavioral Sciences, Unpublished Ph.D. dissertation, Harvard University, 1974.
Widrow, B., Generalization and Information Storage in Networks of ADALINE 'Neurons', in *Self-Organizing Systems,* M. C. Yowitz et al. (Eds.), Spartan Books, Washington, D.C., pp. 435–461, 1962.
Widrow, B. and Hoff, M. E., "Adaptive Switching Circuits," IRE WESCON Convention Record, pp. 96–104, 1960.
Widrow, B. and Sterns, S. D., *Adaptive Signal Processing,* Prentice-Hall, Englewood Cliffs, NJ, 1985.
Wiener, N., *Cybernetics: Or, Control and Communication in the Animal and the Machine,* John Wiley & Sons, New York, 1948.
Zurada, J. M., *Introduction to Artificial Neural Systems,* West Publ., New York, 1992.

3 Preprocessing

3.1 GENERAL

In this chapter we deal with the preprocessing of image data. Image compression, skeletonization, and edge detection all belong to the preprocessing stage of the overall recognition engine. Their relevance to pattern recognition lies in their ability to make the raw input data palatable to the neural network recognizer.

We begin now with a brief discussion about obtaining image data from a scanner.

3.2 DEALING WITH INPUT FROM A SCANNED IMAGE

Scanners are capable of producing image representation in a variety of formats. One of the most common of these is the bit map (.BMP) format. We briefly describe here how to convert a BMP file to the pixel map representation to be used with the recognition techniques that follow. Figure 3.1 below shows a sample scanned image represented as an 8-bit per pixel or 256 level gray scale. Figure 3.2 illustrates a sample image represented as 1-bit per pixel or monochrome.

FIGURE 3.1 Gray-scale image.

FIGURE 3.2 Monochrome image.

BMP files can be thought of as consisting of three main parts. First, a header provides essential information regarding what is to follow. Such information includes the width and depth of the pixel map, the number of bits per pixel, and a pointer to the beginning of the pixel data. The structure of a BMP header file is shown in listing 3.1 below. Second, the header file is followed by a color palette. This is typically represented in intensities in red, blue, and green (RGB), the composite of which designated the actual color. The size of the palette depends on the number of bits per pixel since this, the pixel value, serves as an index into the palette. For gray-scale images the RGB values will generally be equal and will serve as an intensity.

Listing 3.1

```
struct sHeaderBMP{
    char    id1,id2;              // "BM" identifies bitmap
    long    FileSize              // size of the file
    int     reserved[2];          // normally 0
    long    HeaderSize;           // offset to pixel data
                                  // in row X column format
    long    InfoSize;             // normally should be 0x28
    long    Width;                // # of columns
    long    Depth;                // # of rows
    int     BitPlanes;            // # of bit planes
    int     BitsPerPel;           // number of bits per pixel
    long    Compression;          // normally 0, not compressed
    long    ImageSize;            // size of the image.
    long    PelsPerMeterX;        // Resolution in X direction
    long    PelsPerMeterY;        // Resolution in Y direction
    long    ColorsUsed;           // # of colors used in image
    long    ImportantColors;      // # of colors that are important
}; //Palette (if any follows this)
```

A detailed discussion of file formats is beyond the scope of this book; however, a code has been provided on the companion diskette providing a limited read and write capability for BMP files. A number of books are available which are dedicated to discussions of the commonly used graphical file representation schemes. Among these are Rimmer [1992] and Levine [1994], both of which provide much greater detail on bitmap files and in addition discuss in depth many of the other possible file formats.

3.3 IMAGE COMPRESSION

Image compression is useful and often necessary to reduce the input image to a manageable size both for the recognizer and for subsequent preprocessor stages. The amount of compression needed is application specific. In context of the character recognition problem, experiments have shown that a 16×16 representation is sufficient to preserve the shape of the input image [Darwiche et al., 1992]. Many methods exist to perform image compression. These methods are frequently dichotomized into lossless and lossy methods. Image compression algorithms include the Huffman algorithm, run length code algorithms such as CCIT4, and sliding window compression algorithms such as LZ77. A comprehensive and useful discussion of these techniques may be found in Nelson [1992].

The following method [Darwiche et al., 1992] uses run length coding (RLC). Compression is performed first horizontally and then vertically. The image is assumed to be composed of a matrix of binary-valued pixels. We will have more to say about gray-scale images later in the discussion of edge detection. Run length coding builds a list by counting sequence of consecutive ones and zeroes. For example, the sequence 0 0 0 0 1 1 1 0 0 1 1 would be coded as 4, 3, 2, 2. In the horizontal pass this is done for each row and in the vertical pass, for each column. A compression ratio, C, must then be defined in both the horizontal and vertical directions. In the horizontal direction:

$$C = C_{\text{Horizontal}} = \frac{N_{\text{Horizontal}}}{M_{\text{Horizontal}}} = \frac{\text{Number of columns in input matrix}}{\text{Number of columns desired output}} \tag{1}$$

and in the vertical direction:

$$C = C_{\text{Vertical}} = \frac{N_{\text{Vertical}}}{M_{\text{Vertical}}} = \frac{\text{Number of rows in input matrix}}{\text{Number of rows desired output}} \tag{2}$$

The image is reconstructed from this modified RLC. Some care must be exercised in the application of this algorithm. In particular, round-off errors when calculating (run length)/C can result in an uneven number of pixels in the direction of compression. This effect is easily overcome by accumulating round-off errors as shown in the code sample in section 3.3.1 on the following two pages.

3.3.1 Image Compression Example

The listing below shows the class definition for the run length compression class. Its data members are a run length matrix as described above and an index giving the largest x index.

Listing 3.2

```
class RLCX {
 private:
  unsigned int RLmatrix[RLCXMAXX][RLCXMAXY];
   int       LargestX;
 public:
   RLCX();
  void Setup(unsigned char *data, int width, int height );
   void Clear();
  void Compress(int width, int height, double Scale);
  unsigned int Query(int x, int y){return RLmatrix[x][y];}
  unsigned int QueryIndx(){return LargestX;}
 };
```

The method of greatest interest here is "Compress". This is where the compression or scaling actually takes place. This method is shown below in listing 3.3. Note that the result of applying the scale factor is first rounded to the nearest integer value and that subsequently an aggregate error is accumulated to be applied when it has grown to be sufficiently large. In this case, sufficiently large is defined as being in excess of 1.0. The setup method although trivial is shown in the listing as well for context.

Figures 3.3(a)-(e) below illustrate the compression process on character "A":

```
000000000000111111000000000000000000
000000000011111111110000000000000000
000000000111111111111000000000000000
000000001111110011111100000000000000
000000011110000001111100000000000000
000000111000000000111110000000000000
000001111000000000111111000000000000
000011111000000000011111000000000000
000111100000000000111111000000000000
000111100000000000001111100000000000
001111100000000000000111111000000000
011111100000000000000111111000000000
011111100000000000000111111100000000
011111111111111111111111111111110000
111111111111111111111111111111111110
111111111111111111111111111111111110
111111011111111111111111111111111100
111110000111111111111111111111110000
111110000000000001011101111111100000
111110000000000000000000111111100000
111110000000000000000000011111100000
111110000000000000000000001111110000
111111000000000000000000001111110000
111110000000000000000000000111110000
111111000000000000000000000111110000
111111000000000000000000000111111000
111111000000000000000000000111111000
111111000000000000000000000011111000
111110000000000000000000000011111000
011110000000000000000000000001111000
011100000000000000000000000000000110
```

FIGURE 3.3(a) Input character.

```
12, 6, 18
10, 10, 16
9, 12, 15
8, 6, 2, 6, 14
7, 4, 6, 5, 14
6, 3, 9, 5, 13
5, 4, 9, 6, 12
4, 5, 10, 5, 12
3, 4, 11, 6, 11
3, 4, 13, 5, 11
2, 5, 14, 6, 9
1, 6, 14, 6, 9
1, 6, 14, 7, 8
1, 31, 4
0, 35, 1
0, 35, 1
0, 6, 1, 27, 2
0, 5, 4, 23, 4
0, 5, 12, 1, 1, 3, 1, 8, 5
0, 5, 19, 7, 5
0, 5, 20, 6, 5
0, 5, 21, 6, 4
0, 6, 20, 6, 4
0, 5, 22, 5, 4
0, 6, 21, 5, 4
0, 6, 21, 6, 3
0, 6, 21, 6, 3
0, 6, 22, 5, 3
0, 5, 23, 5, 3
1, 4, 24, 4, 3
1, 3, 26, 2, 4
```

FIGURE 3.3(b) RLC of lines.

```
5, 3, 8
4, 5, 7
4, 5, 7
3, 3, 1, 3, 6
3, 2, 3, 2, 6
3, 1, 4, 2, 6
2, 2, 4, 3, 5
2, 2, 5, 2, 5
1, 2, 5, 3, 5
1, 2, 6, 3, 4
1, 2, 6, 3, 4
0, 3, 6, 3, 4
0, 3, 6, 3, 4
0, 14, 2
0, 16
0, 16
0, 15, 1
0, 2, 2, 10, 2
0, 2, 5, 1, 1, 5, 2
0, 2, 9, 3, 2
0, 2, 9, 3, 2
0, 2, 9, 3, 2
0, 3, 9, 2, 2
0, 2, 10, 2, 2
0, 3, 9, 2, 2
0, 3, 9, 3, 1
0, 3, 9, 3, 1
0, 3, 10, 2, 1
0, 2, 10, 2, 2
0, 2, 11, 2, 1
0, 1, 12, 1, 2
```

FIGURE 3.3(c)
Division by compression ratio $C_{Horizontal}$.

```
0000011100000000
0000111110000000
0000111110000000
0001110111000000
0001100011000000
0001000011000000
0011000011000000
0011000001100000
0110000011100000
0110000001110000
0110000001110000
1110000001110000
1110000001110000
1111111111111100
1111111111111111
1111111111111111
1111111111111110
1100111111111100
1100000101111100
1100000000011100
1100000000011100
1100000000011100
1110000000001100
1100000000001100
1110000000001100
1110000000001110
1110000000001110
1110000000001100
1100000000001100
1100000000000110
1000000000000100
```

FIGURE 3.3(d)
Horizontal compressed character.

```
0000111110000000
0001111110000000
0001000011000000
0011000011100000
0110000001100000
0110000000110000
1111101011110000
1111111111111111
1111111111111110
1100000101111100
1100000000011100
1110000000011100
1110000000001100
1110000000001110
1110000000001110
1000000000000110
```

FIGURE 3.3(e)
Fully compressed character.

Listing 3.3

```
void RLCX::Compress(int width, int height, double Scale){
  int i,j,bpl,k;
  double erf;
  double r,frac,ipart;
  bpl=width>>3;
  for (j=0; j<height; j++) {
    erf=0;
    while ((RLmatrix[i][j] & 0x8000) != 0) {
      r=(double)(RLmatrix[i][j] & 0x7fff);
      r=Scale*r;
```

```
      frac=modf(r,&ipart);
      if (frac>0.5) {
        k=(int)ipart+1;
        /* erf = erf- (1.0-frac); */
      } else {
        k=(int)ipart;
        /* erf += frac; */
      } /* endif */
      erf = erf + (r-(double)k);
      if (erf >= 1.0) {
        k=k+1;
        erf=erf-1.0;
      } /* endif */
      if (erf <= -1.0) {
        k=k-1;
        erf=erf+1.0;
      } /* endif */
/*    RLmatrix[i][j] = (2*(RLmatrix[i][j] & 0x7fff))/3;  *
      RLmatrix[i][j] = k;
      RLmatrix[i][j] = RLmatrix[i][j] | 0x8000;
      i++;
    } /* endwhile */
  } /* endfor */
}

void RLCX::Setup(unsigned char *data, int width, int height ) {
  unsigned int Target,indx,Count,Query,i,j,x,Bit;
  unsigned char mask;

  Clear();
  x=0;
  for (j=0; j<height; j++) {
    Target=1;
    i=0;
    indx=0;
    Count=0;
    while (i<width) {
      //x=j*(width>>3)+ (i>>3);
      Bit=i&0x07;
      mask=0x80>>Bit;
      if ((data[x] & mask)!=0) Query=1;
                    else Query=0;
      if (Query==Target) {
        Count++;
        i++;
        if ((i&0x07)==0) x++;
      } else {
        RLmatrix[indx][j]=Count | 0x8000;
        if(indx>LargestX)LargestX=indx;
        indx++;
        if (Target==0) Target=1;
              else Target=0;
        Count=0;
      } /* endif */
    } /* endwhile */
    RLmatrix[indx][j]=Count | 0x8000;
  } /* endfor */
}
```

3.4 EDGE DETECTION

In this section we discuss the segmentation of gray-scale images into points, lines, and most importantly edges. This is often a crucial first step in the analysis of a gray-scale image. The approach in all three cases lies in detecting discontinuities in the digital image. To begin the discussion, consider a template defining an 8-neighborhood surrounding the pixel z_5 as follows:

$$\begin{pmatrix} Z_1 & Z_2 & Z_3 \\ Z_4 & Z_5 & Z_6 \\ Z_7 & Z_8 & Z_9 \end{pmatrix} \tag{3}$$

where the z_i's represent the gray level for the pixel.

The discontinuities representing features such as points and lines may be easily obtained by running a mask consisting of a set of weights over the image. This mask is as follows:

$$\begin{pmatrix} W_1 & W_2 & W_3 \\ W_4 & W_5 & W_6 \\ W_7 & W_8 & W_9 \end{pmatrix} \tag{4}$$

The response of the mask to any point in the image (centered at w_5) can then be computed as:

$$R = \sum_{i=1}^{9} w_i z_i \tag{5}$$

Clearly an isolated point may be detected by using a mask with the following weights:

$$\begin{pmatrix} -1 & -1 & -1 \\ -1 & 8 & -1 \\ -1 & -1 & -1 \end{pmatrix} \tag{6}$$

The point is detected when the response exceeds a predetermined threshold, T, that is, when:

$$|R| = T \tag{7}$$

In a similar fashion the following templates can be used to detect lines:

$$\begin{pmatrix} -1 & -1 & -1 \\ 2 & 2 & 2 \\ -1 & -1 & -1 \end{pmatrix} \begin{pmatrix} -1 & -1 & 2 \\ -1 & 2 & -1 \\ 2 & -1 & -1 \end{pmatrix} \begin{pmatrix} -1 & 2 & -1 \\ -1 & 2 & -1 \\ -1 & 2 & -1 \end{pmatrix} \begin{pmatrix} 2 & -1 & -1 \\ -1 & 2 & -1 \\ -1 & -1 & 2 \end{pmatrix}$$

Horizontal 45° Vertical –45° (8)

We now proceed to the more interesting and more important subject of edge detection. An edge is a boundary between two regions with relatively distinct gray levels. That is, we are looking for areas in which the gray level changes relatively abruptly. Since we are considering the rate of change of the gray level, it is natural to think of the derivative. Figure 3.4 illustrates the behavior of the first and second derivative at the edges of relatively homogeneous gray level regions.

Notice that the first derivative is positive at transitions from low to high transitions (dark to light), negative on high to low transitions, and zero elsewhere. Thus the magnitude of the first derivative is itself an edge detector. The second derivative has the property of a zero crossing at the midpoint of transitions between levels. We will say more about edge detection based on this property of the second derivative later in this section.

Now let us generalize the discussion to the more meaningful domain of two-dimensional images. In this case we proceed as before but we must now use the gradient in place of the first derivative and the Laplacian in place of the second derivative. The gradient of a function $f(x, y)$ is defined as follows:

$$\nabla f = \begin{bmatrix} G_x \\ G_y \end{bmatrix} = \begin{bmatrix} \frac{\partial f}{\partial x} \\ \frac{\partial f}{\partial y} \end{bmatrix} \tag{9}$$

The magnitude of the gradient is given by:

$$\nabla f = |\nabla \mathbf{f}| = \left[G_x^2 + G_y^2\right]^{1/2} \tag{10}$$

and the direction of the gradient vector is then

$$\alpha(x, y) = \tan^{-1}\left(\frac{G_x}{G_y}\right) \tag{11}$$

In order to apply these operators, the function $f(x, y)$ must be a closed form expression which we can differentiate analytically. Clearly, we must now be able to approximate the partial derivatives $\partial/_{\partial x}$ and $\partial/_{\partial y}$ over an array of constant data representing

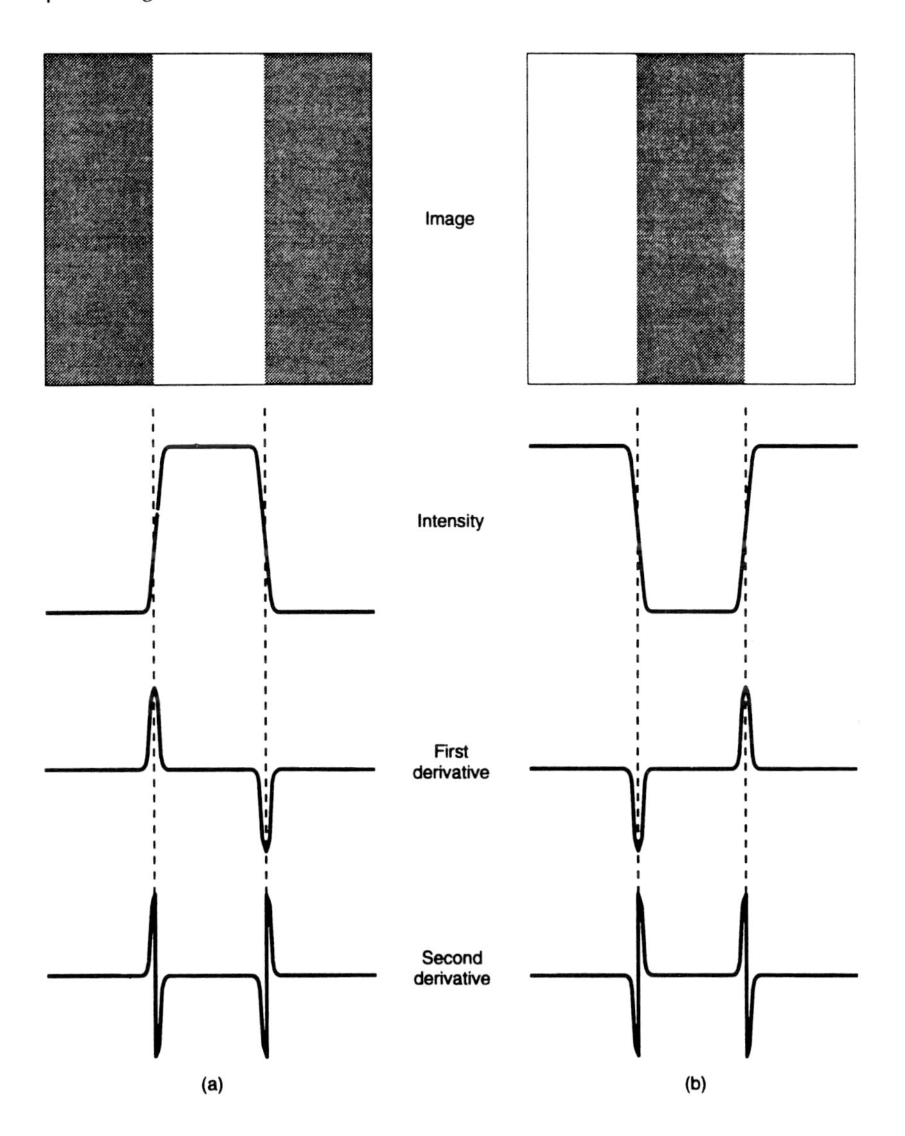

FIGURE 3.4 Edge detection using first- and second-order derivatives. (Adapted from Gonzalez and Woods [1992].)

the intensities of an image. Referring back to Figure 3.4, the most obvious and straightforward way to accomplish this is simply as follows:

$$G_x = Z_5 - Z_8$$
$$G_y = Z_5 - Z_6 \tag{12}$$

A variation of the above method which can be formulated as the response to a 3 × 3 mask is as follows:

$$G_x = Z_7 + Z_8 + Z_9 - (Z_1 + Z_2 + Z_3)$$

$$G_y = Z_3 + Z_6 + Z_9 - (Z_1 + Z_4 + Z_7) \tag{13}$$

The corresponding masks (known as Prewitt operators) are as follows:

$$\begin{pmatrix} -1 & -1 & -1 \\ 0 & 0 & 0 \\ 1 & 1 & 1 \end{pmatrix} \& \begin{pmatrix} -1 & 0 & 1 \\ -1 & 0 & 1 \\ -1 & 0 & 1 \end{pmatrix} \tag{14}$$

Another approach to approximation called Sobel operators may be useful due to their smoothing effect. The discreet operators corresponding to the partial derivatives are given by:

$$G_x = Z_7 + 2Z_8 + Z_9 - (Z_1 + 2Z_2 + Z_3)$$

$$G_y = Z_3 + 2Z_6 + Z_9 - (Z_1 + 2Z_4 + Z_7) \tag{15}$$

and the corresponding masks as:

$$\begin{pmatrix} -1 & -2 & -1 \\ 0 & 0 & 0 \\ 1 & 2 & 1 \end{pmatrix} \& \begin{pmatrix} -1 & 0 & 1 \\ -2 & 0 & 2 \\ -1 & 0 & 1 \end{pmatrix} \tag{16}$$

Finally the second-order derivative, the Laplacian, of a two-dimensional function $f(x, y)$ is given by:

$$\nabla^2 f = \frac{\partial^2 f}{\partial x^2} + \frac{\partial^2 f}{\partial y^2} \tag{17}$$

In discreet form the Laplacian can be calculated as:

$$\nabla^2 f = 4z_5 - (z_2 + z_4 + z_6 + z_8) \tag{18}$$

with a corresponding mask as follows:

$$\nabla^2 f = \begin{pmatrix} 0 & -1 & 0 \\ -1 & 4 & -1 \\ 0 & -1 & 0 \end{pmatrix} \tag{19}$$

Having described in depth some methods for calculating derivatives, the gradient and the Laplacian over discrete data, we now proceed to our first edge detection example. Note from Figure 3.4 that the magnitude of the first derivative effectively identifies the edge in the one-dimensional example. It can then be seen that where the magnitude of the first derivative exceeds some certain threshold, a point belonging to an edge has been detected. Generalizing to two dimensions, we will consider a point as belonging to an edge when the magnitude of its gradient exceeds a threshold, T. That is:

$$\sqrt{\left(\frac{\partial f}{\partial x}\right)^2 + \left(\frac{\partial f}{\partial y}\right)^2} = \sqrt{G_x^2 + G_y^2} > T \tag{20}$$

Figures 3.5(a) and (b) shown below illustrate the application of this technique. Figure 3.5 (a) shows the original image and Figure 3.5 (b) shows the result when only pixels satisfying the equation above are kept.

FIGURE 3.5(a) Original image. **FIGURE 3.5(b)** Image after edge detection.

Another method of edge detection involves taking the convolution of the image with the Laplacian of a two-dimensional Gaussian, $h(x, y)$. The two-dimensional Gaussian has the form:

$$h(x,y) = e^{-\left(\frac{x^2 y^2}{2\sigma^2}\right)} \tag{21}$$

The Laplacian then is

$$\nabla^2 h = \frac{x^2 + y^2 - \sigma^2}{\sigma^4} e^{-\left(\frac{x^2+y^2}{2\sigma^2}\right)} \tag{22}$$

The Laplacian is shown graphically in Figure 6 below:

In one dimension the continuous convolution of two functions $f(x)$ and $g(x)$ is defined as:

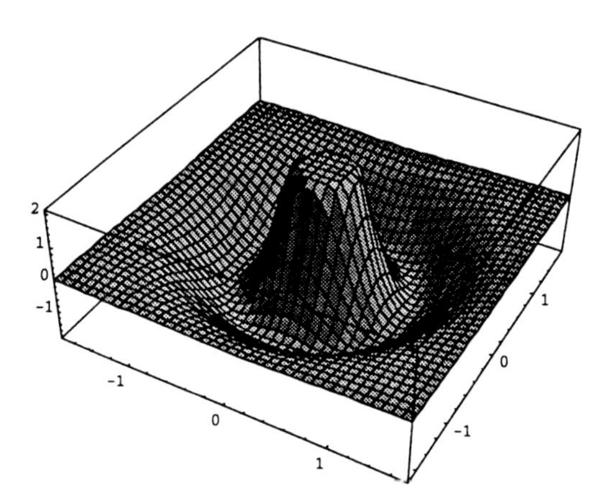

FIGURE 3.6(a) Three-dimensional view of $\nabla^2 h$.

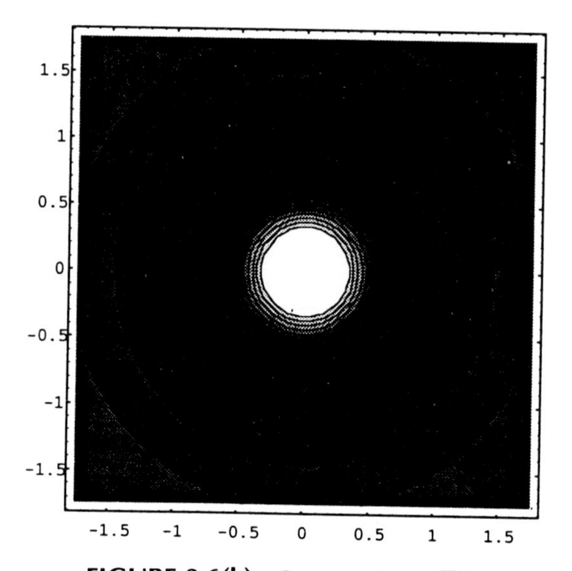

FIGURE 3.6(b) Contour view of $\nabla^2 h$.

$$f(x) \bullet g(x) = \int_{-\infty}^{\infty} f(\alpha) g(x - \alpha) \, d\alpha \tag{23}$$

For the discrete case this becomes:

$$f(x) \bullet g(x) = \sum_{m=0}^{M} f(m) g(x - m) \tag{24}$$

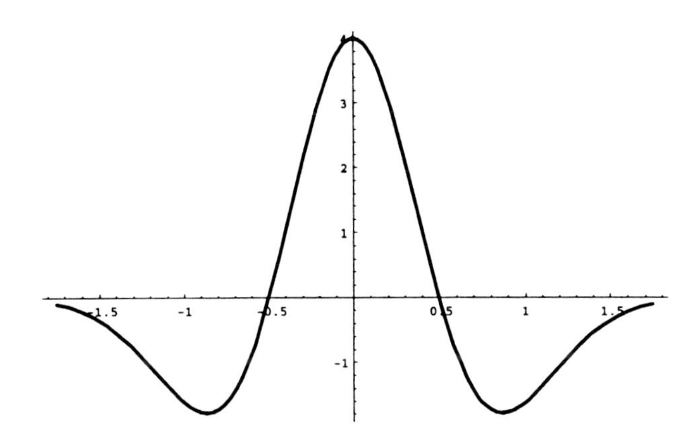

FIGURE 3.6(c) Two-dimensional view of $\nabla^2 h$.

Clearly, in the continuous case, equation 22 can be extended to two dimensions as follows:

$$f(x,y) \bullet g(x,y) = \int_0^\infty \int_0^\infty f(\alpha,\beta)\, g(x-\alpha, y-\beta)\, d\alpha\, d\beta \tag{25}$$

The corresponding generalization of equation to two dimensions for the discrete case is

$$f(x,y) \bullet g(x,y) = \sum_{m=0}^{M} \sum_{n=0}^{N} f(m,n)\, g(x-m, y-n) \tag{26}$$

As stated earlier, we may detect edges by taking the convolution of the image with the Laplacian $\nabla^2 h(x, y)$. Once the convolution matrix is obtained, negative values are set to black and positive values are set to white. The edges are then easily obtained by discarding all but the perimeter pixels where black pixels are adjacent to white ones.

3.5 SKELETONIZING

Skeletonizion is an important approach to representing the shape of a plane region. The objective of skeletonizing is to reduce the representation of a region to a chain of single pixel width while preserving all other relevant features. Notice that after skeletonizing we may easily represent the region as a graph. Below, Figure 3.7(a) shows a binary valued image of a tree and Figure 3.7(b) shows the same tree after skeletonization.

FIGURE 3.7(a) Original image. **FIGURE 3.7(b)** Image after skeletonization.

The width of strokes within characters provides little useful information for the recognizer and may even serve to obscure the classification. Thus there is a sense in which skeletonization, sometimes termed character thinning in this context, can be thought of as removing unnecessary or redundant portions of the input. Additionally the extraction of geometric features (i.e., intersections, endpoints, and loops) is facilitated by this process. Character thinning will ideally reduce the character representation to a single pixel width while preserving all other relevant features. One method for doing this involves eroding the edges of the image until only the skeleton of the character remains. This can be done by raster scanning the image and checking with templates stored in memory as discussed in LeCun et al. [1989]. Each template provides a set of conditions under which the center pixel should be deleted. Sample templates and the results of the character thinning are shown in Figure 3.8 below.

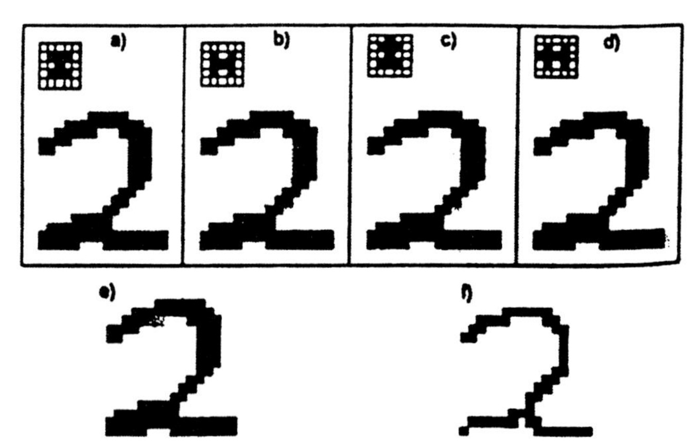

FIGURE 3.8 Four sample templates used by LeCun and the results of character thinning. (From LeCun, Y. et al., *IEEE Commun. Mag.,* vol. 27, no. 11, pp. 41–46, 1989. With permission.)

Darwiche et al. [1992] describe a method of character thinning in detail. In order to highlight the thinning process with an example, we shall follows the presentation found in this reference. Four pixel bands are defined as follows:

$$\begin{pmatrix} X & 0 & X \\ X & 1 & X \\ X & X & X \end{pmatrix} \begin{pmatrix} X & X & X \\ X & 1 & 0 \\ X & X & X \end{pmatrix} \begin{pmatrix} X & X & X \\ X & 1 & X \\ X & 0 & X \end{pmatrix} \begin{pmatrix} X & X & X \\ 0 & 1 & X \\ X & X & X \end{pmatrix}$$

$$\text{NORTH} \qquad \text{EAST} \qquad \text{SOUTH} \qquad \text{WEST} \qquad (27)$$

A "1" pixel in the character bit map matches the band if it has a "0" pixel in the specified direction. That "1" pixel will then be stripped off if and only if doing so would not cause any of the neighboring pixels (*X*s) that are 1's to become disconnected.

The subject of character thinning is somewhat larger than the above discussion may suggest as seen in the work of Jang and Chin [1990]. Further we may note that caution must be exercised when applying thinning techniques. Pattern thinning algorithms may introduce distortion into the pattern resulting in spurious features [Liao et al., 1992].

3.5.1 Thinning Example

Below in listing 3.4 is the class definition for the thinner code along with the most significant methods. A sample main routine is also shown in the listing. The thinning algorithm is a variation of that discussed in section 3.5. Note that when thinner is used it is actually invoked twice. The first invocation uses the Thin1 method which is slightly less aggressive than the Thin2 method that is invoked subsequently. We have found that taking this approach reduces a tendency sometimes found in thinning algorithms to erode away details in the target image that are salient to recognition. The complete code is as usual available on the companion diskette.

Listing 3.4

```
class cThinner {
private:
 int   xMax, yMax;                 // Size of the bitmap
 int   PelxMax, PelyMax;
 int   PelxMin, PelyMin;
 int   M[32][32];                  // Space for the whole bitmap
 int   M1[3][3];                   // 3x3 receptive field — Template sized window
 tMap  Map[7];
 tgrid ZeroPosn[4];
 int   ApplyTemplate(int, int, int);
```

```
    void   Rotate(int);
    int    isConnected();
    void   buildMap();
    int    isPath(int, int);
    int    WeakCond(int x, int y);
    int    StrongCond(int x, int y);

  public:
    cThinner();
    void   SetGridSize(int Mx, int My){xMax=Mx; yMax=My;}
    void   Store(int x,int y,int Pel){M[x][y]=Pel;}
    int    QueryPel(int x,int y){return M[x][y];}
    void   Thin1();
    void   Thin2();
    int    doScan(int);
    void   ShowM();
    void   fShowM(FILE *);
    void   ShowMap();
    void   LocateEndPts();
    void   LocateIntersects();
    void   SetCharExtents();
    int    doScanLR();
    int    doScanInOut();
    int    ApplyAllTemplates(int, int);
  };

  cThinner::cThinner(){
   ZeroPosn[0].x=1;   ZeroPosn[0].y=0;            // SOUTH
   ZeroPosn[1].x=2;   ZeroPosn[1].y=1;            // EAST
   ZeroPosn[2].x=1;   ZeroPosn[2].y=2;            // NORTH
   ZeroPosn[3].x=0;   ZeroPosn[3].y=1;            // WEST
   Map[0].Next1= 1;   Map[0].Next2=-1;
   Map[1].Next1= 3;   Map[1].Next2= 2;
   Map[2].Next1= 3;   Map[2].Next2=-1;
   Map[3].Next1= 5;   Map[3].Next2= 4;
   Map[4].Next1= 5;   Map[4].Next2=-1;
   Map[5].Next1= 6;   Map[5].Next2=-1;
   Map[6].Next1=-1;   Map[6].Next2=-1;
  }

  void cThinner::Thin1(){
   int Tid;
   int dirty;
  SetCharExtents();
  dirty=1;
  while (dirty) {
   //dirty=doScanLR();
   dirty=doScanInOut();
   } /* endwhile */
  }

  void cThinner::Thin2(){
   int dirty;
  dirty=1;
  while (dirty) {
   dirty =doScan(1);
   dirty+=doScan(3);
   dirty+=doScan(0);
```

```
  dirty+=doScan(2);
} /* endwhile */
}

int cThinner::WeakCond(int x, int y){
  int i,sum;

sum=0;
if ((x==PelxMax)||(x==PelxMin)) {
  for (i=0; i<=PelyMax; i++) {
    sum+=M[x][i];
    } /* endfor */
  if (sum<2) return 0;
  } /* endif */
sum=0;
if ((y==PelyMax)||(y==PelyMin)) {
  for (i=0; i<=PelyMax; i++) {
    } /* endfor */
  if (sum<2) return 0;
  } /* endif */
return 1;
}


int cThinner::StrongCond(int x, int y){
if ((x>PelxMin) && (x<PelxMax) && (y>PelyMin) &&(y<PelyMax))
  return 1;
else
  return 0;
}

int cThinner::doScanLR(){
 int x1,y1,x2,y2;
 int dirty;
 char c;

dirty=0;
for (x1=1; x1<xMax-1; x1++) {
  x2= xMax-x1;
  for (y1=1; y1<yMax-1; y1++) {
    //if ((x1>PelxMin) &&(x1<PelxMax) && (y1>PelyMin) &&(y1<PelyMax)) {
    if (WeakCond(x1,y1)) {
     if ((ApplyTemplate(x1,y1,3)) || (ApplyTemplate(x1,y1,0))|| (ApplyTemplate(x1,y1,2)))
           dirty=1;
      } /* endif */
    y2=y1;
    //if ((x2>PelxMin) &&(x2<PelxMax) && (y2>PelyMin) &&(y2<PelyMax)) {
    if (WeakCond(x2,y2)) {
      if (ApplyTemplate(x2,y2,1))
           dirty=1;
      } /* endif */
    } /* endfor */
  } /* endfor */
return dirty;
}
```

```
int cThinner::doScanInOut(){
  int Q1x,Q1y,Q2x,Q2y;
  int Q3x,Q3y,Q4x,Q4y;
  int CenterX1, CenterX2;
  int CenterY1, CenterY2;
  int dirty;
  int done;
  char c;

 CenterX1= PelxMin+(PelxMax-PelxMin)/2;
 CenterX2= CenterX1+1;
 CenterY1= PelyMin+(PelyMax-PelyMin)/2;
 CenterY2=CenterY1+1;
 Q1x=CenterX2; Q2x=CenterX2; Q3x=CenterX1; Q4x=CenterX1;
 Q1y=CenterY2; Q2y=CenterY1; Q3y=CenterY1; Q4y=CenterY2;

 dirty=0; done=0;
 while (!done) {
 // Apply the templates
 if (WeakCond(Q1x,Q1y) && (Q1x<=PelxMax))
   dirty |= ApplyAllTemplates(Q1x, Q1y);
 if (WeakCond(Q2x,Q2y) && (Q2x<=PelxMax))
   dirty |= ApplyAllTemplates(Q2x, Q2y);
 if (WeakCond(Q3x,Q3y)&& (Q3x>=PelxMin))
   dirty |= ApplyAllTemplates(Q3x, Q3y);
 if (WeakCond(Q4x,Q4y)&& (Q4x>=PelxMin))
   dirty |= ApplyAllTemplates(Q4x, Q4y);
 // Move the templates
 Q1y++;
 if (Q1y>PelyMax) {
   Q1y=CenterY2;
   Q1x++;
   }
 Q2y—;
 if (Q2y<PelxMin) {
   Q2y=CenterY1;
   Q2x++;
   }
 Q3y—;
 if (Q3y<PelyMin) {
   Q3y=CenterY1;
   Q3x—;
   }
 Q4y++;
 if (Q4y>PelyMax) {
   Q4y=CenterY2;
   Q4x—;
   }

 if ((Q1x>PelxMax) && (Q2x>PelxMax) && (Q3x<PelxMin) && (Q4x<PelxMin))
   done=1;
 } /* endwhile */
return dirty;
}

int cThinner::doScan(int Tid){
 int x,y;
 int dirty;
```

```
dirty=0;
for (y=1; y<yMax-1; y++) {
  for (x=1; x<xMax-1; x++) {
    if (ApplyTemplate(x,y,Tid))
          dirty=1;
    } /* endfor */
  } /* endfor */
return dirty;
}

int cThinner::ApplyAllTemplates(int x, int y){
int rc;

rc=0;
if (ApplyTemplate(x, y, 1)) rc=1;
if (ApplyTemplate(x, y, 3)) rc=1;
if (ApplyTemplate(x, y, 0)) rc=1;
if (ApplyTemplate(x, y, 2)) rc=1;
return rc;
}
/*****************************************************************************
* int cThinner::ApplyTemplate(int x, int y, int Tid)                        *
*                                                                           *
*  Removes center pixel when template is applicable                         *
*                                                                           *
*  x,y locates center of 3x3 template placement                             *
*  Tid is the templates identification                                      *
*****************************************************************************/
int cThinner::ApplyTemplate(int x, int y, int Tid){
 int i,j,sum;

sum=0;
for (j=0; j<3; j++) {
  for (i=0; i<3; i++) {
    M1[i][j] = M[x+i-1][y+j-1];
    sum+=M1[i][j];
  } /* endfor */
} /* endfor */

if(M1[1][1] != 1) return 0;                                    //Template mismatch
if (M1[ZeroPosn[Tid].x][ZeroPosn[Tid].y] != 0) return 0;       //Template mismatch
if(sum<=2) return 0;                                           // never delete an isolated pixel
if(sum==8) return 0;                                           // don't hollow out the core of fat lines
Rotate(Tid);
if (isConnected()) {
  M[x][y]=0;                                                   // Strip the pixel
  return 1;                                                    // DIRTY
  }
 else {
  return 0;                                                    // NOT DIRTY
  } /* endif */
return 0;
}

void cThinner::Rotate(int Tid){
 int t,i,j,m,n;
 int Temp[3][3];
```

```
for (t=0; t<Tid; t++) {
  for (i=0; i<3; i++) {
     for (j=0; j<3; j++) {
       Temp[i][j] = M1[2-j][i];
       } /* endfor */
    } /* endfor */
  for (m=0; m<3; m++) {
    for (n=0; n<3; n++) {
      M1[m][n] = Temp[m][n];
      } /* endfor */
    } /* endfor */
  } /* endfor */
}

void cThinner::buildMap(){
  Map[0].Pel=M1[0][0];
  Map[1].Pel=M1[0][1];
  Map[2].Pel=M1[0][2];
  Map[3].Pel=M1[1][2];
  Map[4].Pel=M1[2][2];
  Map[5].Pel=M1[2][1];
  Map[6].Pel=M1[2][0];
}

int cThinner::isPath(int i, int j) {
  int k;

k=Map[i].Next1;
while ((k<j) && (k>-1)) {
  if (Map[k].Pel) {
     if (Map[k].Next1==j) return 1;
     if (Map[k].Next2==j) return 1;
    k=Map[k].Next1;
     }
   else {
     return 0;
     } /* endif */
  } /* endwhile */
printf("ERROR:Problem in isPath");
return 1;
}

int cThinner::isConnected(){
  int i,j,rc;
rc=1;
buildMap();
for (i=0; i<=4; i+=2) {
  for (j=i+2; j<=6; j++) {
    if ((Map[i].Pel==1) && (Map[j].Pel==1)) {
       if (!isPath(i,j)) rc=0;
       }
     else {
       } /* endif */
    } /* endfor */
  } /* endfor */
for (i=1; i<=3; i+=2) {
  for (j=i+3; j<=6; j++) {
    if ((Map[i].Pel==1) && (Map[j].Pel==1)) {
       if (!isPath(i,j)) rc=0;
       }
```

```
        else {
          } /* endif */
        } /* endfor */
      } /* endfor */
    return rc;
    }

    int main(int argc, char *argv[])
    {
    .
    cThinner      Thinner;                      // create an instance
          .
          .
      // setup thinner data here.
      // Then Invoke.....
          .
      Thinner.Thin1();           // run first w/ weak condition
      Thinner.Thin2();           // run with strong condition
          .
          .
          .
    }
```

3.6 DEALING WITH INPUT FROM A TABLET

Data from a pen tablet is seen in units of strokes. Characters are typically composed of one to four such strokes. Each stroke is represented as an ordered sequence of Cartesian coordinates, *x, y* pairs, moving through time. Figures 3.9(a) and (b) shown below illustrate a typical sequence of stroke data along with its graphical representation.

```
=S 19
 988 2836 986 2842 984 2848 984 2852 984 2854 982 2856 982 2858 980 2858
 982 2852 990 2826 998 2776 1010 2714 1024 2650 1030 2610 1036 2578 1040 2556
1040 2542 1040 2540 1034 2544
=S 15
862 2838 864 2842 866 2842 868 2844 876 2844 888 2844 904 2844 930 2844
986 2854 1034 2864 1078 2872 1112 2876 1140 2876 1154 2876 1162 2876
=S 44
1150 2786 1150 2790 1148 2790 1148 2792 1150 2792 1148 2792 1148 2790 1150 2786
1150 2782 1148 2786 1146 2788 1146 2792 1144 2794 1144 2796 1142 2796 1144 2794
1150 2780 1164 2740 1174 2702 1182 2664 1192 2634 1206 2600 1214 2584 1216 2568
1216 2552 1216 2540 1216 2538 1216 2540 1216 2542 1214 2544 1214 2546 1214 2554
1214 2564 1216 2580 1220 2604 1236 2636 1254 2650 1274 2650 1290 2640 1308 2608
1316 2586 1322 2570 1330 2562 1332 2556
=S 21
1404 2624 1414 2624 1438 2622 1458 2624 1478 2630 1488 2642 1494 2650 1494 2656
1490 2660 1478 2666 1456 2668 1434 2664 1416 2652 1404 2626 1406 2592 1418 2572
1438 2558 1468 2550 1506 2548 1528 2550 1546 2552
=E
```

FIGURE 3.9(a) Sample stroke data for "The".

The following code demonstrates how to convert pen data to a bit map representation suitable for use with the techniques discussed in later sections.

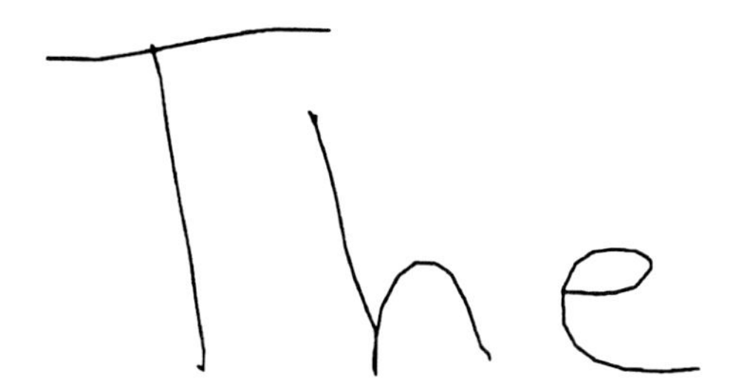

FIGURE 3.9(b) Pixel map representation of the "The".

Listing 3.5

```
struct aPoint {
  int x;                       // x coord of point in pixel map
  int y;                       // y coord of point in pixel map
  int id;                      // id of stroke to which point belongs
  };

struct aPoint point[1024];
char Grid[1024][1024];

int ReadStrokeFile(char *InName,long& maxx, long& maxy,int InvFlag){
  FILE *fpIn;
  char Name[180];
  char command[180];
  int count,dun,sid;
  int i,x,y,xyindx;
  long minx,miny;
strcpy(Name,InName);
strcat(Name,".RAW");
minx=65535; miny=65535;
if((fpIn = fopen(Name,"r")) == NULL){
  printf("Unable to open input file:%s",InName);
  return 0;
  }
sid=0; xyindx=0; dun=0;
while (!dun) {
  fscanf(fpIn,"%s",command);
  if (strcmp("=S",command)==0) {
     fscanf(fpIn,"%d",&count);
     for (i=0; i<count; i++) {
       fscanf(fpIn,"%d%d",&x,&y);
       point[xyindx].x=x;
       point[xyindx].y=y;
       point[xyindx].id=sid;
       xyindx++;
       minx=min(minx,x);
       miny=min(miny,y);
       } /* endfor */
     sid++;
     }
```

```
    else {
     if (strcmp("=E",command)==0) dun=1;
     if (feof(fpIn)) dun=1;
     } /* endif */
  } /* endwhile */
maxx=0; maxy=0;
for (i=0; i<xyindx; i++) {                    //shift image to remove
  if(sid != point[i].id)                      //excess space
    sid = point[i].id;
  point[i].x=point[i].x-minx;
  point[i].y=point[i].y-miny;
  maxx=max(maxx,point[i].x);
  maxy=max(maxy,point[i].y);
  } /* endfor */
if(InvFlag) //optionally invert for screen coord system.
  for (i=0; i<xyindx; i++) {
    point[i].y=maxy-point[i].y; }
fclose(fpIn);
return xyindx; //return number of entries in point array
}

void PlotOnGrid(int x0, int y0, int x1, int y1){
  double m,b,rx,ry,rx0,rx1,ry0,ry1,rlasty;
  int xStart,yStart,xEnd,yEnd;
  int x,y;
//plot a line between points [x0,y0] & [x1,y1]
Grid[x0][y0]=1;
Grid[x1][y1]=1;
if (x1==x0) { //handle vert line as special case
  if (y0>y1) {
    yStart=y1; yEnd=y0; }
    else {
    yStart=y0; yEnd=y1; }
  for (y=yStart; y<=yEnd; y++) {
    Grid[x0][y]=1; } /* endfor */
  }
  else {
   ry0=(double)y0;
   rx0=(double)x0;
   ry1=(double)y1;
   rx1=(double)x1;
   xStart=min(x0,x1);
   xEnd=max(x0,x1);
   m=(ry1-ry0)/(rx1-rx0);                          //calc slope
   b=ry0-m*rx0;                                    //calc intercept
   rlasty=-1.0;
   for (x=xStart; x<=xEnd; x++) {
     rx= (double)x;
     ry=m*rx+b;
     if (rlasty>=0.0){
       if (ry<rlasty) {
         for (y=(int)ry; y <= rlasty; y++) {
           Grid[x][y]=1;
           } /* endfor */
```

```
            }
           else {
            for (y=(int)rlasty; y <= ry; y++) {
               Grid[x][y]=1;
               } /* endfor */
             } /* endif */
         rlasty=ry;
          }
        else {
         Grid[x][(int)ry]=1;
         rlasty=ry;
          }
       } /* endfor */
    } /* endif */
}

void PlotStrokes(int xyPoints){
  int sid,xprev,yprev,i;
sid=-1;
//Now map all strokes onto a 2 dim grid
for (i=0; i<xyPoints; i++) {
  if(sid != point[i].id)                          //seed. must wait til a prev stroke exists
     sid  = point[i].id;
   else
    PlotOnGrid(xprev,yprev,point[i].x,point[i].y);  //plot it
  xprev=point[i].x;                               //curr stroke becomes previous stroke
  yprev=point[i].y;
  } /* endfor */
}

int main(int argc, char *argv[]) {
  int xyPoints;
  long Maxx,Maxy;
xyPoints=ReadStrokeFile(argv[1],Maxx,Maxy,1);     //Read in strokes from stroke file
PlotStrokes(xyPoints);                            //Plot strokes on grid.
//Pixel map is now in Grid Array
   .
   .
   .
return 1;
}
```

3.7 SEGMENTATION

As was discussed in the introduction, character segmentation is a significant task in the recognition of handwritten characters when presented in ruled and in unruled mode. Word segmentation can be tricky in that interword spacings may not always even be larger than intercharacter spacing. Clustering characters, multiple strokes may be connected while belonging to different characters, and conversely disconnected strokes may comprise a single character as in Figure 3.10 below:

The sample in Figure 3.10 (a) might be recognized as an "1", "c" and Figure 3.10 (b) as an "1", "2". There is perhaps some potential to bring a time delay neural network to bear on this problem. Conventional techniques are only partially successful

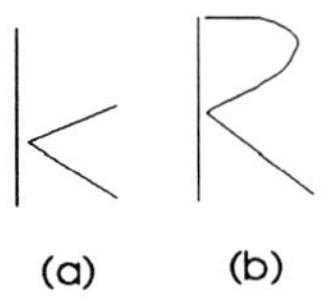

FIGURE 3.10 Character segmentation ambiguity.

in dealing with this problem. Spatial histograms for character segmentation are typically only about 60% accurate. Segmentation based on distance algorithms do somewhat better [Seni and Cohen, 1992].

Such algorithms include:

1. Bounding box methods
2. Euclidean distance methods
3. Minimum run length distance method
4. Average run length distance method
5. Run length with bounding box heuristic
6. Run length with Euclidean heuristic

Accuracy of the above techniques is reported to seem like good results but before becoming too pleased it must be remembered that segmentation occurs at the front end of the process. Any errors that occur at the segmenter are propagated through the rest of this system. Put more plainly: if the segmenter makes a mistake, the recognition system lying downstream does not have a hope of making a correct classification. The error is additive.

The reason why distance classifiers cannot do better is that they discard a great deal of relevant information. Put bluntly, distance measures do not contain all necessary information to perform segmentation accurately. The shapes of the strokes must be considered in combination with the Euclidean distance between the strokes. Neural network classifiers are suitable for this task. Garris and Wilson [1992] propose two neural network implementations both of which achieved 100% accuracy in their test. Unlike many other reports the results incorporate a large database (1104 images) and thus may be taken with some trust. The two methods reported on were

1. A self-organizing multimap network
2. A feed-forward multilayer perceptron network using Gabor coefficients as inputs and trained by the conjugate gradient technique

We also need to mention here that a form of the neocognitron which utilizes selective attention is also capable of character segmentation as part of its recognition scheme. All of these techniques are discussed in considerable detail in later chapters.

Before leaving the topic of segmentation it is worth noting that if the segmentation system is cognizant of stroke boundaries as atomic units (as would be the case for an on-line system), it has an advantage that does not accrue to systems that experience only the bit map image. In this case strokes can be scored when taken individually or when clustered into n-stroke groupings. The maximum score would then win and therefore define the character cluster.

REFERENCES AND BIBLIOGRAPHY

Darwiche, E., Pandya, A. S., and Mandalia, A. D., "Automated optical recognition of degraded characters," *SPIE,* vol. 1661, pp. 579-84, 1992.

Garris, M. D. and Wilson, C. L., "A neural approach to concurrent character segmentation and recognition," *Proc. IEEE Southcon,* pp. 154+159, 1992.

Gonzalez, R. C. and Woods, R. E., *Digital Image Processing,* Addison-Wesley, Reading, MA, 1992.

Jang, B. K. and Chin, R. T., "Analysis of thinning algorithms using mathematical morphology," *IEEE PAMI,* vol. 12, no. 6, pp. 541-555, 1990.

LeCun, Y., Le, L. D., Jackel, L. D., Boser, J. B., Denker, J. S., Graf, H. P., Guyon, J., Henderson, D., Howard, R. E., and Hubbard, W., "Handwritten digit recognition: applications of neural network chips and automatic learning," *IEEE Commun. Mag.,* vol. 27, no. 11, pp. 41-46, 1989.

Levine, J., *Programming for Graphics Files in C and C++,* John Wiley & Sons, New York, 1994.

Liao, H. Y., Huang, J. S., and Huang, S.-T., "Two-dimensional networks for handwritten Chinese character recognition," *IEEE Conf., Baltimore,* vol. 3, pp. 579-84, 1992.

Nelson, M., *The Data Compression Book,* M & T Books, San Mateo, CA, 1992.

Rimmer, S., *Supercharged Bitmapped Graphics,* Windcrest/McGraw-Hill, New York, 1992.

Seni, G. and Cohen, E., "Segmenting handwritten text lines into words using distance algorithms," *Proc. SPIE,* vol. 1661, pp. 61-72, 1992.

4 Feed-Forward Networks with Supervised Learning

4.1 FEED-FORWARD MULTILAYER PERCEPTRON (FFMLP) ARCHITECTURE

Multilayer perceptron (MLP) networks trained by back propagation [Rumelhart et al., 1986] are among the most popular and versatile forms of neural network classifiers. It has been shown that multilayer perceptron networks with a single hidden layer and a nonlinear activation function are universal classifiers [Funahashi, 1989; Cybenko, 1989, Hartman et al., 1990; Hornik et al., 1989]. That is, such networks can approximate decision boundaries of arbitrary complexity. Thus given an input vector consisting of a set of features representing an input pattern, this network is known to have the inherent discriminative power necessary to serve as a strong recognition engine. This does not mean that we can achieve zero error. Classes are often overlapping so that classification by minimizing error due to misclassification cannot produce zero error. Even so, this is more an artifact of ambiguity in the problem domain that a deficiency in the network capability. Multilayer perceptron networks remain among the strongest tools in the neural arsenal and as such they represent a reasonable place to start the discussion of neural approaches to pattern recognition. The structure of an unadorned multilayer perceptron network is shown below in Figure 4.1.

The input vector representing the pattern to be recognized is incident on the input layer and distributed to subsequent hidden layers and finally to the output layer via weighted connections. Each neuron in the network operates by taking the sum of its weighted inputs and passing the result through a nonlinear activation function. This is shown mathematically as:

$$out_i = f(net_i) = f\left(\sum_j W_{ij} out_j + \theta_i\right) \tag{1}$$

Here out_i is the output of the ith neuron in the layer under consideration; out_j is the output of the jth neuron in the preceding layer. There are several conventionally used choices for the nonlinear activation function, *f*. One of those most frequently employed is the sigmoid:

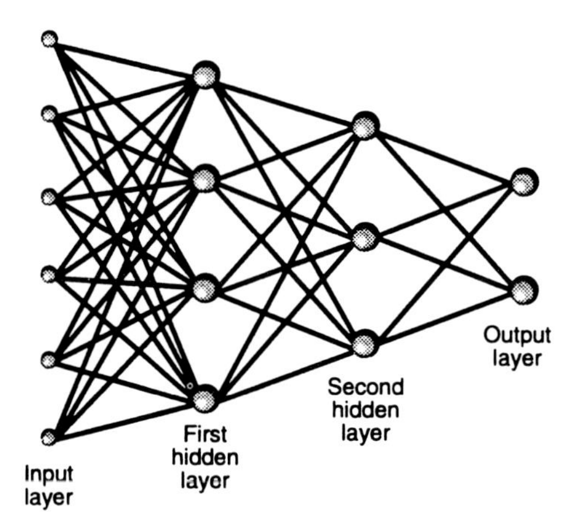

FIGURE 4.1 Feed-forward multilayer perceptron architecture.

$$f(net_i) = \frac{1}{1+e^{\frac{-net_i}{Q_0}}} \tag{2}$$

The term Q_0 in equation 2 is referred to as the temperature of the neuron. The higher the temperature the more gently the sigmoid changes. At very low temperatures it approaches a step function. Figure 4.2 shows the sigmoidal function for a variety of temperatures.

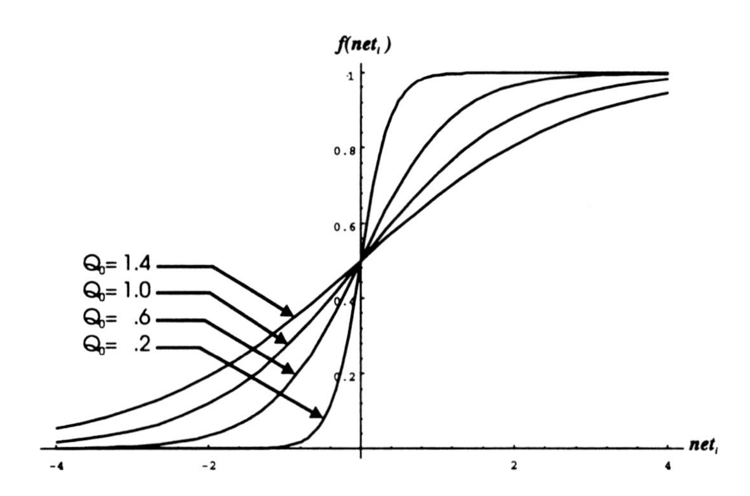

FIGURE 4.2 Sigmoid function.

Note that there is a nearly linear region surrounding $x = 0$ and that the extent of this region depends on the temperature.

It has been suggested that an antisymmetric activation function may have some advantages with regard to training a network. A function is antisymmetric when equation 3 below holds true.

$$f(-x) = -f(x) \tag{3}$$

One such antisymmetric function that is often used as an activation function is the hyperbolic tangent. For the hyperbolic tangent equation 2 becomes:

$$f(net_i) = \tanh(net_i) = \frac{1 - e^{-net_i}}{1 + e^{-net_i}} \tag{4}$$

The antisymmetric property is evident in Figure 4.3 which shows the hyperbolic tangent activation function.

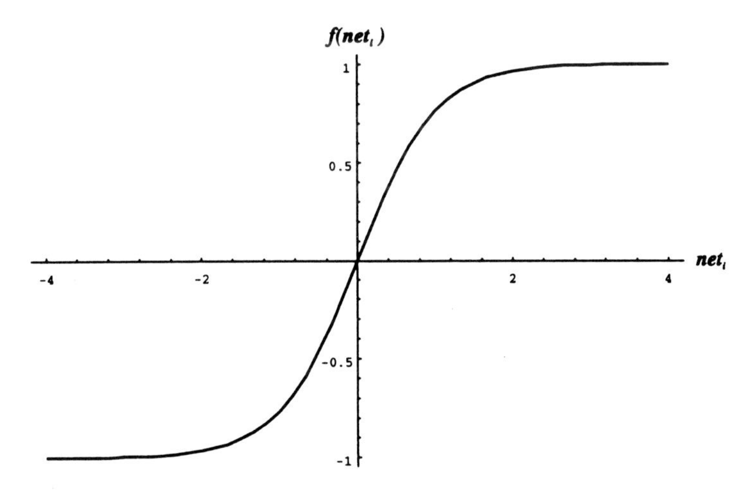

FIGURE 4.3 Hyperbolic tangent.

The knowledge required to map input patterns into an appropriate classification is embodied by the weights. Initially the weights appropriate to a given problem domain are unknown. Until a set of applicable weights is found the network has no ability to deal with the problem to be solved. The process of finding a useful set of weights is called training. Training begins with a training set consisting of specimen inputs with associated outputs that represent a correct classification. The existence of a specific desired output for each of the training set vectors is what differentiates supervised from unsupervised learning.

Training the network involves moving from the training set to a set of weights which correctly classifies the training set vectors at least to within some defined error limit. In effect the network learns what the training set has to teach it. If the training set is good and the training algorithm is effective, the network should then be able to correctly classify inputs not belonging to the training set. This phenomenon is sometimes termed generalization. We will have a great deal more to say about what makes a training set "good" in later sections.

Thus we see that the application of a neural network to a recognition problem involves two distinct phases. During the training phase the network weights are adapted to reflect the problem domain as shown in Figure 4.4(a) at the left. In the second or operational phase, the weights have been frozen and the network when presented with test data or real-world data will predict a classification. This is illustrated in Figure 4.4(b).

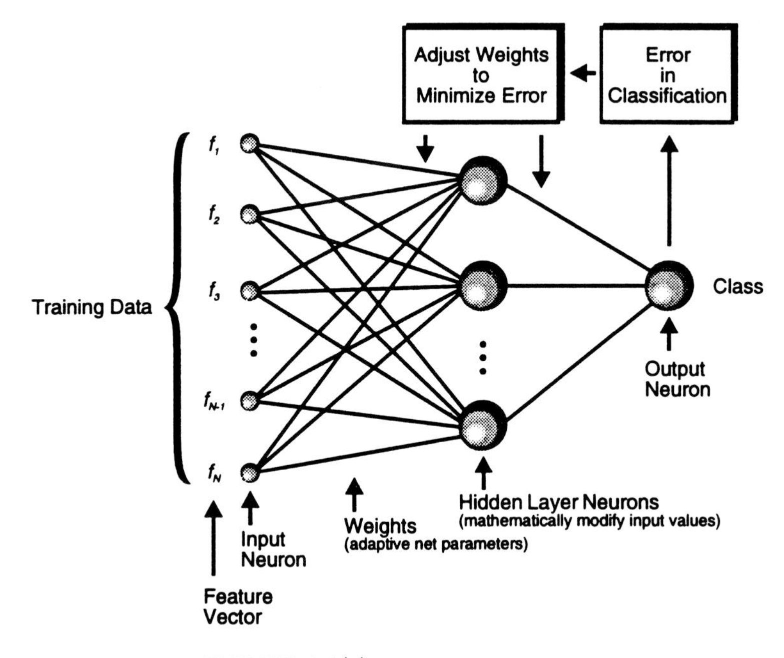

FIGURE 4.4(a) The training phase.

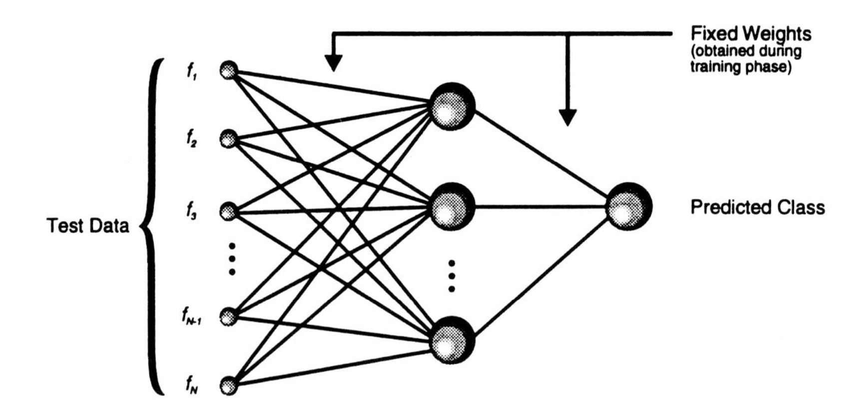

FIGURE 4.4(b) The prediction phase.

4.2 FFMLP IN C++

Now that we have seen some of the theory, we can see how we might implement a feed-forward network. To begin, let us look at the header file so we can see the overall structure of the network.

Listing 4.1

```
/*-------------------------------------------------*
*                                                  *
* Code listing PNET.HPP                            *
*                                                  *
*-------------------------------------------------*/
#define     NEURONHIGH   1.0         // Neuron's high output value
#define     NEURONLOW    0.0         // Neuron's low output value
#define     TRUE         1
#define     FALSE        0

class       WEIGHT                   // Forward reference

struct      WEIGHTIMAGE {
 double     data;                    // Weight value
 int        sneuron;                 // Source neuron for this weight
 int        dneuron;                 // Dest neuron for this weight
 WEIGHTIMAGE *next;
};

struct NETRESULTS {
 int        index;                   // Neurons identification number
 double     value;                   // Neurons output value
 char       character;               // char representation of digit
};

class NETFILEDATA {
private:
 double     temperature;             // Neurons temperature
 double     threshold;               // Neurons firing threshold
 int        Nlayers;                 // Number of layers in the net
 int        neurons[MAXLAYERS];      // Number of neurons per layer
 int        status;                  // Error status ( 0 = OK)
 WEIGHTIMAGE *weights[MAXLAYERS - 1]; //Temp weight storage area
 void       ADDweights(int l, int d, int s, double w);
 double     GETweights(int l, int d, int s);

public:
 NETFILEDATA(void);
 int  SetupNet(char *);
 double GetTemp(void) { return temperature; }
 double GetThresh(void) { return threshold; }
 int  GetNlayers(void) { return Nlayers; }
 int  GetLayerSize(int layer) { return neurons[layer]; }
 double GetWeight(int l, int d, int s) { return GETweights(l-1, d, s); }
 int  GetStatus(void) { return status; }
};
```

```
//**************************************************************
//   THE NEURON CLASS                                          *
//**************************************************************
class NEURON {
private:
 static double temperature;       //Holds a single copy for all neurons
 static double threshold;         // Holds a single copy for all the neurons
 int         id;                  // Holds a neuron identification number
 double      out;                 // Holds a neurons output value
  WEIGHT     *weight1;            // Pointer to list of weights (head)
  WEIGHT     *weightL;            // Pointer to list of weights (tail)
 int         BiasFlg;             // 1 = Bias Neuron,  0 otherwise
 NEURON      *next;               // Hook to allow neurons to be list members

public:
 NEURON(void) { id = 0; out = 0;
          weight1 = (WEIGHT *)NULL; next = (NEURON *)NULL; }
 NEURON(int ident, int bias=0) { id = ident; out = 0; BiasFlg=bias;
          weight1 = (WEIGHT *)NULL; next = (NEURON *)NULL; }
 void calc(void);                 // Update out based on weights/inputs
 void SetNext(NEURON *N) { next = N; }
 NEURON *GetNext(void) { return next; }
 void SetWeight(double Wght, NEURON *SrcPtr);
 int GetId(void) { return id; }
 double GetOut(void) { return out; }
 void SetTemperature(double tmpr) { temperature = tmpr; }
 void SetThreshold(double thrsh) { threshold = thrsh; }
 void SetOut(double val) { out = val; }
 int IsBias(){return BiasFlg;}
};
double NEURON::temperature = 0.0;          // REQUIRED for static data elements
double NEURON::threshold  = 0.0;

//**************************************************************
//   THE WEIGHT CLASS                                          *
//**************************************************************
class WEIGHT {
private:
 NEURON   *SRCneuron;       // Source neuron for this weight
 double   WtVal;            // Magnitude of weight
  WEIGHT  *next;            // Hook so weights can be list members
public:
 WEIGHT(double W, NEURON *SN) { next = (WEIGHT *)NULL;
              SRCneuron = SN; WtVal = W; }
 NEURON *GetSRCNeuron(void) { return SRCneuron; }
 double getWeight(void) { return WtVal; }
 void   SetNext(WEIGHT *W) { next = W; }
 WEIGHT *GetNext(void) { return next; }
};
```

```
//***********************************************************
//   THE  LAYER CLASS                                       *
//***********************************************************
class LAYER {
private:
  int           Layer ID;          // 0 for input, 1 for 1st hidden,...
 unsigned int Ncount;              // # of neurons in layer;
 NEURON    *Neuron1;               // Pointer to 1st neuron in layer
 NEURON    *NeuronL;               // Pointer to last neuron in layer
  LAYER     *next;                 // Hook so layers can be list members

public:
 LAYER(int layer_id, NETFILEDATA *netdata);
 int SetWeights(NEURON *PrevNeuron, NETFILEDATA *netdata);
 void SetNext(LAYER *Nlayer) { next = Nlayer; }
 int GetLayerID(void) { return LayerID; }
 NEURON *GetFirstNeuron() { return Neuron1; }
 LAYER *GetNext(void) { return next; }
 unsigned int getCount(void) { return Ncount; }
 void calc(int);
};

//**************************************************************************
// THE NETWORK CLASS                                                       *
//**************************************************************************
class NETWORK {
private:
 int                 Alive;              // True when weights are valid
 NETFILEDATA         netdata;            // Class to load saved weights
 int                 Nlayers;            // Number of layers in the network
 LAYER               *INlayer;           // Pointer to the input layer
 LAYER               *OUTlayer;          // Pointer to the output layer
  NETWORK            *next;              // Need more than 1 network ... OK

public:
 NETWORK(void) { Alive = 0; next=(NETWORK *)NULL; }
 int Setup(char *);
 void ApplyVector(unsigned char *, int);
 void RunNetwork(void);
 int RequestNthOutNeuron(int, NETRESULTS *);
 double RequestTemp(void);
 double RequestThresh(void);
 int RequestLayerSize(int);
 int GetAlive() { return Alive; }
 NETWORK *GetNext() {return next;}
 void SetNext(NETWORK *netptr) {next=netptr;}
private:
 int SetWeights(void);
};
```

Looking at the class structure it is evident that the number of layers and the number of neurons in each layer are limited only by available memory. The network class consists of a linked list of dynamically allocated layer objects as shown in Figure 4.5:

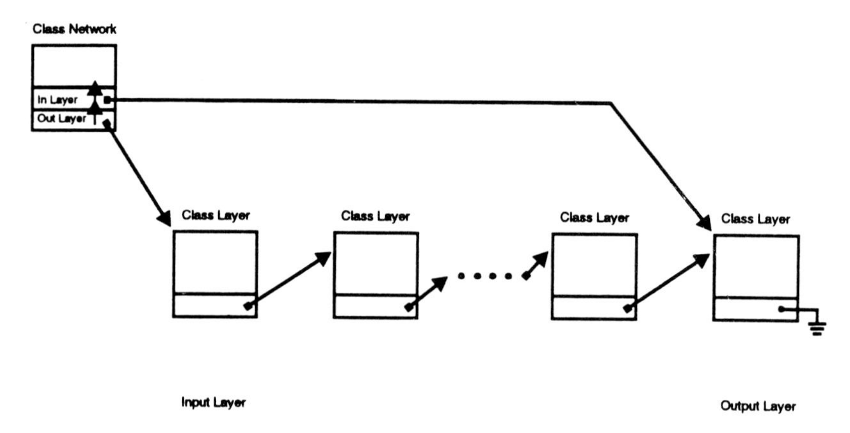

FIGURE 4.5 The network class.

In like manner the layer class consists of a linked list of neuron objects. Each neuron object in turn points to a linked list of weight objects. Finally, each weight object contains both the weight value and a pointer to the source neuron in the previous layer. The overall structure of the layer class is illustrated in Figure 4.6:

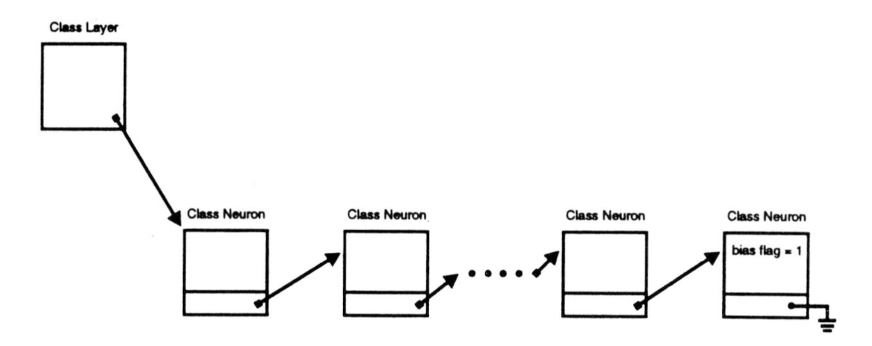

FIGURE 4.6 The layer class.

The weight class in relation to both the layer and the neuron classes is illustrated in Figure 4.7.

Only the network class needs to be visible to the calling program. The method **ApplyVector** in the network class placed an input vector into the input layer neurons. The **RunNetwork** method invokes each of the layer **calc** methods starting at the first hidden layer and ending with the output layer. In turn, when a layer **calc** method is invoked, it causes each neuron **calc** method to run. All of this is transparent to the caller. In effect the network class asks each of the layers to calculate its own values and the layers respond by asking each of the neurons to calculate its values. Finally, each neuron, in cooperation with its associated list of weight class objects, calculates its activation level. Here is the implementation of the more important methods. The complete source may be found on the companion disk.

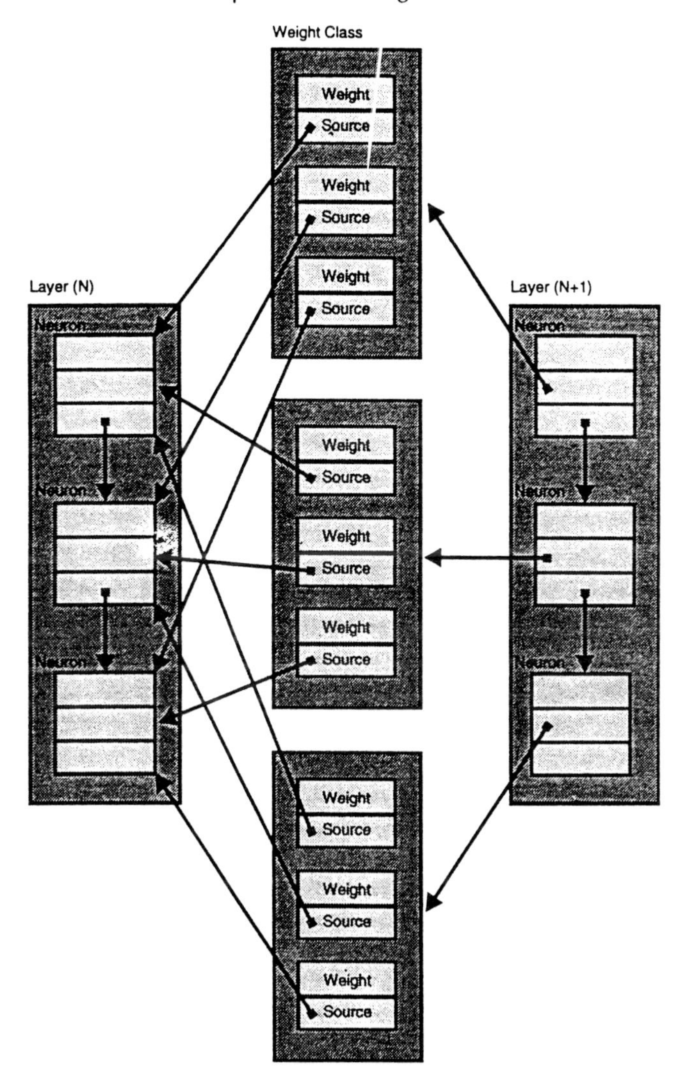

FIGURE 4.7 The weight class.

Listing 4.2

```
#include "pnet.hpp"

//***************************************************************************
// METHODS FOR CLASS NETFILEDATA                                           *
//***************************************************************************

NETFILEDATA::NETFILEDATA(){
  int i;
for (i=0; i<(MAXLAYERS-1); i++) {
  weights[i]= (WEIGHTIMAGE *)NULL;
  } /* endfor */
}
```

```
int NETFILEDATA::SetupNet(char *wgt_file_name) {
  FILE  *wgt_file_ptr;                           // Pointer to a weight file
  double AWeight;                                // Used to hold a temporary weight
  if((wgt_file_ptr = fopen(wgt_file_name, "r")) == NULL)
    return 1;
  fscanf(wgt_file_ptr, "%lg", &threshold);       // Read in threshold value
  fscanf(wgt_file_ptr, "%lg", &temperature);     // Read in temperature value
  fscanf(wgt_file_ptr, "%d", &Nlayers);          // Read in # of layers
  for(int j = 0; j < MAXLAYERS; j++)
    neurons[j] = 0;                              // Initialize array to zero
  for(int i = 0; i < Nlayers; i++)
    fscanf(wgt_file_ptr, "%d", &neurons[i]);     // Read # neurons in each layer
  for(int lyr = 1; lyr < Nlayers; lyr++)         // Traverse all layers
   {
    for(int dn = 0; dn < neurons[lyr]; dn++)     // Traverse dest layer nodes
      {                                          // Pick up bias neuron
       for(int sn = 0; sn <= neurons[lyr-1]; sn++) // Traverse src lyr nodes
         {
        //Read in the weight from the weight file
         fscanf(wgt_file_ptr, "%lg", &AWeight);
         ADDweights(lyr-1, dn, sn, AWeight);     // Add new weight to the net
         }
      }
   }
  fclose(wgt_file_ptr);                          // Close the weight file
  return 0;
}

void NETFILEDATA::ADDweights(int l, int d, int s, double w){
 WEIGHTIMAGE *WI   = weights[l],                 //Point to weights for this layer
          *Wnew = new WEIGHTIMAGE,               //Create a new weight
          *cursor,
          *trailer;
 Wnew->data    = w;                              // Assign Wnew with
 Wnew->dneuron = d;                              //     the information
 Wnew->sneuron = s;                              //         that was passed in
 Wnew->next    = (WEIGHTIMAGE *)NULL;
 if(WI) {
    cursor  = WI;
    trailer = (WEIGHTIMAGE *)NULL;
    while(cursor) {
       trailer = cursor;
       cursor  = cursor->next;
      }
    trailer->next = Wnew;
  }
 else
    weights[l] = Wnew;
}

double NETFILEDATA::GETweights(int l, int d, int s) {
 WEIGHTIMAGE *WI = weights[l];                   // Point to 1st weight in the current layer

 while(WI) {
    if((WI->sneuron == s) && (WI->dneuron == d)) {
       status = 0;
       return WI->data;
```

```
      }
     Wl = Wl->next;
    }
  status = 1;
  return 0.0;
}

//***************************************************************************
// METHODS FOR CLASS NEURON                                                 *
//***************************************************************************
void NEURON::SetWeight(double Wght, NEURON *SrcPtr) {
  WEIGHT *W = new WEIGHT(Wght, SrcPtr);
  if(weight1 == NULL) {
    weight1 = weightL = W;
    }
  else {
    weightL->SetNext(W);
    weightL = W;
    }
  }

  void NEURON::calc(void) {
   NEURON *Nptr;                          // Pointer to neuron
   WEIGHT *Wptr = weight1;                // Pointer to the first weight in the layer
   double NET  = 0.0,                     // Accumulates the sum
          PLNout,                         // Previous layer neuron output
          Weight;                         // Connection strength

   if (!BiasFlg) {
     while(Wptr) { // Traverse src layer & weights
        Weight = Wptr->getWeight();
      Nptr   = Wptr->GetSRCNeuron();      // Get weight between prev/curr layer
      PLNout = Nptr->GetOut();            // Get the previous layer output (out)
      NET   += Weight * PLNout;           // Sum(weight * out) over the curr layer
      Wptr   = Wptr->GetNext();           // Get the next weight in the weight list
      }
    // Calculate a neuron output using a sigmoid
    out = 1 / (1 + exp(- (NET + threshold)/temperature));
     }
   else {
    out = 1.0;                            // force output on for bias neuron
     }
  }

  //***************************************************************************
  // METHODS FOR CLASS LAYER                                                  *
  //***************************************************************************
  LAYER::LAYER(int layer_id, NETFILEDATA *netdata) {
   NEURON *Nptr;
   LayerID = layer_id;
   Neuron1 = NeuronL = (NEURON *)NULL;
   next   = (LAYER *)NULL;
   //Get # of neurons in layer #layer_id
```

```
  Ncount = netdata->GetLayerSize(layer_id);
  for(int i = 0; i <= Ncount; i++) { // include bias
      if (i==Ncount) {
        Nptr = new NEURON(i,TRUE);          // This is a bias Neuron
        }
      else {
        Nptr = new NEURON(i);               // This is a normal Neuron
        } /* endif */
    // Attach neuron to the list
    if(Neuron1 == NULL) {
      Neuron1 = NeuronL = Nptr;
      }
    else {
      NeuronL->SetNext(Nptr);
      NeuronL = Nptr;
      }
    }
}

void LAYER::calc(int out_layer)
{
 NEURON *Nptr = Neuron1;
  while(Nptr) {                          // Traverse the layer
   Nptr->calc();                         // Ask the neuron to calculate its own value
   Nptr = Nptr->GetNext();               // Move to next neuron in the layer
   }
}

int LAYER::SetWeights(NEURON *PrevNeuron, NETFILEDATA *netdata) {
 NEURON *CurNeuron = Neuron1, *PrevPtr;
 double ZWeight = 0.0;
 int   curx = 0,                          // Current layer neuron index
       prevx,                             // Previous layer neuron index
       status = 0,                        // Error status ( 0 = ok)
 while(CurNeuron != NULL) {               //Traverse curr lyr starting at 1st neuron
    if ( !CurNeuron->IsBias()) {          // Bias neurons dont have incoming wgts
      PrevPtr = PrevNeuron;               // Pointer to 1st neuron in prev layer
      prevx   = 0;                        // Index of 1st neuron in prev layer
      while(PrevPtr) {                    // Read a weight from the file
        ZWeight = netdata->GetWeight(LayerID, curx, prevx++);
         status = netdata->GetStatus();
          if(status > 0)
            return status;
         CurNeuron->SetWeight(ZWeight, PrevPtr);
         PrevPtr = PrevPtr->GetNext();
        }
      }
    curx++;                               // Bump index to next for current layer
    CurNeuron = CurNeuron->GetNext();     // Set pointer to next for current layer
   }
 return status;
}
```

```
//***************************************************************************
// METHODS FOR CLASS NETWORK                                                *
//***************************************************************************
int NETWORK::Setup(char *wgt_file_name) {
 LAYER *Lptr;                            // Pointer to a layer
 NEURON N;                               // Use to set statics (temp. & thres. for
NEURON class)
 int status = 0;                         // Error status > 0 if error occured
 char *tbl_file_name;                    // holds the table file name
 // Setup network using info in the weight file
 status = netdata.SetupNet(wgt_file_name);
 if(status > 0)                          // error occurred opening weight file
   return 1;
 Nlayers = netdata.GetNlayers();         // Get the number of layers
 N.SetTemperature(netdata.GetTemp());    // Set the temp for all the neurons
 N.SetThreshold(netdata.GetThresh());    // Set the thresh for all the neurons
 for(int i = 0; i < Nlayers; i++) {
   Lptr = new LAYER(i, &netdata);
    if(i == 0) {
      INlayer = OUTlayer = Lptr;
     }
    else {
      OUTlayer->SetNext(Lptr);
      OUTlayer = Lptr;
     }
  }
 status = SetWeights();                  // Setup connection strengths
 if(status > 0)                          // If > 0 then a Weight linked list
   return 2;                             //      out-of-bounds occured
 Alive = 1;                              // The network is in working condition
 return 0;                               // Return 0 on successful operation
}

int NETWORK::SetWeights(void) {
 LAYER *L1ptr = INlayer,                 // Pointer to the input layer
      *L2ptr = INlayer->GetNext();       // pointer to the second layer
 int status = 0;
 while(L2ptr) {                          // Start at the hidden layer and traverse
   status = L2ptr->SetWeights(L1ptr->GetFirstNeuron(), &netdata);
   L1ptr = L1ptr->GetNext();             // Used to mark the current layer
   L2ptr = L2ptr->GetNext();             // Get the next layer in the layer list
  }
 return status;
}

void NETWORK::ApplyVector(unsigned char *InVecPtr, int size) {
 LAYER *Lptr = INlayer;                  // Start at the 1st layer
 NEURON *Nptr = Lptr->GetFirstNeuron();  // Start at 1st neuron in 1st layer
 double value;                           // Holds pixel value (on/off)
 unsigned char mask;                     // Holds the mask value
 while(Nptr && ( !Nptr->IsBias() ) ) {   // traverse the list or neurons applying
   for(int i = 0; i < size; i++) {              // the inputs (But not to bias neuron)
      mask = 0x80;
      for(int j = 0; j < 8; j++) {       // Cycle thru bit positions
```

```
          if((InVecPtr[i] & mask) != 0)
            value = NEURONHIGH;
          else
            value = NEURONLOW;
          Nptr->SetOut(value);              // Set output of current neuron to value
          Nptr = Nptr->GetNext();           // Get  next neuron in the input layer
          mask = mask >> 1;
        }
      }
    }
}

void NETWORK::RunNetwork() {
  LAYER *Lptr = INlayer->GetNext();        // Traverse layers starting w/ 1st hidden
  int out_layer=0;                         // Use to indicate that you are on output layer
  while(Lptr){                             //       and ending with the output layer
    if(Lptr->GetNext() == NULL)            // If NULL, then end of output layer so
      out_layer = 1;                       //   build sorted list of output values
    // Ask layer to calculate its values
    Lptr->calc(out_layer);
    Lptr = Lptr->GetNext();                // Move on to next layer
    }
}
int NETWORK::RequestNthOutNeuron(int neuron_num, NETRESULTS *results) {
  if(Alive == 0)                                    // If Alive = 0 then
    return 2;                                       //   declare failure
  NEURON *Nptr = OUTlayer->GetFirstNeuron();        // Start at 1st neuron in out lyr
  while(Nptr) {   // Traverse the list of neurons
    if(neuron_num == Nptr->GetId()) {
       results->index    = neuron_num;
       results->value    = Nptr->GetOut();
       results->character = '1';
       return 0;
      }
    Nptr = Nptr->GetNext();
    }
  //  OUT OF BOUNDS ERROR on link-list traversal
  return 1;
}

double NETWORK::RequestTemp(void) {
  if(Alive == 0)                           // If Alive = 0 then no network data is avail and
    return -1.0;                           //   network is not in working condition
  else
    return netdata.GetTemp();              // Return the networks temperature value
}

double NETWORK::RequestThresh(void) {
  if(Alive == 0)
    return -1.0;
  else
    return netdata.GetThresh();  // Return the networks threshold value
}
```

```
int NETWORK::RequestLayerSize(int layer_num) {
  if(Alive == 0)
    return 0;
  else if(layer_num >= MAXLAYERS)         // If layer_num is an invalid
    return 0;                             //  layer number then return 0
  else
    return netdata.GetLayerSize(layer_num); // return # of neurons in
}                                         //  layer number layer_num
```

4.3 TRAINING WITH BACK PROPAGATION

To begin our discussion of training the network we must first recognize the need for a measure of how close the network has come to an established desired value. This measure is the network error. Since we are dealing with supervised training, the desired value is known to us for the given training set. We will see later that the proper selection of a training set will be a crucial factor in any successful network application. The training set must be of an appropriate size and it must be reasonably well representative of the problem space. For now we assume that such a training set exists so we may interrogate how to use it to train a network.

Therefore we begin by defining an error measure. Typically for the back propagation [Rumelhart et al., 1986] training algorithm, an error measure known as the mean square error is used. This is in fact not a requirement. Any continuously differentiable error function can be used, but the choice of another error function does add additional complexity and should be approached with a certain amount of caution. Remember also that whatever function is chosen for the error function must provide a meaningful measure of the "distance" between desired and actual outputs of the network. The mean square error is defined as follows:

$$E_p = \frac{1}{2}\sum_{j=1}^{N}\left(t_{pj} - o_{pj}\right)^2 \tag{5}$$

where E_p is the error for the pth presentation vector; t_{pj} is the desired value for the jth output neuron (i.e., the training set value); and o_{pj} is the actual output of the jth output neuron.

Each term in the sum is the error contribution of a single output neuron. Notice that by taking the square of the absolute error, the difference between desired and actual, we cause outputs that are distant from the desired value to contribute most strongly to the total error. Increasing the exponent, if we chose to do that, would augment this effect.

Back propagation is one of the simpler members of a family of training algorithms collectively termed gradient descent. The idea is to minimize the network total error by adjusting the weights. Gradient descent, sometimes known as the method of steepest descent, provides a means of doing this. Each weight may be thought of as a dimension in an N-dimensional error space. In error space the weights act as independent variables and the shape of the corresponding error surface is determined by the error function in combination with the training set.

The negative gradient of the error function with respect to the weights then points in the direction which will most quickly reduce the error function. If we move along this vector in weight space, we will ultimately reach a minimum at which the gradient becomes zero. Unfortunately this may be a local minimum, but we will have more to say about that a bit later. Figure 4.8 illustrates the operation of the gradient in the context of a two-dimensional cross section of the error space.

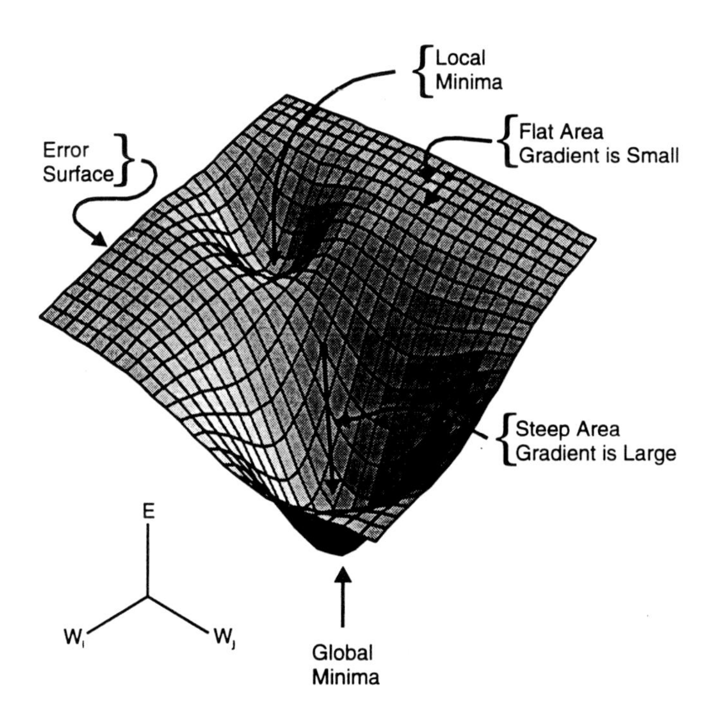

FIGURE 4.8 Surface gradient diagram.

We can express the above observations mathematically as:

$$\Delta_p w_{ji} \propto -\frac{\partial E_p}{\partial w_{ji}} \tag{6}$$

The term, $\Delta_p W_{ji}$, designates the change in the weight connecting a *source* neuron, *i*, in layer L-1 and a *destination* neuron, *j*, in layer L. This change in the weight results in a step in the weight space (shown in Figure 4.8) toward lower error.

The objective is to determine how we must adjust each weight to achieve convergence for the network. Equation 6 states that the change in each weight w_{ij} will be along the negative gradient leading to a steepest descent along the local error surface.

The task now is to convert equation 6 into a difference equation suitable for use in a computer implementation. To accomplish this we must evaluate the partial derivative, $\partial E_p/\partial w_{ji}$. We begin by applying the chain rule:

$$\frac{\partial E_p}{\partial w_{ji}} = \frac{\partial E_p}{\partial \text{net}_{pj}} \frac{\partial \text{net}_{pj}}{\partial w_{ji}} \tag{7}$$

However, we know that net_{pj} is given by:

$$\text{net}_{pj} = \sum_l w_{jl} O_{pl} \tag{8}$$

where the sum in equation 8 is taken over the output, O_{pl}, of all neurons in the L-1 layer. We may therefore evaluate $\partial \text{net}_{pj}/\partial w_{ji}$, the second term in equation 7, as follows:

$$\frac{\partial \text{net}_{pj}}{\partial w_{ji}} = \frac{\partial}{\partial w_{ji}} \sum_l w_{jl} O_{pl} \tag{9}$$

By expanding equation 9 we obtain:

$$\frac{\partial \text{net}_{pj}}{\partial w_{ji}} = \frac{\partial}{\partial w_{ji}} \left(\sum_{l' \neq i} w_{jl'} O_{pl'} + w_{ji} O_{pi} \right) = O_{pi} \tag{10}$$

Substituting equation 10 into equation 7 we obtain:

$$\frac{\partial E_p}{\partial w_{ji}} = O_{pi} \frac{\partial E_p}{\partial \text{net}_{pj}} \tag{11}$$

Now we define the error signal δ_{pj} as:

$$\delta_{pj} = -\frac{\partial E_p}{\partial \text{net}_{pj}} \tag{12}$$

By combining equations 11 and 12 we have:

$$-\frac{\partial E_p}{\partial w_{ji}} = \delta_{pj} O_{pi} \tag{13}$$

We may rewrite equation 6 by substituting equation 13 and supplying a constant of proportionality, η.

$$\Delta_p w_{ji} = \eta \delta_{pj} O_{pi} \tag{14}$$

The constant η is known as the learning rate. As its name implies, it governs the distance traveled in the direction of the negative gradient when a step in weight space is taken.

In order to achieve a usable difference equation, the task of evaluation δ_{pj} still remains. Once again we must apply the chain rule:

$$\delta_{pj} = -\frac{\partial E_p}{\partial \text{net}_{pj}} = -\frac{\partial E_p}{\partial O_{pj}} \frac{\partial O_{pj}}{\partial \text{net}_{pj}} \tag{15}$$

Now recall that the output O_{pj} is directly a function of net_{pj} as follows:

$$O_{pj} = f(\text{net}_{pj}) \tag{16}$$

$$\frac{\partial O_{pj}}{\partial \text{net}_{pj}} = f'(\text{net}_{pj}) \tag{17}$$

where *f*() is the squashing function.

To evaluate $\partial E_p / \partial O_{pj}$ (the first term of equation 15), we must consider two cases individually:

1. The destination neuron *j* is an output neuron.
2. The destination neuron *j* is a hidden layer neuron.

For a destination neuron *j* in the output layer we have direct access to the error E_p as a function of O_{pj}. Therefore we write:

$$\frac{\partial E_p}{\partial O_{pj}} = \frac{\partial}{\partial O_{pj}} \left(\frac{1}{2} \sum_{j'} \left(t_{pj'} - O_{pj'} \right)^2 \right) = -\left(t_{pj} - O_{pj} \right) \tag{18}$$

Notice that with equation 18 we have specialized the algorithm to the specific error function. An alternate choice of error function will lead to a different difference equation. Substituting equations 17 and 18 into equation 15 we may now write δ_{pj} (for destination neurons in the output layer) as:

$$\delta_{pj} = \left(t_{pj} - O_{pj} \right) f'(\text{net}_{pj}) \tag{19}$$

For destination neurons that reside in hidden layers we cannot differentiate the error function directly. Therefore we must once again apply the chain rule to obtain:

$$\frac{\partial E_p}{\partial O_{pj}} = \sum_k \frac{\partial E_p}{\partial \text{net}_{pk}} \frac{\partial \text{net}_{pk}}{\partial O_{pj}} \tag{20}$$

In equation 20 the sum k is over all neurons in the L + 1 layer. Recalling the definition of net_{pk}, we may evaluate the second factor in equation 20 as follows:

$$\begin{aligned}\frac{\partial \text{net}_{pk}}{\partial O_{pj}} &= \frac{\partial}{\partial O_{pj}}\left(\sum_l w_{kl} O_{pl}\right) \\ &= \frac{\partial}{\partial O_{pj}}\left(\sum_{l' \neq j} w_{kl'} O_{pl'} + w_{kj} O_{pj}\right) \\ &= w_{kj}\end{aligned} \tag{21}$$

Substituting equation 21 back into equation 20 yields:

$$\frac{\partial E_p}{\partial O_{pj}} = \sum_k \frac{\partial E_p}{\partial \text{net}_{pk}} w_{kj} \tag{22}$$

Now we have it by definition that:

$$\delta_{pk} = p \frac{\partial E_p}{\partial \text{net}_{pk}} \tag{23}$$

Substituting equation 23 into equation 22 yields:

$$\frac{\partial E_p}{\partial O_{pj}} = \sum_k \delta_{pk} w_{kj} \tag{24}$$

Finally combining equation 15, 17, and 24 we can represent the error signal d_{pj} for hidden layers as:

$$\delta_{pj} = f'(\text{net}_{pj}) \sum_k \delta_{pk} w_{kj} \tag{25}$$

To summarize the results so far, equation 14 provides the difference equation in terms of δ_{pj}. This is valid for both hidden and output layer weights. Equations 19 and 25 specify δ_{pj} for the output layer and hidden layer weights, respectively. Equation 18 particularized our solution to the mean square error. Therefore to use an alternative error function equation 18 would require modification. To obtain a difference equation suitable for use on a digital computer it now only remains to evaluate $f(\text{net}_{pj})$. To do this we must again particularize our solution by choosing a specific squashing function $f'(\text{net}_{pj})$. We now proceed using the sigmoid function as follows:

$$O_{pj} = f\left(\text{net}_{pj}\right) = \frac{1}{1+e^{-\text{net}_{pj}+\theta}} \tag{26}$$

From equations 17 and 26 we may write $f'(\text{net}_{pj})$ as:

$$f'\left(\text{net}_{pj}\right) = \frac{\partial}{\partial \text{net}_{pj}}\left(\frac{1}{1+e^{-\text{net}_{pj}+\theta}}\right) \tag{27}$$

Evaluating the derivative in equation 27 leads to:

$$f'\left(\text{net}_{pj}\right) = \left(\frac{-1}{\left(1+e^{-\text{net}_{pj}+\theta}\right)^2}\right)\frac{\partial}{\partial \text{net}_{pj}}\left(1+e^{-\text{net}_{pj}+\theta}\right) \tag{28}$$

We continue the evaluation of $f'(\text{net}_{pj})$ as follows:

$$f'\left(\text{net}_{pj}\right) = \left(\frac{-1}{\left(1+e^{-\text{net}_{pj}+\theta}\right)^2}\right)e^{-\text{net}_{pj}+\theta}\frac{\partial}{\partial \text{net}_{pj}}\left(-\text{net}_{pj}+\theta\right) \tag{29}$$

$$= \left(\frac{1}{1+e^{-\text{net}_{pj}+\theta}}\right)\left(\frac{e^{-\text{net}_{pj}+\theta}}{1+e^{-\text{net}_{pj}+\theta}}\right) \tag{30}$$

$$= \left(\frac{1}{1+e^{-\text{net}_{pj}+\theta}}\right)\left(\frac{1+e^{-\text{net}_{pj}+\theta}}{1+e^{-\text{net}_{pj}+\theta}} - \frac{1}{1+e^{-\text{net}_{pj}+\theta}}\right) \tag{31}$$

$$= \left(\frac{1}{1+e^{-\text{net}_{pj}+\theta}}\right)\left(1 - \frac{1}{1+e^{-\text{net}_{pj}+\theta}}\right) \tag{32}$$

We may now express $f'(\text{net}_{pj})$ in term of O_{pj} by substituting equation 26 into equation 32. We then obtain:

$$f'\left(\text{net}_{pj}\right) = O_{pj}\left(1 - O_{pj}\right) \tag{33}$$

Taken together equations 14, 19, 25, and 33 provide all that is necessary to write the difference equation needed to implement training by back propagation on a digital computer where the error function is the mean square error and the squashing function is the sigmoid. As we have proceeded through this derivation we have taken pains to show the points at which modifications would be required for alternative error or activation functions.

To summarize, the difference equation required for back-propagation training is

$$\Delta w_{ji} = \eta \delta_{pj} O_{pi} \tag{34}$$

where η refers to the learning rate; δ_{pj} refers to the error signal at neuron j in layer L; and O_{pi} refers to the output of neuron i in layer L-1.
With the error signal, δ_{pj}, given by:

$$\delta_{pj} = \left(t_{pj} - O_{pj}\right) O_{pj}\left(1 - O_{pj}\right) \quad \text{for output neurons} \tag{35}$$

$$\delta_{pj} = O_{pj}\left(1 - O_{pj}\right) \sum_k \delta_{pk} w_{kj} \quad \text{for hidden neurons} \tag{36}$$

where O_{pj} refers to layer L; O_{pi} refers to layer L – 1; and δ_{pk} refers to layer L + 1.

True gradient descent would proceed in infinitesimal steps along the direction established by the gradient. Since this is obviously impractical for our purposes, the learning rate, η, is defined (see equation 34). It can be seen that equation 34 results in a finite step size in the direction of the gradient. Here η is a constant which acts like a gain to determine the step size. The idea is to choose η large enough to cause the network to converge quickly without introducing overshoot and therefore oscillations. Later we will look at the conjugate gradient technique which in effect uses the rate at which the gradient is changing to establish a step size. Understanding the effect of the learning rate can help to choose its value judiciously. Even so, a certain amount of experimental tuning is generally required to optimize this parameter.

For clarity, the application of equations 34, 35, and 36 is shown in Figure 4.9. In particular, this figure is useful to clarify which layers are involved when calculating the various components of the difference equation. The top half of the figure delineates the training of the output layers. The bottom half depicts training in hidden layers. A cautionary reminder is that the above difference equation is valid only for mean square error with a sigmoidal activation function (given in equation 2). To use alternate error or activation functions with back propagation the difference equation must be modified as shown in the derivation.

In practice a momentum term is frequently added to equation 34 as an aid to more rapid convergence in certain problem domains. The momentum takes into account the effect of past weight changes. The momentum constant, α, determines the emphasis to place on this term. Momentum has the effect of smoothing the error-surface in weight space by filtering out high-frequency variations. The weights are adjusted in the presence of momentum by:

$$\Delta w_{ji}(n+1) = \eta\left(\delta_{pj} O_{pi}\right) + \alpha \Delta w_{ji}(n) \tag{37}$$

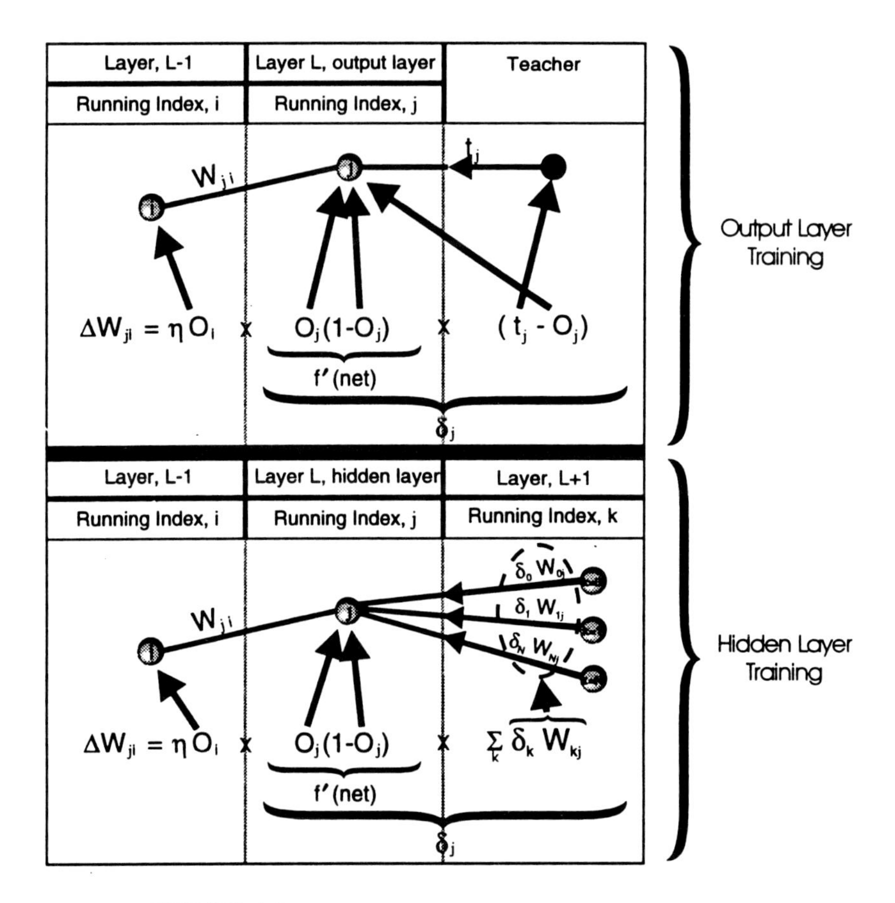

FIGURE 4.9 Back-propagation interaction between layers.

The momentum term is but the first of several departures from what might be described as pure gradient descent that is intended to augment the algorithm with respect to its ability to converge more rapidly. At this point we remark that there is much in back propagation and its variations that is largely empirical.

The overall process of back-propagation learning including both the forward and backward pass is presented in Figure 4.10 above. To apply the back-propagation algorithm the network weights must first be initialized to small random values. It is important to make the initial weights "small". Choosing initial weights too large will make the network untrainable for reasons we will discuss later. After initialization, training set vectors are then applied to the network. Running the network forward will yield a set of actual values. Back propagation can then be utilized to establish a new set of weights. The total error should decrease over the course of many such iterations. If it does not, an adjustment to the training parameters, α and η, may be required. (Contradictory data, a training vector duplicated with an oppositely

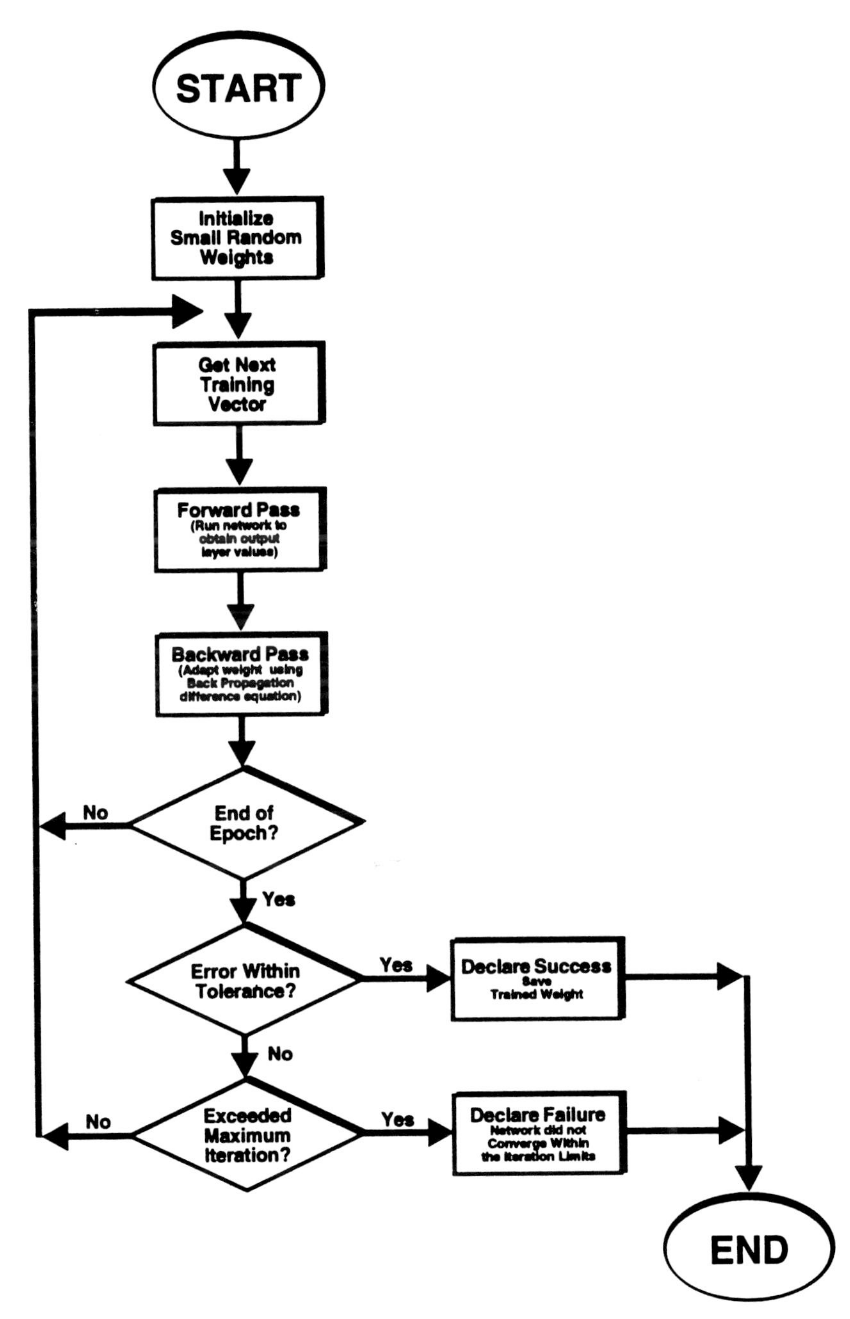

FIGURE 4.10 Back-propagation flow chart.

sensed desired value, will inhibit the network ability to converge irrespective of training parameter. In the event that extreme difficulty is encountered, verification of the training data set can prove to be worthwhile.)

One full presentation of all the vectors in the training set is termed an epoch. When the weights approach values such that the total network error, over a full epoch, falls below a preestablished threshold, the network is said to have converged. The error does not fall necessarily uniformly. Local fluctuations in the total network error are normal and expected, especially early in the training cycle. It is useful to look at error profiles as a function of iteration to gain insight into the convergence. Figure 4.11 shown below illustrates convergence behavior during a typical training cycle.

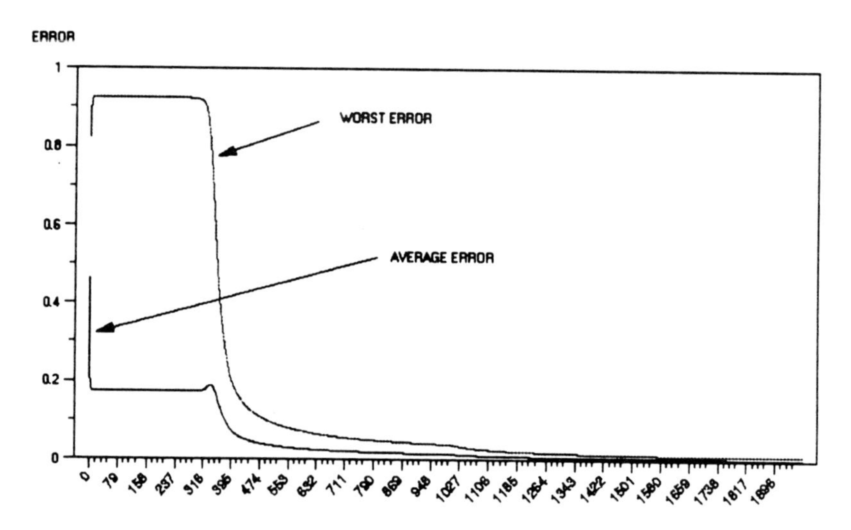

FIGURE 4.11 Back-propagation convergence curves. The worst error is the maximum error over a full epoch. The average error is the average over a full epoch.

4.3.1 Back Propagation in C++

Now we will look at the companion code implementing back propagation. The representation of the network in this code is much more basic than that presented in section 4.2. This simplicity will allow us to focus on the subject at hand, training with back propagation. The astute reader will observe that the network above would require additional backward pointers to be used effectively with back propagation; thus we avoid this distraction. The weights calculated by the code we now show (listing 4.3) can, however, be mapped onto the network of section 4.2 without difficulty.

The back-propagation algorithm has been implemented as a single class the structure of which is:

Listing 4.3

```
/****************************************************************************
*                                                                          *
* 0-0 BACK-PROP                                                            *
*                                                                          *
****************************************************************************/
// DEFINES

#define MAXLAYERS   4    // MAX NUMBER OF LAYERS IN NET (IN OUT & HIDDEN)
#define MAXNEURONS 120 // MAX NUMBER OF NEURONS PER LAYER
#define MAXPATTERNS 80 // MAX NUMBER OF PATTERNS IN A TRAINING SET
#define SIGMOID     0      // Choose Squashing fn
#define STEPFN      1      // Choose Squashing fn
#define ARCGRAN     1      // ARCHIVE GRANULARITY
#define NONE      0        // DO NOT ARCHIVE ERROR DATA
#define ALL       1        // ARCHIVE ALL ERROR DATA FOR EACH ITERATION
#define AVERAGE   2      // ARCHIVE AVERAGE ERROR DATA FOR EACH EPOCH
#define WORST     3        // ARCHIVE WORST ERROR FOR EACH EPOCH
#define TRUE      1
#define FALSE     0
#define SHUFFLE   0        // O TO DEFEAT RAND PRESENTATION ORDER
                         // 1 TO RANDOMIZE PRESENTATION OF INPUT VECTORS
//-----------------------------------------------------------------------

class BackProp {
private:
 double W[MAXLAYERS][MAXNEURONS][MAXNEURONS],       // WEIGHTS MATRIX
 double Wprev[MAXLAYERS][MAXNEURONS][MAXNEURONS]; // previous WEIGHTS
for momentum
 double Neuron[MAXLAYERS][MAXNEURONS];
 double DELTAj[MAXLAYERS][MAXNEURONS];
 double DELTA_Wij[MAXLAYERS][MAXNEURONS][MAXNEURONS];
 double ERROR[MAXNEURONS];
 double WorstErr;
 double AvgErr;
 double LastAvgErr;
 double LastWorstErr;
//Topology
 int NumLayers;              // Number of layers
 int OutLayerIndx;           // array index of last layer
 int LayerSize[MAXLAYERS];  // Number of neurons in each layer

 // Pattern data
 double InPattern[MAXPATTERNS][MAXNEURONS];   // Input values for each pattern
 double Desired[MAXPATTERNS][MAXNEURONS];     // desired value for each
                             //  pattern/output
 unsigned int PatPresOrder[MAXPATTERNS];          // Order to present patterns to net
 unsigned long int CurrIter;    // Current  iterations number.
 long int Epoch;
 int NumPatterns;            // Total patterns in training set
 int CurrPat;                // Current pattern used in training
 int ConvCount;              // The number of consecutive
                             // patterns within tolerance
 int ConvergeFlg;            // Flag indicates convergence has
                             // occurred
```

```
//PARMS
 double Temperature;          // For sigmoid
 double ETA;                  // Learning rate
 double ALPHA;                // Momentum
 double ERRTOL;               // min error tolerance required for convergence
 unsigned long int MAXITER;  // Max iterations to do before stopping

public:
 BackProp(void);
 void LoadTrainingSet(char *Fname);
 void  GetInputs(void);
 void  RunNet(void);
 void  GetParms(char *);
 void  GetWeights(void);
 double Sigmoid(double Net, double Tempr);
 void  SetRandomWeights(void);
 double HCalcDelta(int lyr,int j,double dNETj);
 int   CalcErrors(void);
 void  AdaptWeights(void);
 void  SaveWeights(char *);
 int   train(char *, char *);
 long int QueryEpoch(){return Epoch;}
 double QueryTemperature(){return Temperature;}
 double QueryEta(){return ETA;}
 double QueryAlpha(){return ALPHA;}
};


//MISC FLAGS
int ArchOn  = AVERAGE;        // Set to 1= ALL, NONE, AVERAGE or WORST to control
                              //  data sent to ARCHIVE file for graphical
                              //  analysis
FILE *ARCHIVE;                // Archive training sequence
```

Now we will look at the implementation of the methods. Only the **train** method needs to be invoked to train the network. **RunNet** accomplishes the forward pass. **AdaptWeights** and **HCalcDelta** cooperate to perform the backward pass. Notice that we can choose whether to present patterns in random or original order by specifying the define **SHUFFLE**. The convergence criteria requires that the mean square for all patterns be less than the error tolerance over a full epoch.

Listing 4.4

```
//-----------------------------------------------------------------------

//METHOD DEFINITIONS

BackProp::BackProp() {
 LastAvgErr= 1.0;
 LastWorstErr=1.0;
```

```
 WorstErr = 0.0;
 AvgErr   = 0.0;
 CurrIter=0;
 Epoch=0;
 CurrPat = -1;                          // Current pattern used in training
 ConvCount=0;
 ConvergeFlg=FALSE;                     // Flag indicates convergence
}

void BackProp::LoadTrainingSet(char *Fname) {
FILE *PFILE;
int PGindx;
int x,mask;
int pat,i,j;
double inVal;
int NumPatBytes;

 PFILE = fopen(Fname,"r");              // batch
 if (PFILE==NULL){
   printf("\nUnable to open file \n");
   exit(0);
   }
 fscanf(PFILE,"%d",&NumPatterns);
 NumPatBytes= LayerSize[0] / 8;         // # of Input lyr neurons must be divisible by 8
 for (pat=0; pat<NumPatterns; pat++) {
   PGindx=0;
   for (i=0; i<NumPatBytes; i++) {
     fscanf(PFILE,"%x",&x);
     mask = 0x80;
     for (j=0; j<8; j++) {
       if ((mask & x) > 0) {
         InPattern[pat][PGindx]=1.0;
         }
        else {
         InPattern[pat][PGindx]=0.0;
         } /* endif */
       mask=mask/2; //printf("{%x}",mask);
        PGindx++;
       } /* endfor */
     } /* endfor */
   // Now get desired / expected values
   for (i=0; i<LayerSize[OutLayerIndx]; i++) {
     fscanf(PFILE,"%lf",&inVal);
     Desired[pat][i]=inVal;
     } /* endfor */
   } /* endfor */
  fclose(PFILE);
  //init pattern presentation order
  for (i=0; i<NumPatterns; i++) {
    PatPresOrder[i]=i;                  // Start with unmodified order
    } /* endfor */
  }
```

```
/*****************************************************************************
* Function GetParms                                                          *
*    Loads topology from file spec'd file if avail.            *    otherwise tries
DEFAULT.PRM                                    *
*
*****************************************************************************/
void BackProp::GetParms(char *PrmFileName){
FILE *PRMFILE;
PRMFILE = fopen(PrmFileName,"r");        // batch
if (PRMFILE==NULL){
  printf("\nUnable to open Parameter file: %s \n",PrmFileName);
  printf("\nAttempting to open default file: PARMS1");
  PRMFILE = fopen("PARMS1","r");
  if (PRMFILE==NULL){
    printf("\nUnable to open Parameter file\n");
     exit(0);
     }
  }
printf("\nLoading Parameters from file: %s\n",PrmFileName);
fscanf(PRMFILE,"%lf",&Temperature);
fscanf(PRMFILE,"%lf",&ETA);
fscanf(PRMFILE,"%lf",&ALPHA);
fscanf(PRMFILE,"%ld",&MAXITER);
fscanf(PRMFILE,"%lf",&ERRTOL);
fscanf(PRMFILE,"%d",&NumLayers);
printf("\nNumber of layers=%d\n",NumLayers);
for (int i=0; i<NumLayers; i++) {
  fscanf(PRMFILE,"%d",&LayerSize[i]);
  printf("Number of neurons in layer %d = %d\n",i,LayerSize[i]);
  }
OutLayerIndx = NumLayers-1;                 // accommodate 0 org'd arrays
fclose(PRMFILE);
}

/*****************************************************************************
* FUNCTION GetWeights                                                        *
*    Loads a set of previously stored weights from file                      *
     (Weights must match current net parms.  No advance                      *
     error checking or validation is done)                                   *
*****************************************************************************/
void BackProp::GetWeights(){
FILE *WTFILE;
char WtFileName[30];
int i,j,k;
double zWT;

WTFILE = fopen("DEFAULT.WGT","r");
if (WTFILE==NULL){
  printf("Unable to open weight file");
  exit(0);
  }
for (i=0; i<NumLayers-1; i++) {
  for (k=0; k<LayerSize[i+1]; k++) {
    for (j=0; j<=LayerSize[i]; j++) {          // One extra for bias neuron
```

```
        fscanf(WTFILE,"%lf",&zWT);          // read weights from file
        W[i][j][k] =zWT;
        }
      }
    }
fclose(WTFILE);
}

void BackProp::SetRandomWeights(){
int i,j,k;
double zWT;
  //randomize();
  srand(6);
  for (i=1; i<NumLayers; i++) {
    for (k=0; k<LayerSize[i]; k++) {
      for (j=0; j<=LayerSize[i-1]; j++) {     // One extra for bias neuron
        zWT=(double)rand();
         zWT=zWT/2.0;
        W[i][j][k] =zWT/65536.0;             // random weight normalized to [0,0.2]
        Wprev[i][j][k]=0;;
        }
      }
    }
}

/*****************************************************************************
* GetInputs                                                                  *
*  Loads training vectors into input layer neurons                           *
*  Keeps track of iteration and epoch                                        *
*****************************************************************************/
void BackProp::GetInputs(){
int i,j,k,a,b,c;
unsigned int PatID;
CurrIter++;
CurrPat++;                                   // Update the current pattern
if (CurrPat>=NumPatterns){
  CurrPat=0;
  Epoch++;
  }
 if ((CurrPat==0) &&(SHUFFLE>0)) {          // shuffle

    for (j=0; j<=5; j++) {
      a=NumPatterns*rand()/32767;
      b=NumPatterns*rand()/32767;
       c=PatPresOrder[a];
      PatPresOrder[a]=PatPresOrder[b];
       PatPresOrder[b]=c;
       } /* endfor */
       if ( 100*(Epoch/100) ==Epoch) {
          printf("Epoch:%d",Epoch);
          printf(" Worst=%lf Avg=%lf\n",LastWorstErr, LastAvgErr);
          }
    } /* endif */
  PatID=PatPresOrder[CurrPat];
```

```
for (i=0; i<LayerSize[0]; i++) {
  Neuron[0][i]=InPattern[PatID][i];          // Show it to the neurons
   }
}

/*****************************************************************************
*
* FUNCTION RunNet                                                            *
*   Back-Propagations forward pass
*    Calculates outputs for all network neurons                              *
*****************************************************************************/
void BackProp::RunNet(){
int lyr;    // layer to calculate
int dNeuron; // dest layer neuron
int sNeuron; // src layer neuron
double SumNet;
double out;
for (lyr=1; lyr<NumLayers; lyr++) {
  Neuron[lyr-1][LayerSize[lyr-1]]=1.0;       //force bias neuron output to 1.0
  for (dNeuron=0; dNeuron<LayerSize[lyr]; dNeuron++) {
    SumNet=0.0;
    for (sNeuron=0; sNeuron <= LayerSize[lyr-1]; sNeuron++) {              //add 1 for bias
     SumNet += Neuron[lyr-1][sNeuron] * W[lyr][sNeuron][dNeuron];
       }
    out=Sigmoid(SumNet,Temperature);
    Neuron[lyr][dNeuron] = out;
     }
  }
}

/*****************************************************************************
* HCalcDelta
*  Calculate backward error signal for hidden nodes
*****************************************************************************/

double BackProp::HCalcDelta(int lyr,int j,double dNETj){
int k;
double Delta, SUMk;
SUMk=0.0;
for (k=0; k<LayerSize[lyr+1]; k++) {
  SUMk += DELTAj[lyr+1][k] * W[lyr+1][j][k];
  } /* endfor */
Delta = dNETj * SUMk;
return Delta;
}

/*****************************************************************************
* CalcErrors                                                                 *
*   Determines convergence & archives convergence behavior                   *
*                                                                            *
*****************************************************************************/
int BackProp::CalcErrors(){
double dNETj;
double MOMENTUM;
int i, j;
int LocalConvFlg=TRUE;
unsigned int PatID;
```

```
PatID = PatPresOrder[CurrPat];
for (j=0; j<LayerSize[OutLayerIndx]; j++) {
  ERROR[j] = (Desired[PatID][j] - Neuron[OutLayerIndx][j]);
  if (fabs(ERROR[j])>=ERRTOL) LocalConvFlg=FALSE;   // Any Nonconverged error kills
  AvgErr +=fabs(ERROR[j]);
  if (fabs(ERROR[j])  > WorstErr) WorstErr= fabs(ERROR[j]);
  if (CurrPat==(NumPatterns-1)) {
    AvgErr=AvgErr/NumPatterns;
     if (ArchOn==AVERAGE) {
       if ((ARCGRAN*(Epoch/ARCGRAN)) == Epoch) {  // only save ARCGRANth epochs
         fprintf(ARCHIVE,"%ld %lf %lf\n",Epoch,AvgErr, fabs(WorstErr) );
         } /* endif */
       }
     if (ArchOn==WORST) {
       if ((ARCGRAN*(Epoch/ARCGRAN)) == Epoch) {         // only save alternate epochs
         fprintf(ARCHIVE,"%ld %lf\n",Epoch, fabs(WorstErr) );
         } /* endif */
       }
     LastAvgErr=  AvgErr;
     LastWorstErr=WorstErr;
     AvgErr=0.0;
     WorstErr=0.0;
     }
   else {
     } /* endif */
  if (ArchOn==ALL) fprintf(ARCHIVE,"%ld %lf\n",CurrIter,fabs(ERROR[j]) );
  } /* endfor */
if (LocalConvFlg) {
  ConvCount++; //Record that another consec pattern is within ERRTOL
  if (ConvCount==2*NumPatterns) {
    ConvergeFlg=TRUE;
    printf("Epoch:%d",Epoch);
    printf(" Worst=%lf Avg=%lf\n",LastWorstErr, LastAvgErr);
     } /* endif */
  }
  else {
  ConvCount=0; //Start over. This pattern had an error out of tolerance
   } /* endif */
return(ConvergeFlg);
}
/*******************************************************************************
* AdaptWeights
*   Back-propagations backward pass
*******************************************************************************/
void BackProp::AdaptWeights(){
double dNETj;
double MOMENTUM;
int lyr, i, j;
unsigned int PatID;
PatID = PatPresOrder[CurrPat];
for (lyr=OutLayerIndx; lyr>0; lyr—) {
  for (j=0; j<LayerSize[lyr]; j++) {
    dNETj=Neuron[lyr][j] * (1 - Neuron[lyr][j]);
    if (lyr==OutLayerIndx) {
      ERROR[j] = (Desired[PatID][j] - Neuron[lyr][j]);
       DELTAj[lyr][j] = ERROR[j] * dNETj;
       }
```

```
      else {
       DELTAj[lyr][j] = HCalcDelta(lyr,j,dNETj);
        }
     } /* endfor */
   } /* endfor */
for (lyr=OutLayerIndx; lyr>0; lyr—) {
  for (j=0; j<LayerSize[lyr]; j++) {
    for (i=0; i<=LayerSize[lyr-1]; i++) {  // include bias
       DELTA_Wij[lyr][i][j] = ETA * DELTAj[lyr][j] * Neuron[lyr-1][i];
       MOMENTUM= ALPHA*(W[lyr][i][j] - Wprev[lyr][i][j]);
       Wprev[lyr][i][j]=W[lyr][i][j];
       W[lyr][i][j] = W[lyr][i][j]+ DELTA_Wij[lyr][i][j] + MOMENTUM;
        } /* endfor */
     } /* endfor */
   } /* endfor */
}

/*****************************************************************************
*                                                                           *
* SaveWeights                                                               *
*                                                                           *
*****************************************************************************/
void  BackProp::SaveWeights(char *WgtName) {
int lyr,s,d;
double zWT;
FILE *WEIGHTFILE;
  WEIGHTFILE = fopen(WgtName,"w");
  if (WEIGHTFILE==NULL){
   printf("Unable to open weight file for output:%s\n",WgtName);
    exit(0);
    }
  printf("SAVING CALCULATED WEIGHTS:\n\n");
  fprintf(WEIGHTFILE,"0.00\n");                    // Threshold always 0
  fprintf(WEIGHTFILE,"%lf\n",Temperature);         // Temperature
  fprintf(WEIGHTFILE,"%d\n",NumLayers);            // Number of layers
  for (lyr=0; lyr<NumLayers; lyr++) {              // Save topology
    fprintf(WEIGHTFILE,"%d\n",LayerSize[lyr]);     // Number of neurons/layer

    }
  for (lyr=1; lyr<NumLayers; lyr++) {              // Start at 1st hidden
    for (d=0; d<LayerSize[lyr]; d++) {
      for (s=0; s<=LayerSize[lyr-1]; s++) {        // One extra for bias
         zWT=W[lyr][s][d];
         fprintf(WEIGHTFILE,"%lf\n",zWT);
         }
      }
    }
  fclose(WEIGHTFILE);
}

/*****************************************************************************
* Sigmoid                                                                   *
*   This is the activation function                                         *
*                                                                           *
*****************************************************************************/
double BackProp::Sigmoid(double Net, double Tempr){
return 1.0/(1.0 + exp(-Net/Tempr));
}
```

```
/*************************************************************************
* train                                                                  *
*   Top level control to train the network with back-propagation        *
*                                                                        *
*************************************************************************/
int  BackProp::train(char *TrnFname, char *ParmFname){
if (ArchOn) ARCHIVE=fopen("archive.lst","w");
GetParms(ParmFname);
LoadTrainingSet(TrnFname);
SetRandomWeights();
int Converged=0;
while ((!Converged) && (CurrIter < MAXITER) ) {
  GetInputs();
  RunNet();
  Converged=CalcErrors();
  AdaptWeights();
  }
if (ArchOn) fclose(ARCHIVE);
return Converged;
}
```

Finally, the following code illustrates the use of the **BackProp** class in a text mode program:

Listing 4.5

```
/*************************************************************************
*
* FUNCTION ShowResults                                                  **
*                                                                        *
*************************************************************************/

void ShowResults(int ConvergeFlg, long int CurrEpoch,
           double temp, double alpha, double eta){
printf("\n------------------------------------------------------------\n");
if (ConvergeFlg) {
  printf("SUCCESS: Convergence has occured at iteration %ld\n",CurrEpoch);
  }
  else {
  printf("FAILURE: Convergence has NOT occured!!\n");
  } // endif
printf("Temperature = %lf\n",temp);
printf(" ETA  = %lf\n Alpha= %lf\n",eta,alpha);
printf("\n------------------------------------------------------------\n");
}

BackProp BPnet;

int main(int argc, char *argv[]) {
int Converged ;
if (argc>3) {
  Converged = BPnet.train(argv[1],argv[2]);
  }
  else {
```

```
  printf("USAGE: BPROP TRAINING_FILE PARMS_FILE WEIGHT_FILE\n");
  exit(0);
  }
// Show how we did
ShowResults(Converged,BPnet.QueryEpoch(),
      BPnet.QueryTemperature(),BPnet.QueryAlpha(),BPnet.QueryEta());
if (Converged)
  BPnet.SaveWeights(argv[3]);                    // Save the weights for later use
return 0;
}
```

Now that we have established a methodology for obtaining the weights for a trained network it remains to put them to use. We will do this in the next section where we will present a very basic interactive character recognition system based on what we have done so far. More typically the available data are divided into a training set for use with back propagation (or another training algorithm) and a test set which does not participate in the training process. In general it will be the performance of the recognizer on the test set that is most salient. In chapter 13 we will discuss additional metrics (in particular, the reliability and the confusion matrices) by which to judge the overall goodness of a neural recognition system. For now a simple and intuitive metric is the accuracy:

$$\text{Accuracy} = \frac{\text{Number of correct classifications}}{\text{Total number of patterns}} \tag{38}$$

Something else that we may wish to do with the weights that we have worked so hard to obtain is to look at them in some visually informative way. One of the most common means to present the trained weights graphically is to use the Hinton diagrams. To see how these diagrams are obtained and interpreted, let us consider the network shown below in Figure 4.12:

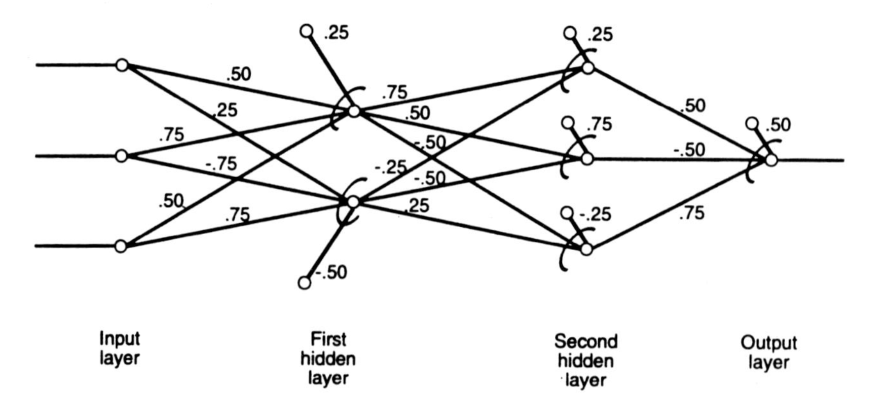

FIGURE 4.12 The Hinton network.

The corresponding Hinton diagram is shown below in Figure 4.13:

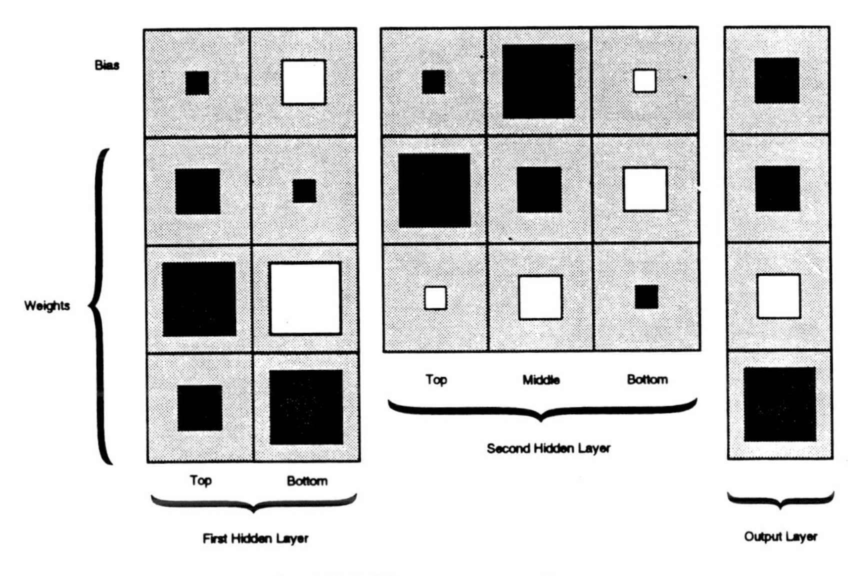

FIGURE 4.13 The Hinton diagram.

The magnitude of weight in the diagram corresponds to the size of the box. The sign of the weight corresponds to the shading. Dark boxes represent positive weights while white boxes represent negative weights. Notice that each neuron is represented as a column in the diagram and that the weights shown are those incident on the inputs of the neuron. Thus input neurons do not appear in the diagram since they have no such incoming weights.

4.4 A PRIMITIVE EXAMPLE

Perhaps the most simplistic and obvious approach to the pattern recognition problem would be to present a feed-forward network, trained by back propagation, with a vector consisting of the rasterized bit map formed from the image. In our later discussion of feature extraction (see chapters 6 and 7) we will describe several more sophisticated techniques which can be used to transform the bit-map input space into another sort of representation having some more desirable properties when presented to the neural recognition system. However, despite the existence of other and probably more effective techniques, raw bit-map inputs have been utilized with some success and should perhaps be considered where substantial preprocessing may be too computationally expensive for the given application. The presentation of the bit-mapped image to the network is shown schematically in Figure 4.14:

Results from just such implementations, those using direct bit maps as inputs, can be found in the neural network literature [Baker and McCartor, 1991]. Most frequently, however, they appear as a benchmark to which other more sophisticated strategies are compared.

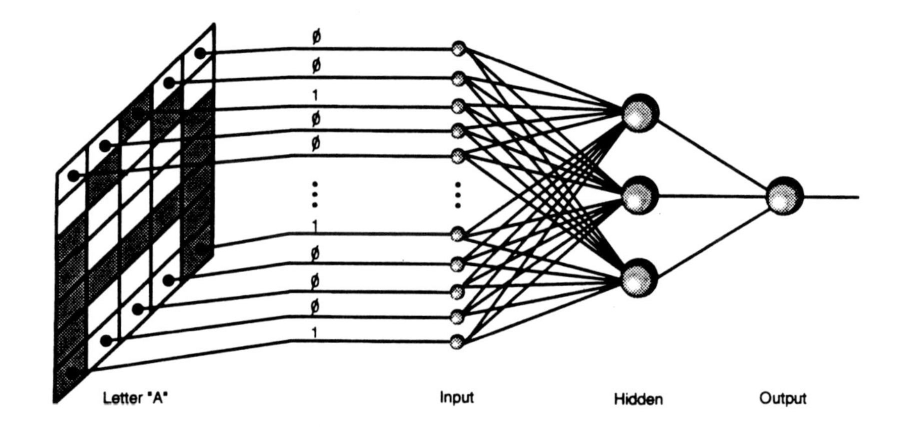

FIGURE 4.14 Character bit map "A" feeding into network.

To get a feel for scale we present one such implementation [Zurada et al., 1991]. In this application a feed-forward neural network was utilized to classify printed alphanumeric characters. More specifically, characters were represented as a 7×10 pixel bit map (from a standard dot matrix printer). Since the 7×10 pixel map was used directly as input to the network, 70 neurons were required in the input layer. This study evaluated several network topologies for the given problem domain from a point of view of training effectiveness and accuracy over the training set. No results for an independent test set were reported. Two variations of the output layer were assessed:

1. Ninety-six neurons each representing an individual digit, upper- or lower-case α or punctuation mark
2. Eight neurons which collectively encoded the ASCII representation of the character

Networks having one and two hidden layers were considered. The number used in the hidden layers ranged from 25 to 80. The best classification results were achieved for the two-hidden-layer topology in which both hidden layers contained 70 neurons.

In the companion code to this volume we have taken a somewhat different approach. Our recognizer system has one network for each of the 10 digits that it recognizes. That is, we use fewer hidden neurons but more networks. So that we can see how this works in terms of our class structure let us look at the class definition for the recognizer:

Listing 4.6

```
//*************************************************************************
// THE RECOLIST CLASS                                                     *
//*************************************************************************
```

```
struct RecoDat{
double Val;
int    NetID;
char   digit;
RecoDat *next;
};

class RecoList{
private:
 RecoDat      *head;
public:
 RecoList(){head = (RecoDat *)NULL;}
 void kill();                              //delete entire list
 void AddSorted(double, int);              //Add value + net id to list
 RecoDat QueryNth(int);                    //Get nth element in sorted list
 ~RecoList();                              //destructor
};

//***************************************************************************
// THE RECOSYS CLASS                                                        *
//***************************************************************************
class RECOSYS {
private:

  NETWORK      *net;
  RecoList     rList;
public:
  RECOSYS(void) ;
  int   Setup(char *);                          // parm is filename of file with
                                                // weight filenames for all networks
  void  ApplyVector(unsigned char *, int);      // same vect for -> all nets
  void  RunReco(void);                          // run all nets, sort results
  int   QueryNth(int, NETRESULTS *);            // results from Nth net
  int   QueryNthBest(int, NETRESULTS *);        // results from Nth best net
  double QueryTemp() {return net->RequestTemp();}
  double QueryThresh() {return net->RequestThresh();}
  int  QueryLayerSize(int l){return net->RequestLayerSize(l);}
  int  QueryNetCount() { return 10; }
  int  QueryAlive() { return net->GetAlive(); }
};
```

As can be seen from listing 4.6 the **RECOSYS** class consists of two parts. The first and most important is a linked list of **NETWORK** objects (as illustrated in Figure 4.15). Each **NETWORK** object in the list is specialized to recognize a specific digit and thus has only a single output. **ApplyVector** loads identical input vectors into each of the **NETWORK** objects. RunReco invokes each of the **NETWORK** objects in the list causing each to evaluate the input pattern. The second part of **RECOSYS** is a **RecoList** object whose function is to provide access to the individual **NETWORK** results sorted by activation level.

The implementation of the class **RECOSYS** and **RecoList** methods can be found in listing 4.7.

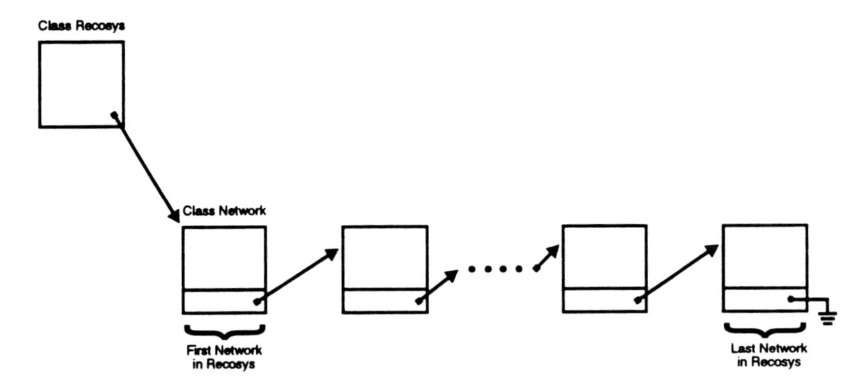

FIGURE 4.15 The recognizer class.

Listing 4.7

```
//***************************************************************************
// METHODS FOR CLASS RECOLIST                                               *
//***************************************************************************
void RecoList::kill() {                                   //delete entire list
RecoDat *p1;
RecoDat *p2;
p1=head;
while (p1) {
  p2=p1;
  delete p1;
  p1=p2->next;
  } /* endwhile */
head = (RecoDat *)NULL;
}

void RecoList::AddSorted(double V, int id) {              //Add value + net id tp list
  RecoDat *Itm;
  RecoDat *cur;
  RecoDat *prev;
  int got1;
Itm = new RecoDat;
Itm->Val=V;
Itm->NetID=id;
Itm->next=(RecoDat *)NULL;
Itm->digit=id+0x30;
if (head) {
  if (V > head->Val) {
    Itm->next=head;                                       // add as 1st item on list
    head = Itm;
    }
   else {
    cur=head->next;
    prev=head;
    got1 =0;
    while (cur && !got1) {
      // if we find 1 here its in the middle
```

```
    if (V > cur->Val) {
      got1=1;                               // found the spot...Add BEFORE cur
      prev->next=Itm;
      Itm->next=cur;
      }
     else {
      prev=cur;
      cur=cur->next;
      } /* endif */
    } /* endwhile */
    if (!got1) {
      // add at end using prev
      prev->next=Itm;
      } /* endif */
  } /* endif */
}
else {
  head=Itm;                                 // add as only item on list
  } /* endif */
}

RecoDat RecoList::QueryNth(int n) {         //Get nth element in sorted list
  RecoDat *cur;
  RecoDat rv;
  int found,cnt;
rv.Val=0;
rv.NetID=-1;
rv.digit='*';
cur=head;
found =0;
cnt=0;
while (cur && !found) {
  if (n==cnt) {
    found=1;
    rv.Val=   cur->Val;
    rv.NetID= cur->NetID;
    rv.digit= cur->digit; }
  else {
    cnt++;
    cur=cur->next;
    } /* endif */
  } /* endwhile */
return rv;
}

RecoList::~RecoList() {                     //destructor
kill();
}

//*****************************************************************************
// METHODS FOR CLASS RECOSYS                                                  *
//*****************************************************************************
RECOSYS::RECOSYS() {
  int i;
  NETWORK *N;
for (i=0; i<10; i++) {                      // create 10 generic networks
```

```
  N = new NETWORK;
  if (net) {
    N->SetNext(net);                          // point to old 1st net
    net=N;
    }
   else {
    net=N;
    } /* endif */
  } /* endfor */
}

int RECOSYS::Setup(char *Zname) {
  FILE *WFL;                                  //weight file list pointer
  char Wname[40];
  int i;
  NETWORK *N;
  int rv=0;
if((WFL = fopen(Zname, "r")) == NULL) return 1;
N=net;                                        // set equal to head
for (i=0; i<10; i++) {                        //init each of the nets
  fscanf(WFL,"%s",Wname);
  if (N->Setup(Wname)) rv=1;
  N=N->GetNext();
  } /* endfor */
return rv;
}

void RECOSYS::ApplyVector(unsigned char *Vect, int Sz) {
  int i;
  NETWORK *N;
N=net;                                        // set equal to head
for (i=0; i<10; i++) {                        // send same vector to each of the nets
  N->ApplyVector(Vect, Sz);
  N=N->GetNext();
  } /* endfor */
}

void RECOSYS::RunReco(void) {
  int i;
  NETWORK *N;
  NETRESULTS results;
N=net;                                        // set equal to head
for (i=0; i<10; i++) {                        // run each of the nets successively
  N->RunNetwork();
  N=N->GetNext();
  } /* endfor */
//BUILD THE SORTED LIST HERE!!!!
//BUT 1st INIT LIST TO EMPTY
rList.kill();
N=net;                                        // reset to head
for (i=0; i<10; i++) {
 N->RequestNthOutNeuron( 0, &results);
 rList.AddSorted(results.value, i);
 N=N->GetNext();
 } /* endfor */
}
```

```
int RECOSYS::QueryNth(int n, NETRESULTS *rv)  {
  int i;
  NETWORK *Nptr;
  NETRESULTS results;
rv->value   = 0.0;
rv->index   = -1;
rv->character= '*';
if (n>9) return 1;                              // out of bounds
Nptr=net;                                       // set equal to head
for (i=0; i<n; i++) {
  Nptr=Nptr->GetNext();
  } /* endfor */
if (!Nptr->RequestNthOutNeuron(0, &results)) {
  rv->value   = results.value;
  rv->index   = n;
  rv->character= n+0x30;
  return 0;
  }
return 2;                                       // Nth net not found
}

int RECOSYS::QueryNthBest(int n, NETRESULTS *rv) {
  RecoDat RD=rList.QueryNth(n);
  rv->value = RD.Val;
  rv->index = RD.NetID;
  rv->character = RD.digit;
return 0;
}
```

The **recosys** method **ApplyVector** causes an input vector to be loaded into the input layer of all member networks. A single reference to the **recosys** method **RunReco** causes all member networks to be run. Upon completion the results from running the suite of networks against the input vector are sorted by activation level. They are available by use of the method **QueryNthBest.** The highest activation level wins. Complete source code for both the recognizer system and the graphical user interface (GUI) is available on the diskette. In addition two sets of pretrained weights are available on the diskette: one for 8 hidden units and another for 16 hidden units.

The companion disk provides a program using the recognizer class with a graphical user interface as an aid to experimentation with the network implementation we have described. Now we briefly describe the operation of user interface for this program. To run the program simply type GRID1 at any OS/2 command prompt. Figure 4.16 shows a sample screen after all required steps to recognize a character have taken place.

The action bar menu selections are self-explanatory. Network weights can be loaded from a saved set of weights by selecting **LOAD WEIGHTS**. **LOAD WEIGHTS** will bring up a dialog box containing existing weight files in the current directory. The network topology and temperature coefficient will appear in the ***NETWORK PARAMETERS*** window upon successful completion.

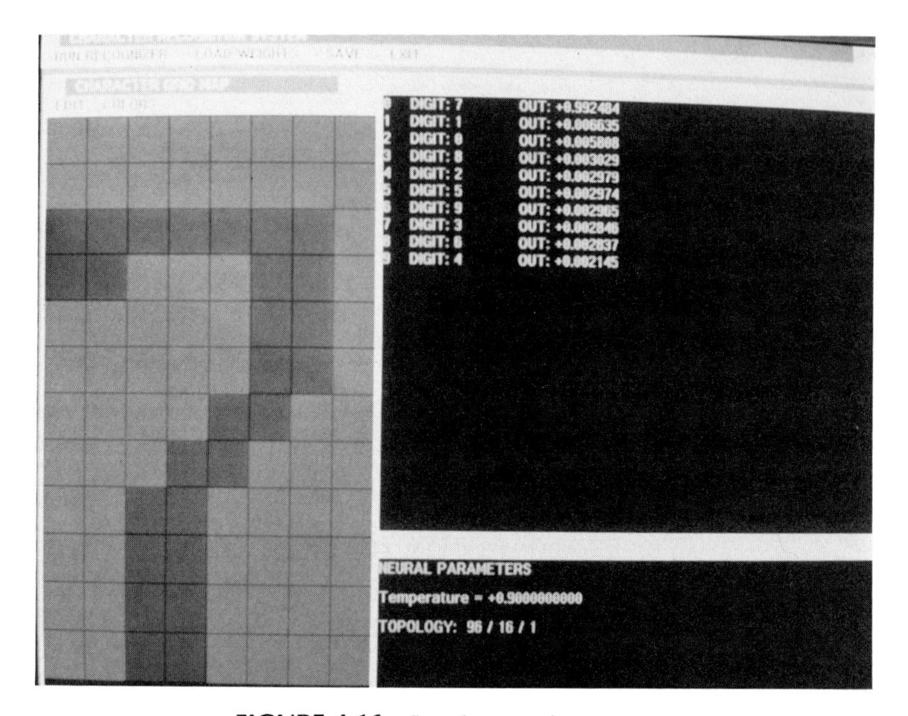

FIGURE 4.16 Sample recognizer screen.

Once weights have been established as above, the next step is to load a bit map representation of the image to be recognized. From **EDIT** in the ***CHARACTER GRID MAP*** window, specimen images of all the upper- and lowercase characters as well as the digits can be loaded. These character maps are the same ones provided in the training set. Clicking on individual squares in the grid with the left mouse button will turn pixels or on off. In this manner the exemplar characters can be modified in any way desired. Another option under **EDIT** will clear the grid to create patterns without referring to the exemplars. The remaining action bar item in the ***CHARACTER GRID MAP*** window simply provides the means to select foreground and background colors for the grid.

Now that we have established a set of weights to be used and a bit map to be recognized all that remains is to tell the recognition system to go ahead and recognize it. Predictably this is done by selecting **RUN RECOGNIZER** from the top level action bar. Recognition results appear in the ***NETWORK — ALL OUTPUTS*** window, sorted by activation level.

A number of issues related to multilayer perceptron networks trained by back propagation are discussed in the literature. Although perhaps such issues are tangential to the broader overall subject of pattern recognition, they remain important to any real implementation.

Such issues include

1. The choice of network training parameters and strategies
2. The selection of an appropriate network topology
3. The issue of overtraining
4. The characteristics and size of the training set

4.5 TRAINING STRATEGIES AND AVOIDING LOCAL MINIMA

The choice of training parameters and avoidance of local minima can be something of a delicate art. No rule for selecting optimal parameters currently exists. Slow convergence or even nonconvergence is one of the problems encountered with back propagation. In addition to the problem of local minima, another difficulty, sometimes termed false minima, may occur when the error space surface contains an expansive shallow plane with a near zero gradient (see Figure 4.8). When such a plane is long enough, convergence may be painfully slow. If the error surface is shallow enough, round-off errors become an enemy which may take a gradient descent algorithm in an entirely spurious direction.

Premature saturation is another problem to which the back-propagation algorithm is subject. Premature saturation occurs when neurons operate beyond their linear region. The error of the network then will stay uniformly high throughout the training process because such neurons cannot be effectively trained. The reason for this is as follows. When the weights associated with a given destination neuron grow sufficiently large, the neuron operates in the region within which the activation function asymptotically approaches its limit (in the case of the sigmoid 1 or 0). Within this region the derivative, f' (net), will be extremely small. Referring to equations 14, 19, and 25 it is evident that when f' (net) approaches zero, the weight adjustment made through back propagation also approaches zero. The result is that training becomes ineffective. Such neurons are said to be in saturation. If initial weight settings are chosen too large, premature saturation can occur. An inauspicious choice of the training parameters, α and η, can also lead to saturation. We also note that training algorithms such as ALOPEX (see section 4.12) which do not depend on the derivative of the activation function do not exhibit this problem.

One class of training techniques focuses on the manner in which training data are presented to the network. One such training strategy termed Snowball training [Wang and Jean, 1993] involves first presenting with only positive examples to establish a set of weights favorable to them. Gradually, as training proceeds, negative or counterexamples are added to the training set. Other techniques include presenting training patterns in random order during training and ensuring that nearly equal numbers of training vectors are provided for all classes.

Here is a slightly more complex technique for managing the presentation of training data by which the back-propagation algorithm can be made to converge more quickly. The method involves pruning training patterns which may be termed "redundant". Redundant patterns are those which fail to cause an appreciable weight update when presented to the network during the learning process. Input data are divided into two categories — the learning set and the idle set (to which each so-called redundant datum belongs). Once stigmatized as an idle set member, pruning is accomplished by presenting that vector to the network only every N epochs. As much as a two-thirds reduction in training time has been reported for this technique [Tsay et al., 1992].

Another class of training strategy involves manipulation of the network learning rate, η. The later layers in the feed-forward network tend to have larger local gradients and thus tend to be faster learners. Making the learning rate specific to a given layer, neuron, or even weight can be easily accomplished. All that is needed is to embed the learning rate within the layer, neuron, or weight private area to be used as needed. It is therefore a simple enough matter to allow the earlier layers, those closer to the input layer, to have a relatively larger learning rate. It has also been suggested that neurons having many inputs may do better with the learning rate reduced [Haykin, 1994]. Again the implementation is similar.

Thus far we have seen that modification of the learning rate can be beneficial to training the network. Let us now pursue this line of inquiry further and consider yet further adaptation of the learning rate.

Adaptation of learning rate involves varying the learning rate from one iteration to the next. It has been suggested that one of the reasons why the back-propagation algorithm may be slow to converge is that a single learning rate may not be acceptable for a given weight dimension in a given region of the error surface. At a given point on the error surface one weight dimension may be dropping sharply while another remains virtually level. All this could of course change on the next iteration. To deal with such conditions, the following concept can be applied. If the derivative of the error function with respect to a given weight maintains the same sign for several consecutive iterations, we may try increasing its individual learning rate. Conversely, if the sign is alternating for several iterations, this is a signal that the learning rate should be reduced.

In order to utilize these strategies involving adaptation of the learning rate a suitable algorithm for determining the learning rates at each iteration is needed. Two such algorithms are the delta-bar-delta algorithm of Jacobs [1988] and the decoupled momentum algorithm [Soucek, 1992]). The delta-bar-delta algorithm updates the learning rate at each iteration. This can be expressed mathematically as follows:

$$\eta_{ij}(n+1) = \eta_{ij}(n) + \Delta\eta_{ij}(n) \tag{39}$$

n_{ij} designates the learning rate for the weight connecting the ith and j^{th} neuron.

The change in learning rate $\Delta\varphi_{ij}(n)$ can be calculated as follows:

$$\Delta\eta_{ij}(n) = \begin{cases} \kappa & \text{if } \bar{\delta}(n-1)\delta_{ij}(n) > 0 \\ -\phi\eta_{ij}(n) & \text{if } \bar{\delta}(n-1)\delta_{ij}(n) < 0 \\ 0 & \text{otherwise} \end{cases} \tag{40}$$

where

$$\delta_{ij}(n) = \frac{\partial E(n)}{\partial w_{ij}} \tag{41}$$

and

$$\bar{\delta}_{ij}(n) = (1-\theta)\delta_{ij}(n) + \theta\bar{\delta}_{ij}(t-1) \tag{42}$$

where κ,ϕ, and θ are constant parameters.

A detailed description and derivation of the delta-bar-delta algorithm can to be found in Haykin [1994]. For a discussion of the decoupled angular momentum algorithm see Soucek [1992].

These modifications of the learning rate as we move through iterations modify the back-propagation algorithm in a fundamental way. This probably lies at the fringes of the changes that can be made to back propagation while still remaining somewhat recognizable as back propagation. Nevertheless, there are many useful variations on the theme that are largely empirical and the final arbiter is a converged set of network weights.

As mentioned earlier, avoidance of local minima is an issue in training by back propagation. Two of the methods which are used to avoid local minima are simulated annealing and genetic algorithms. Simulated annealing functions by scheduling a temperature term in the activation function decrease it according to a schedule as training proceeds. Genetic algorithms (GAs) use probabilistic rules to guide their search and consider many points in a search space simultaneously. Therefore they stand a better chance to avoid a local minima. A detailed discussion of their application to avoidance of local minima is found in Masters [1993]. Yet another technique that has been suggested for the avoidance of local minima involves injecting noise into the training pattern. Early in the training, the amount of noise produced by a random generator may be quite large. As training progresses the noise is gradually reduced and ultimately eliminated to achieve convergence. The presence of the noise early in the training cycle is in some sense analogous to simulated annealing.

4.6 VARIATIONS ON GRADIENT DESCENT

Variations on the theme of gradient descent can be broadly dichotomized by frequency of update (block adaptive vs. data adaptive techniques) and direction of update (first order vs. second order).

4.6.1 Block Adaptive vs. Data Adaptive Gradient Descent

Data adaptive techniques update network weights after each training vector is experienced. That is, each individual pattern vector produces a gradient and we follow each of these individual gradients as we proceed through an epoch. The result may resemble a drunkard's walk through the weight space which tends ultimately, we hope, toward the direction that most directly reduces the error function over the ensemble of training patterns. In this sense the data adaptive method can be thought of as stochastic nature. It is also evident that data adaptive updates will tend to be more sensitive to noise effects in individual patterns.

Block adaptive techniques withhold weight adjustments until the effects of an entire block of data has been accumulated. Typically this block size will be the entire training set although smaller block sizes are possible. In some sense the block adaptive method is a purer form a gradient descent in that the direction of update is the true negative gradient over the complete ensemble of training set vectors. As a result, block adaptive techniques can be more robust numerically since the training step is averaged over all training patterns within the block. It is for this reason that second-order techniques tend to use block adaptive updating. It should not necessarily be concluded from this that block adaptive updates are necessarily superior. The stochastic nature of data adaptive techniques can actually be useful in avoiding local minima, one of the most commonly discussed problems attributed to back propagation.

Back propagation can be used with either style of update. The code we examined earlier for back propagation was data adaptive in nature. However, to really take full advantage of the stochastic aspect of data adaptive updates the order of presentation of training vectors should be randomized in between epochs. To make the code block adaptive requires very little work. All that is required is an area to store the accumulated changes during an epoch. This area is cleared when the epoch starts. During training the sum of the individual training vector weight adjustments is accumulated. Subsequently, at the end of the epoch, the networks weights are adjusted.

4.6.2 First-Order vs. Second-Order Gradient Descent

First-order techniques depend only on the gradient. Second-order techniques utilize second-order gradients. In both cases back propagation remains useful for calculating the gradient. Second-order gradients can be sensitive to compute, but in general second-order techniques produce superior numerical performance when used in a block adaptive context. One particularly common second-order technique, the conjugate gradient method, will be discussed in section 4.11.

4.7 TOPOLOGY

The number of layers and number of nodes per lay will affect the decision surface. Again no simple rule exists which indicates how many hidden units are required for a given task. It is desirable to obtain a network with the fewest possible neurons in the hidden layer. Aside from issues of performance both during training and after, an excessive number of hidden layer neurons invite a phenomenon termed overfitting. Overfitting occurs when the network has so much information processing capability that it will learn insignificant aspects of the training set. Overfitting is illustrated in Figure 4.17.

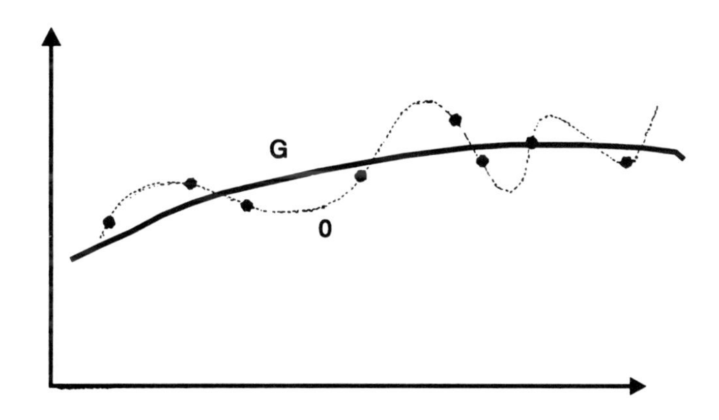

FIGURE 4.17 Overfitting diagram.

One suggestion is to start with a low number of hidden layer nodes, preferably fewer than are required to solve the problem. If the network does not converge satisfactorily, the hidden layer size would then be increased in the belief that the current network lacked the capacity to partition the data. This is done iteratively until a minimum network that can solve the problem is obtained.

Pruning the network hidden nodes is yet another approach to the definition of network topology. Lee et al. [1992] propose a methodology for pruning. In their method nodes which are considered to be candidates for pruning are

1. Those for which the output across all training patterns is constant (within a tolerance)
2. Those for which the magnitude of the weight vector is relatively much smaller than the norms of other hidden node weights and the output layer

4.8 ACON VS. OCON

Another topological issue involves whether to use a single network with multiple outputs (ACON, all classes one network) or multiple networks with a single output (OCON, one class one network). In an ACON network a single neural network has responsibility for recognizing all characters and only this single network needs to run to achieve recognition. An ACON network designed to recognize the digits, for example, would have ten output neurons, one per digit. By contrast, the corresponding OCON implementation would consist of ten networks each with one output. Each of these ten individual networks would specialize in the recognition of a particular digit. A back end resolves the result as shown in Figure 4.18. As specialists, the individual OCON networks may be less complex, faster, and more easily trainable. Kung [1993] asserts that an OCON may even require fewer overall hidden layer nodes. In a side-by-side test, OCON performed favorably in terms of convergence speed and accuracy. This test, over the 36 alphanumeric characters, used 432 training vectors and 396 test vectors. The results are summarized in Table 4.1 below.

TABLE 4.1
Performance Comparisons Between the ACON and OCON Models on Handwritten Character Recognition

	Training accuracy	Generalization accuracy	Training time
ACON (BP)	405/432 = 94%	324/396 = 82%	Normalized = 1.00
OCON (BP)	430/432 = 99.5%	344/396 = 87%	About = 0.25

Tsay et al. [1992] also provide test results for an OCON-based network recognizer for the handwritten digits. They achieved recognition rates of 100% on the training set and 84.3% on the test set. The training set contained 700 digits from 70 writers while the test set size was 500. As a final stage in the recognition process the outputs of the individual OCON networks are input to a final classifier called a MAXNET. This is shown in Figure 4.18.

A novel variation [Bebbis et al., 1992] on the above architecture follows. It is neither purely OCON nor ACON, but lies in between. The process begins with an ACON network. After training and testing, the confusion matrix is evaluated. (We will provide detail about the confusion matrix in chapter 12.) Based on the results, a determination is made about which characters the network is likely to confuse. The network is then broken up into smaller networks which it is reasoned can do better. This process is iterated until acceptable results are obtained. In effect the network is evolved from ACON toward OCON. The problem addressed was the recognition of handwritten digits. Using the process just described, four subnetworks were defined. These subnetworks were specialized as follows: ({4, 1, 7, 2}, {3, 5}, {6, 9} {0, 8}). The manner in which these networks were combined is almost exactly the reverse of the MAXNET approach. A front-end classifier was used to select the particular subnet to be used in the final stage of classification. This is shown in Figure 4.19 below:

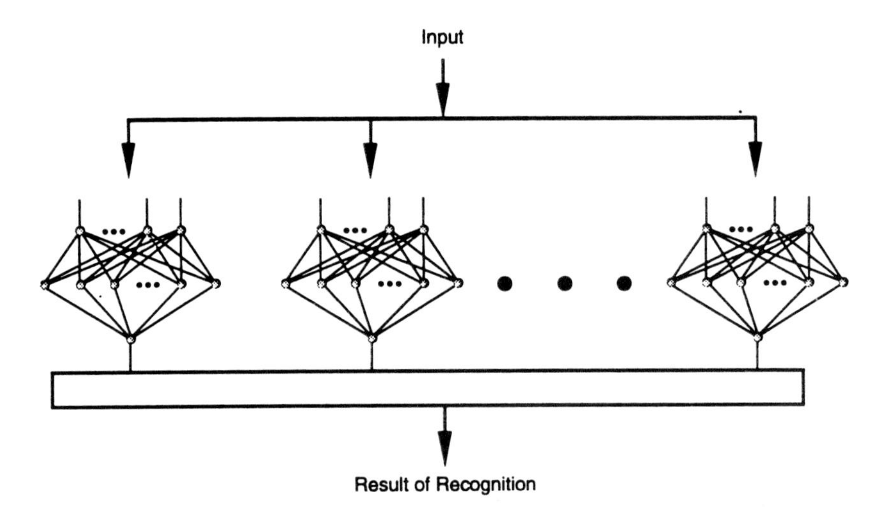

FIGURE 4.18 The MAXNET classifier.

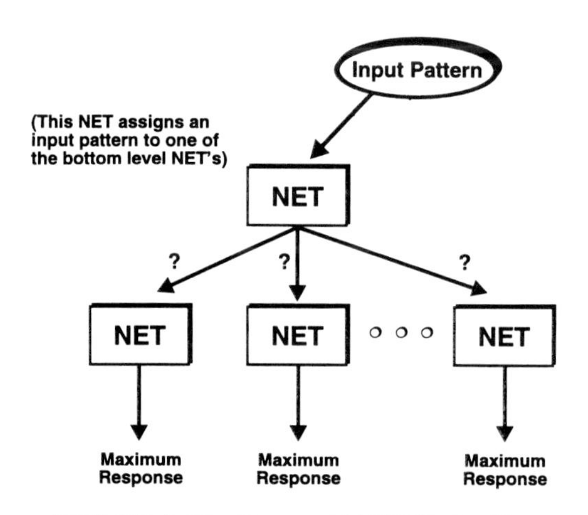

FIGURE 4.19 Reverse MAXNET classifier.

4.9 OVERTRAINING AND GENERALIZATION

Figure 4.20 shows that training to an excessively low error tolerance can result in overall poor performance by the network [Hammerstrom, 1993]. In effect the network begins memorizing minutiae of the training set and loses its ability to generalize. This is similar in nature to the idea of overfitting as illustrated in Figure 4.21.

The point at which test set error begins to rise even as training set error decreases is not known *a priori*. The question becomes: how do we know when productive training has ceased and overtraining is in progress?

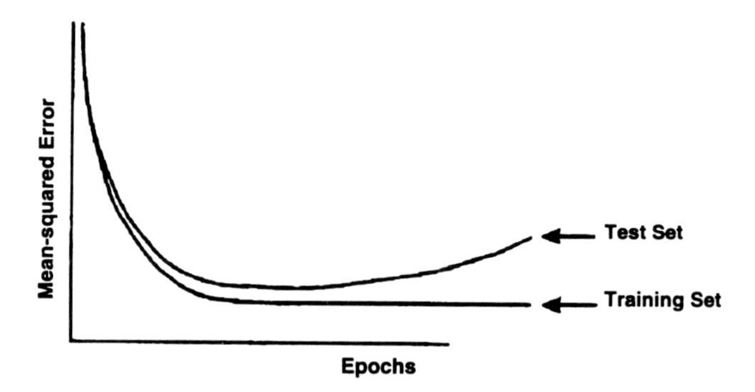

FIGURE 4.20 Overtraining curves.

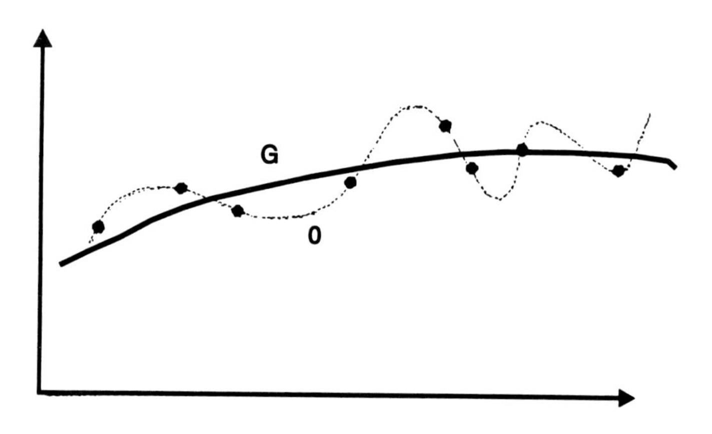

FIGURE 4.21 Overfitting diagram.

The conventional test of convergence requires that the overall error E_t over the ensemble of training vectors be less than some specified minimum value. Suppose that an independent test set ensemble were provided and the error over these test set vectors, E_{test}, is calculated periodically. For example, say every epoch or every K epochs. If $E_{test}(n)<E_{test}(n - K)$, the test set error has decreased. It would be tempting to save the current values for the weights now as $\mathbf{W}_{test\_min}$. We resist this temptation and instead save only the epoch number, n_{test_min}. Note that the error term propagated back to the network to train the weights is based only on the training set vectors so we have done nothing that would alter the composition of the weights themselves. The modification is aimed only at providing a metric to indicate when a problem exists with the topology and/or training set. If n_{test_min} is significantly lower than the epoch at which convergence occurred, then one of the following is probably true:

1. There are too many hidden units in the network and the hidden layer size should be decreased.
2. The training set does not adequately represent the decision class and should be augmented.

Techniques that can improve the generalization capability of back-propagation trained networks include training with noise and weight decay. Noise injection we recall has also been used as a means to avoid local minima.

Another approach toward improving generalization is known as weight decay. Weight decay introduces an additional term to the error function as follows:

$$E_{new} = E_{MSE} + E_{WD} = E_{MSE} + \alpha\|\mathbf{W}\|^2 \quad (43)$$

where α is a constant and $\|\mathbf{W}\|^2$ is given by:

$$\|\mathbf{W}\|^2 = \sum_{i \in C_{\text{total}}} w_i^2 \quad (44)$$

In the above, C_{total} refers to the set of all synaptic weights in the network, E_{MSE} denotes the familiar mean square error, and E_{WD} is the newly added weight decay error term. The new term, E_{WD}, acts to reduce the overall size of weights in the system. Weights that strongly reduce the mean square error term, E_{MSE} will still tend to grow. Excess weights, having little effect on the error term, E_{MSE}, will tend to diminish (assuming values very close to zero) due to the tendency toward overall lowering of weights induced by the weight decay term. Without the weight decay error term, excess weights may result in poor generalization. They may take on arbitrary values without penalty. They may also cause the network to over-fit the data in order to produce a relatively minor improvement in training error. This technique of adding an extra error term, found in weight decay and weight smoothing (which we shall discuss shortly), are regularization techniques which are commonly used to stabilize solutions to ill-posed problems.

In general, patterns will contain some degree of correlation between neighboring points. For a fully connected feed-forward neural network no neighborhood information is implicit in the neural network architecture. Thus the correlation between neighboring points within the image will be strictly an artifact of having done sufficient training. In this sense, we have challenged the network, in effect to recognize what amounts to a (deterministically) scrambled or disconnected version of the pattern. The required training time and pattern set size to achieve adequate training are therefore increased. Deprived of the *a priori* information of a correlation between neighboring points, the network (under the influence of the training algorithm) is free to extract whatever relationships it can. Clearly, a network that is constrained in some fashion will be more successful in generalizing in the pattern recognition problem domain.

Several approaches have been taken to deal with this problem. The neocognitron (see chapter 11) and feature maps (see chapter 6) represent architectural approaches to this problem. Transformations performed on the input data which reflect localized spatial correlation have also been brought to bear on this problem (see chapter 6). In this section we will discuss a method known as weight smoothing which embeds the spatial correlation through an adjustment to the back-propagation training algorithm.

The neural networks used for pattern classification normally contain a great many free weights. The training of the network is a search for a good solution in weight space. Since the weight space is generally huge, many solutions may exist not all of which necessarily reflect spatial correlation. The following technique, termed weight smoothing, represents a method of training the network in a manner that forces spatial correlation to be considered. Like weight decay, the weight smoothing technique introduces an additional error term to the typical mean square error frequently used in back propagation.

Let I and H represent the number of neurons in the input layer and hidden layer, respectively. If $\mathbf{W}_{ji}$ denotes the weight connecting the jth hidden layer neuron and the ith input layer neuron, the weight vector of the jth hidden layer neuron is $(\mathbf{W}_{j1}, \mathbf{W}_{j2}, \ldots, \mathbf{W}_{jI})^T$. The following metric can then be used as a measure for the lack of smoothness in the weight vector of the jth hidden neuron:

$$\sum_{i=2}^{I}\left[\mathbf{W}_{ji} - \mathbf{W}_{j(i-1)}\right]^2 \tag{45}$$

The error term used by the weight smoothing algorithm is as follows:

$$\begin{aligned} E_{new} &= E_{MSE} + E_{WS} \\ &= E_{MSE} + \xi\sum_{j=1}^{H}\sum_{i=2}^{I}\left[\mathbf{W}_{ji} - \mathbf{W}_{j(i-1)}\right]^2 \end{aligned} \tag{46}$$

where ξ is a constant and E_{ws} is the measure for lack of smoothness taken over all hidden layer neurons.

Notice that although equation 46 above clearly applies to a one-dimensional pattern, it could easily be extended to patterns of two or higher dimensions. Further, notice that equation 46 applies a penalty through the error term proportional to the square distance between adjacent weight vectors. (This is in marked contrast to the weight decay error term.) It is assumed that adjacent pattern points are incident on adjacent input layer neurons.

Recall from Section 4.3 that changing the error function utilized by back propagation leads to a new difference equation. Since we have changed the error function, we must determine the new difference equation required by the use of the weight smoothing error function. Taking the derivative of equation 4 we obtain:

$$-\frac{\partial E_{NEW}}{\partial \mathbf{W}_{ji}} = \frac{\partial E_{MSE}}{\partial \mathbf{W}_{ji}} + 2\xi\left[2\mathbf{W}_{ji} - \mathbf{W}_{j(i-1)} - \mathbf{W}_{j(i+1)}\right] \tag{47}$$

Note that the term $(2\mathbf{W}_{ji} - \mathbf{W}_{j(i-1)} - \mathbf{W}_{j(i+1)})$ represents the sum of differences between the weight $\mathbf{W}_{ji}$ and its two (input layer) neighbors. Equation 47 leads to the following difference equation for back propagation when using the weight smoothing error function:

$$\Delta \mathbf{W}_{ji}(t+1) = \alpha \Delta \mathbf{W}_{ji}(t) + \eta \left\{ \Delta \mathbf{W}_{ji}^{S}(t+1) - 2\xi \left[2\mathbf{W}_{ji}(t) - \mathbf{W}_{j(i-1)}(t) - \mathbf{W}_{j(i+1)}(t) \right] \right\} \quad (48)$$

where

$$\Delta \mathbf{W}_{ji}^{S}(t+1) = -\frac{\partial E_{MSE}}{\partial w_{ji}} \quad (49)$$

Two further enhancements may be incorporated into the weight smoothing algorithm. These are (1) overrelaxation and (2) an annealing smoothing factor. The overrelaxation technique approximates $\Delta \mathbf{W}_{ji}(t + 1)$ by means of applying the following two operations (see equations 50 and 51) consecutively:

$$\Delta \mathbf{W}_{ij}^{bp} = \eta \Delta \mathbf{W}_{ji}^{S}(t+1) + \alpha \Delta \mathbf{W}_{ji}(t) \quad (50)$$

$$\Delta \mathbf{W}_{ji}^{sm}(t+1) = \mathbf{W}_{ji}^{bp}(t+1) - 2\eta\xi \left[2\mathbf{W}_{ji}^{bp}(t+1) - \mathbf{W}_{j(i-1)}^{bp}(t+1)\, \mathbf{W}_{j(i+1)}^{bp}(t+1) \right] \quad (51)$$

Notice that equation 50 is nothing more than conventional back propagation with the mean square error. Equation 51 provides the smoothing operation based on the newer weights obtained by equation 50. The technique of updating based on these newer weights is termed overrelaxation and is frequently used in iterative optimization. Equation 50 can be rewritten as follows:

$$\Delta \mathbf{W}_{ji}^{sm}(t+1) = \gamma \mathbf{W}_{ji}^{bp}(t+1) + \frac{(1-\gamma)}{2} \left[\mathbf{W}_{j(i-1)}^{bp}(t+1) - \mathbf{W}_{j(i+1)}^{bp}(t+1) \right] \quad (52)$$

where

$$\gamma = 1 - 4\eta\xi \quad (53)$$

Equation 52 highlights the idea that the smoothing operation can be viewed as a weighted average between the weight $\mathbf{W}_{ji}$ and its two neighboring weights, with the constant γ acting as a weighting factor. Constant γ is chosen to be less than or equal to 1 with smaller values having the strongest smoothing effect. When $\gamma = 1$ the smoothing term has no effect. Notice that through the averaging process the smoothing effect is propagated throughout the network weights.

The smoothing constant may be annealed such that it has a strong effect early in training, but decreases such that in the later stages of training it approaches unity. As a result it has only a negligible effect during the latter stages of training where it may damage refinement of the solution. In the early stages of training γ enforces a strong smoothing effect and thus imposes a course structure on the solution.

A monotonically increasing function can be chosen for γ as follows:

$$\gamma = \gamma(t) = 1 - (1 - \gamma_0)e^{-t/T} \quad (54)$$

where γ_0 and T are constants. Constant T determines the rate at which the smoothing effect will decay and thus should be chosen in accordance with the training set size. Constant γ_0 is chosen fairly close to 1 to avoid causing too strong a smoothing effect compared to the gradient descent based on the mean square error.

4.10 TRAINING SET SIZE AND NETWORK SIZE

For a given network topology with inputs of a given dimensionality, there will be some minimum number of training samples required to successfully train the network weights. The required number of samples grows exponentially [Kneer et al., 1992] with the dimensionality of the input space, a phenomenon that is sometimes called the "curse of dimensionality". This among other reasons will motivate the use of feature extraction techniques. Feature extraction often provides a means by which to reduce the dimensionality of the input vector while retaining information that will be salient to recognition.

The feed-forward network can, of course, be used in conjunction with a wide variety of feature extraction techniques and may be trained by a variety of methods. The discussion of networks whose distinguishing features arise out of a front-end feature extraction, neutral or otherwise, is referred to in chapter 6.

The training set must be both large enough and diverse enough to adequately represent the problem domain. At peril of stating the obvious, each class that will be recognized must be present. Perhaps less obviously, counterexamples even from classes that will not be recognized may also be required. This is especially true when dealing with OCON. If an OCON network is trained to recognize the digit 2, it should also be trained with counterexamples of the other digits that are not 2's.

Within each class sufficient samples must be present to reflect real-world variations within the class. Returning to the digit recognition example, people write their 2's in a great many and sometimes surprising ways. Such variations may even be regional and are almost certainly national.

We have observed earlier that the larger a network becomes, the more likely it is to be overfitted. The remedy lies in providing training data in sufficient abundance that the network does not learn all of the minutiae present in the training vectors and therefore is forced to generalize. Thus, how many training set vectors will be enough for a given topology? If there are L_0 input units and L_1 hidden units in the first hidden layer, the number of weights between the input layer and the first hidden layer is $N = (L_0 + 1) \times L_1$, including the bias weight. Where the input layer is large compared to other layers, these constitute the majority of free parameters in the system. In this case one rough guideline would be to use at least $2N$ and preferably $4N$ training vectors.

In the above a great deal of attention has been given to dealing with the apparent shortcomings of back propagation, in particular, and to gradient descent, in general. It may appear that these techniques are too filled with problems to be useful. If that impression was given, it was entirely unintentional. Gradient descent techniques are both ubiquitous and powerful. The great volume of work devoted to their refinement is only testimony to their great utility. Before leaving the topic of gradient descent we will briefly discuss one of the more important second-order techniques.

4.11 CONJUGATE GRADIENT METHOD

The conjugate gradient method first establishes a direction vector in weight space based on the first-order gradient only and then determines how far to proceed along that vector based on the second-order gradient. As with back propagation, there are many variations of the conjugate gradient. The following discussion follows that in Kung [1993]. Another somewhat less mathematically oriented discussion of conjugate gradient techniques can be found in Masters [1993].

Notationally: $\overline{\mathrm{W}}$ is a vector consisting of all weights

$\mathrm{E}'_{\overline{\mathrm{W}}}$ is the gradient vector

$\mathrm{E}''_{\overline{\mathrm{W}}}$ is the Hessian matrix

After computing the gradient vector (by utilizing the back-propagation algorithm) the direction vector $\mathbf{d}_k$ is updated:

$$\mathbf{d}_k = -\mathrm{E}'_{\overline{\mathrm{W}}}(k) + \beta_{k-1}\mathbf{d}_{k-1} \tag{55}$$

$$\mathbf{d}_0 = \mathrm{E}'_{\overline{\mathrm{W}}}(0) \tag{56}$$

where β_{k-1} may be computed as:

$$\beta_{k-1} = \frac{\mathrm{E}'_{\overline{\mathrm{W}}}(k)^T \mathrm{E}'_{\overline{\mathrm{W}}}(k)}{\mathrm{E}'_{\overline{\mathrm{W}}}(k-1)^T \mathrm{E}'_{\overline{\mathrm{W}}}(k-1)} \tag{57}$$

The step size for the update is now determined by:

$$\eta_k = \frac{\mathrm{E}'_{\overline{\mathrm{W}}}(k)^T \mathrm{E}'_{\overline{\mathrm{W}}}(k)}{\mathbf{d}_k^T \mathrm{E}''_{\overline{\mathrm{W}}}(k)\mathbf{d}_k} \tag{58}$$

and finally the weight vector is updated:

$$\overline{\mathrm{w}}_{k+1} = \overline{\mathrm{w}}_k + \eta_k \mathbf{d}_k \tag{59}$$

From equation 55 we can see that the direction of update is the vector sum of two terms. The first term is the gradient of the error function in weight space, exactly what we experienced in back propagation. The second term is somewhat reminiscent of the momentum term from back propagation. In the back-propagation algorithm the momentum tended to sustain the movement of a particular weight based on its individual history. The term $\beta_{k-1}\mathbf{d}_{k-1}$ operates more globally. Its tendency is toward sustaining the previous overall direction of the last update. We may recall, fondly or otherwise, the gentle art of choosing free parameters for the back-propagation algorithm. We notice now with some relief that although β_{k-1} plays a role that is somewhat analogous to that of the momentum constant α in back propagation, it is not a free parameter. Equation 57 tells us exactly how to choose it as the ratio of the squared magnitude of the current gradient to the squared gradient of the previous cycle.

Referring to equation 59 we observe that the term η_k plays the role of back-propagation learning rate by establishing the step size. It is the distance in weight space that we will travel along the direction vector. Unlike back propagation, this is not a free parameter that must be optimized through intuition and experiment. Let us examine what equation 58 has to tell us about the calculation of step size. In particular, we note the second-order gradient in the denominator. Thus, when the slope of the error surface is changing sharply, the tendency becomes to select a smaller step size. Conversely, where the slope of the error surface is changing, very little the step size is increased.

One major benefit of the conjugate gradient technique (in comparison to back propagation) is elimination of the necessity to find learning parameters by trial and error, a daunting task for large training sets. Another is its tendency to converge in fewer epochs than back propagation although we must bear in mind that each conjugate gradient training cycle is significantly more computationally intensive than its back-propagation alternative. In general, second-order methods do offer superior numerical performance, but should be used in a block adaptive context because the second-order gradients are numerically sensitive to compute.

Gradient descent techniques are a powerful and popular means to train multi-layered perceptron neural networks. They are, however, far from the only class of method which can be used. Especially when we wish to vary the error function or activation function, it is convenient to have a training algorithm that does not depend on being able to calculate the gradient. What happens, for example, if we have chosen an error function that is not differentiable? Several alternate training algorithms will be encountered as various character recognition implementations are discussed. One such algorithm, ALOPEX, is of special interest for its broad applicability over network structures, activation functions, error functions, and neuronal types.

4.12 ALOPEX

ALOPEX is a stochastic parallel algorithm which treats the learning process in a neural network as an optimization problem [Pandya and Szabo, 1991]; Venugopal and Pandya, 1991].

ALOPEX is the acronym for algorithm for pattern extraction. The ALOPEX procedure was first described in connection with the problem of ascertaining the shapes of visual receptive fields [Harth and Tzanakou, 1974; Tzanakou et al., 1979]. Harth [1976] originally proposed a model of visual perception as a stochastic process where sensory messages received at a given level are modified to maximize responses of central pattern analyzers. Later computer simulations were carried out using a feedback generated by a single scalar response, and very simple neuronal circuits in the visual pathway were shown to carry out the ALOPEX algorithms [Harth et al., 1987]. Harth and Pandya [1988] showed that the process represented a new method of approaching a classical mathematical problem, that of optimizing a scalar function $f(x1, \ldots, xN)$ of N parameters xi, $i = 1, \ldots, N$ where N is a large number. Herman et al. [in press] have discussed ALOPEX in the context of pattern classification, in particular using piecewise linear classification. Rosenfeld [1987] refers to it in the survey of picture processing algorithms.

Over the past few years back propagation has become the most popular learning algorithm for neural networks. As Hinton [1989] points out, in spite of its impressive performance on relatively small problems and its promise as a widely applicable, robust algorithm for extracting underlying structure of a domain, back-propagation scales poorly as the learning task becomes complex. Also, it is interesting to note that this most popular learning algorithm (back propagation) is not plausible biologically. There is no evidence that biological synapses can reverse direction to propagate the error or the neurons can compute derivatives [Hinton, 1989]. The general drawbacks of gradient descent approaches apply to the back-propagation algorithm also. The algorithm is sensitive to initial weight distribution [Kolen and Pollack, 1990] and may converge to a local minimum rather than to a global minimum [Rumelhart et al., 1986; Ackley et al., 1985]. Further, the back-propagation algorithm [Rumelhart et al., 1986] requires the nonlinear neuronal transformation to be a monotonically increasing and differentiable function. The existence of the derivative is a necessary requirement. Also a condition can arise in training by back propagation in which the role of the derivative leads to a difficulty that is not encountered using ALOPEX (see section 4.5).

ALOPEX has the advantage that it is computationally simple and can be realized in high-speed very large-scale integration (VLSI) since interconnections between processing elements which update the weights are not required. While back propagation computes new weights by recursively computing the gradient through back-propagating error information, ALOPEX works by broadcasting a measure of global performance (a scalar cost function) to all the weights processors in the network. Weight changes are made by each weight processor stochastically, based on a feedback which is a correlation between changes in its own output and the change in the scalar cost function. Since each weight processor solely depends on the global cost function, no interaction is needed between processors and the algorithm is more amenable to parallel implementation in hardware.

Results from Pandya and Venugopal [1994] demonstrate that a network trained with ALOPEX has better measurement noise rejection capabilities compared to back propagation over a large range of signal-to-noise ratio (SNR).

Monk's problems were initiated by Thrun et al. [1991] to evaluate the performance of various learning techniques. When applied to Monk's problems ALOPEX was observed to generalize better than many of the alternative algorithms (including back propagation) especially over noisy data sets. Note that the training set for

Monk's problem is much smaller than the test set, thus providing a good test of generalization capability.

We will now describe the ALOPEX algorithm as it is used for training neural networks. For context, let the net input to the ith neuron be

$$\text{net}_i = \sum_j w_{ij}\,\text{out}_j + \theta_i \tag{60}$$

and assume a sigmoidal activation function the output of the jth neuron will be

$$\text{out}_i = \frac{1}{1 + e^{-\text{net}_i / Q_0}} \tag{61}$$

(Note, however, that ALOPEX does not assume or require any special choice of activation function.)

The ALOPEX algorithm specifies that the network weights be updated as follows during the nth iteration:

$$w_{ij}(n) = w_{ij}(n-1) + \delta(n) \tag{62}$$

$$\delta_{ij}(n) = \begin{cases} -\delta \text{ with probability } P_{ij}(n) \\ +\delta \text{ with probability } P_{ij}(n) \end{cases} \tag{63}$$

δ is a small positive constant and the probability $P_{ij}(n)$ is given by:

$$P_{ij}(n) = \frac{1}{1 + e^{\frac{\Delta_{ij}(n)}{T}}} \tag{64}$$

with

$$\Delta_{ij}(n) = \Delta w_{ij}(n)\,\Delta E(n) \tag{65}$$

where

$$\Delta w_{ij} = \left[w_{ij}(n-1) - w_{ij}(n-2)\right] \tag{66}$$

$$\Delta E(n) = \left[E(n-1) - E(n-2)\right] \tag{67}$$

If e_k is the error at output neuron k, then the total error, E, taken over all output neurons is given by:

$$E = \sum_k e_k \tag{68}$$

The algorithm takes a biased random walk in the direction of decreasing error, E. The step size, δ, is constant and the temperature T determines the effective randomness of the walk. It is suggested that T be chosen large initially and decreased as training progresses to the average value of the correlation Δ_{ij}.

$$T = |\Delta| \tag{69}$$

It is easily verified that the sense of the probability is always skewed toward lowering the total error, E. To see this more clearly consider Figure 4.22 which illustrates the sense in which the probability will be taken.

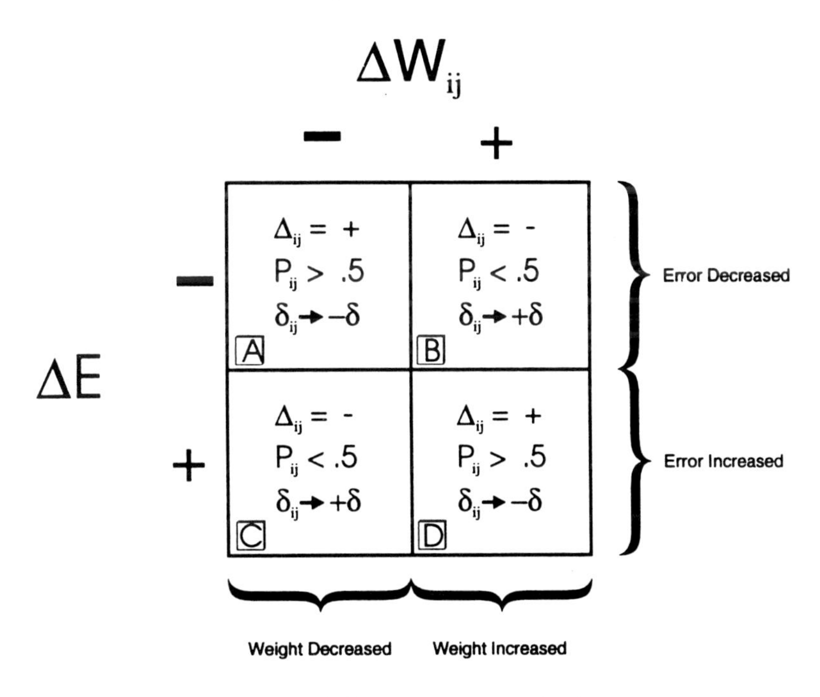

FIGURE 4.22 ALOPEX combinations.

A comparison of the convergence characteristics of ALOPEX training in contrast to back propagation is of interest. It is pointed out that ALOPEX will understandably not converge as quickly as back propagation for a class of problems lacking local minima and having strong global minima. Under such specific circumstances the problems which ALOPEX provides protection against are not present while the specific situation which back propagation is very good at appears in abundance. This is not a serious drawback. There are very few problems indeed for which these conditions are likely to be known to exist in advance, and for other classes of problems ALOPEX performs favorably. Even in those cases in which ALOPEX may not converge as quickly as back propagation, its other advantages may still provide strong motivation for its use.

ALOPEX will be a strong candidate for many possible problems where its strengths can be utilized. In summary, ALOPEX may be chosen for:

1. Its flexibility with respect to network topology and the choice of error function and neuronal types
2. Its tendency to avoid local minima due to its stochastic nature
3. Its independence of the derivatives which mark gradient descent
4. Its ability to generalize well (especially over noisy data)
5. Its applicability to highly parallel implementations

Due to its popularity a great deal of effort has been expended to enhance and extend the back-propagation algorithm. By contrast, ALOPEX being comparatively obscure has received less attention. Modifications which could improve the ALOPEX algorithm include an annealing scheme similar to the "simulated annealing", a temperature perturbation scheme to get out of local minimum, an addition of a momentum term, etc.

One cautionary note, however, needs to be made regarding the implementation of ALOPEX before leaving the topic. Clearly the ALOPEX algorithm will be highly dependent on random numbers for the generation of the probabilities $P_{ij}(n)$. The random number generators available in most compilers use a linear congruential algorithm and thus are subject to some severe limitations [Masters, 1993]. Let p_n be a nonnegative integer in the pseudo-random sequence. By the linear congruential algorithm the next pseudo-random number is generated as: $p_{n+1} = (ap_n + c) \bmod m$. Most C-compilers use $m = 2^{15} = 32{,}768$. As a result there are, at most, that many random numbers available. This problem is aggravated by the fact that the algorithm is periodic. Once a given value is seen, the same sequence will always follow from it. This means there can be, at most, m sequences with no real guarantee that there will be even that many. As a result, a more propitious choice of a random generator to be used in conjunction with ALOPEX is indicated.

The effectiveness of the ALOPEX algorithm for continuous recognition of undersea targets from side-scan sonar returns has been investigated by Pandya and Venugopal (1994). This is a practical problem in undersea explorations and obstacle avoidance (for underwater vehicle navigation), which requires human involvement to a large extent because of the complex and noisy nature of the sonar returns. The ALOPEX algorithm was used for training a network architecture chosen to incorporate the learning of spectral temporal dependencies of the consecutive sonar returns. Figure 4.23 shows the network architecture used in the case of the recognition problem. The power densities of the sonar returns are presented as the inputs to the neural network and each group in the input layer has 31 neurons. The first hidden layer consists of four groups of 62 neurons each, and the second hidden layer consists of four groups having 2 neurons each. At an instant five consecutive returns are presented.

The simulations were conducted with side-scan sonar returns corresponding to two known targets: a wooden box and a metallic drum. The training was performed with 52 returns in each class. Figure 4.24 (a) shows the error convergence characteristic

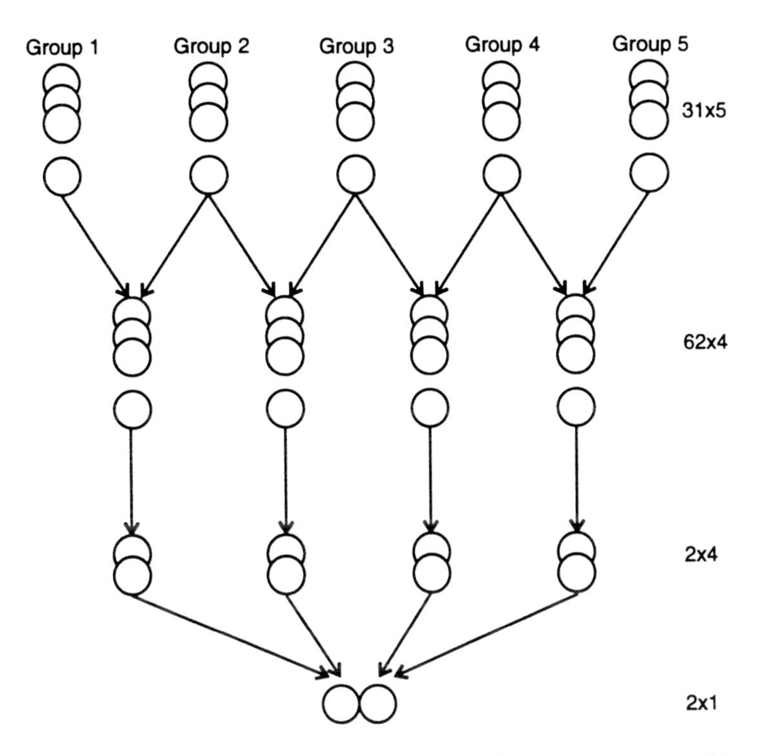

FIGURE 4.23 The neural network architecture for sonar target recognition.

for different numbers of hidden layer neurons. It can be seen that an increase in the number of hidden layer neurons does not increase the rate of convergence. Similar characteristics were obtained during the experiments when the step size was changed (Figure 4.25 [b]). The converged network was presented with 12 test returns and could classify all of them correctly.

The implementation of the ALOPEX algorithm is given in listing 4.8.

Listing 4.8

```
*******************************************************************************
*                                                                             *
* ALOPEX                                                                      *
*                                                                             *
*******************************************************************************/

#include <stdio.h>
#include <stdlib.h>
#include <string.h>
#include <conio.h>
#include <math.h>

// FUNCTION PROTOTYPES
```

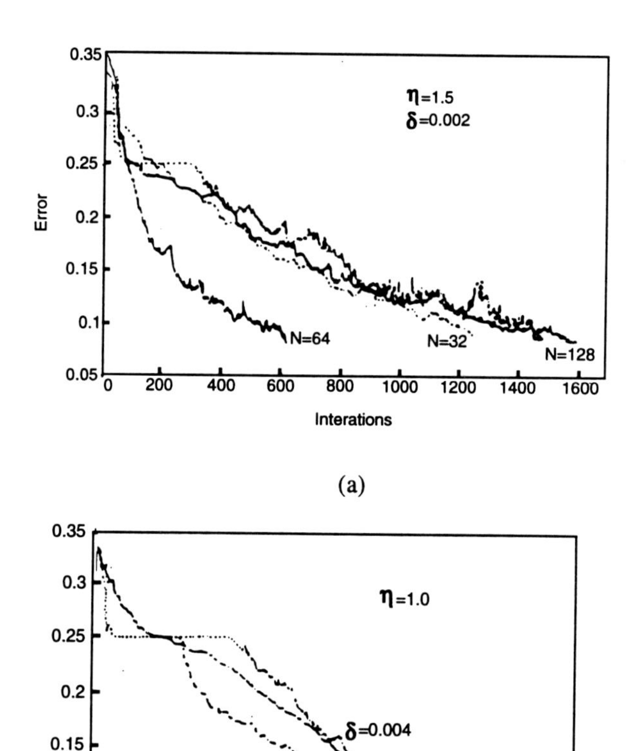

FIGURE 4.24 Change in the error convergence (a) when the number of neurons in the first hidden layer is changed; (b) as the step size is changed.

```
void  ShowResults(int, long int,double temp, double alpha, double eta);

// DEFINES

#define MAXLAYERS 4        // MAX NUMBER OF LAYERS IN NET (IN OUT & HIDDEN)
#define MAXNEURONS 30  // MAX NUMBER OF NEURONS PER LAYER
#define MAXPATTERNS 10  // MAX NUMBER OF PATTERNS IN A TRAINING SET
#define MAXPREV     3        // NUMBER OF PREV EPOCH FOR WHICH WEIGHTS
                              ARE MAINTAINED
#define ARCGRAN   250      // ARCHIVE GRANULARITY
#define NONE        0          // DO NOT ARCHIVE ERROR DATA
#define ALL         1            // ARCHIVE ALL ERROR DATA FOR EACH ITERATION
#define AVERAGE 2            // ARCHIVE AVERAGE ERROR DATA FOR EACH
                              EPOCH
```

```
#define TRUE     1
#define FALSE    0

//-----------------------------------------------------------------------

class ALOPEX
{
private:
 double W[MAXLAYERS][MAXNEURONS][MAXNEURONS];        // WEIGHTS MATRIX
 double Wprev1[MAXLAYERS][MAXNEURONS][MAXNEURONS]; // previous WEIGHTS
for correlation
 double Neuron[MAXLAYERS][MAXNEURONS];               // Output of all net neurons
 double ERROR[MAXPATTERNS][MAXNEURONS];  // Stored error over epoch/outlayer
 double AvgErr;

// Alopex specific data
 double delta;                           // The ALOPEX step size
 double T;                               // Temperature for ALOPEX probability fn
 double Ep1;                             // Sum error over iteration n-1
 int AnnealCount;
 double SumDELTAc;

// Topology
 int NumLayers;                     // Number of layers
 int OutLayerIndx;                  // array index of last layer
 int LayerSize[MAXLAYERS];          // Number of neurons in the each layer
 int NumWeights;                    // Tot number of weights

// Pattern data
 double InPattern[MAXPATTERNS][MAXNEURONS];   // Input values for each pattern
 double Desired[MAXPATTERNS][MAXNEURONS];     // desired value for each
                                              //  pattern/output
 unsigned long int CurrIter;                  // Current  iterations number.
 long int Epoch;                              // Counts number of epochs
 int EpochFlag;                               // True at end of each epoch
 int NumPatterns;                             // Total patterns in training set
 int CurrPat;                                 // Current pattern used in training
 int ConvCount;                               // The number of consecutive
                                              // patterns within tolerance
 int ConvergeFlg;                             // Flag indicates convergance has
                                              // occurred

// PARMS
 double Qzero;                      // For Neuron activation function
 double ERRTOL;                     // min Error tolerance required for convergence
 double ETA;                        // ALOPEX Version of learning rate
 unsigned long int MAXITER;         // Max iterations to do before stopping

// PRIVATE METHODS
 double ErrFunc(double Target, double Actual);        // Calc the log error
 double CalcSumErr();               // Calc tot err over all patterns/outlayer neurons
 void  RecordErrors();              // At each iter, record error results
 void  ArchivePut();                // Keep a record on disk of net progress
```

```
public:
 ALOPEX(void);
 void  GetTrnSet(char *Fname);                  // Load training set from file
 void  GetInputs(void);                         // Load input layer w/ curr pattern
 void  RunNet(void);                            // Run the networks forward pass
 void  GetParms(char *);                        // Load parameters from parm file
 double Sigmoid(double Net, double Tempr);      // The non-linear squashing function
 void  SetRandomWeights(void);                  // Init weights to random values
 int   IsConverged(void);                       // Check for convergence
 void  AdaptWeights(void);                      // Modify weights by ALOPEX method
 void  SaveWeights(char *);                     // Save trained weights to file
 int   train(char *, char *);                   // Top level control of training the net
 long int QueryEpoch(){return Epoch;}
 double QueryQzero(){return Qzero;}
};


// MISC FLAGS
int ArchOn  = AVERAGE;      // Set to 1= ALL, NONE, AVERAGE or WORST to control
                            //  data sent to ARCHIVE file for graphical analysis
FILE *ARCHIVE;              // Archive Training sequence
//---------------------------------------------------------------------------
```

```
// METHOD DEFINITIONS

/**************************************************************************
* Function GetTrnSet                                                      *
*    Class constructor                                                    *
**************************************************************************/

ALOPEX::ALOPEX() {
 AvgErr  = 0.0;
 CurrIter=0;
 Epoch=0;
 AnnealCount=0;
 CurrPat = -1;                          // Current pattern used in training
 ConvCount=0;
 ConvergeFlg=FALSE;                     // Flag indicates convergence
 SumDELTAc=0.0;
 ETA=2.0;
}

/**************************************************************************
* Function GetTrnSet                                                      *
*    Load training set                                                    *
**************************************************************************/

void ALOPEX::GetTrnSet(char *Fname) {   // for small training sets
FILE *PFILE;
int  x;
int  pat,i,j;
double inVal;

PFILE = fopen(Fname,"r");
if (PFILE==NULL){
  printf("\nUnable to open file \n");
  exit(0);
  }
fscanf(PFILE,"%d",&NumPatterns);

for (pat=0; pat<NumPatterns; pat++) {
  for (i=0; i<LayerSize[0]; i++) {
    fscanf(PFILE,"%lf",&inVal);
    InPattern[pat][i]=inVal;
    } /* endfor */

  // Now get desired / expected values
  for (i=0; i<LayerSize[OutLayerIndx]; i++) {
    fscanf(PFILE,"%lf",&inVal);
    Desired[pat][i]=inVal;
    } /* endfor */
  } /* endfor */
fclose(PFILE);
}
```

```
/*****************************************************************************
*                                                                           *
* Function GetParms                                                         *
*   Loads topology from file spec'd file if avail.                          *
*   otherwise tries DEFAULT.PRM                                             *
*                                                                           *
*****************************************************************************/

void ALOPEX::GetParms(char *PrmFileName){
FILE *PRMFILE;
PRMFILE = fopen(PrmFileName,"r");
if (PRMFILE==NULL){
  printf("\nUnable to open Parameter file: %s \n",PrmFileName);
  printf("\nAttempting to open default file: PARMS1");
  PRMFILE = fopen("PARMS1","r");
  if (PRMFILE==NULL){
    printf("\nUnable to open Parameter file\n");
    exit(0);
    }
  }
printf("\nLoading Parameters from file: %s\n",PrmFileName);
fscanf(PRMFILE,"%lf",&Qzero);
fscanf(PRMFILE,"%lf",&ETA);
fscanf(PRMFILE,"%lf",&delta);
fscanf(PRMFILE,"%lf",&T);
fscanf(PRMFILE,"%ld",&MAXITER);
fscanf(PRMFILE,"%lf",&ERRTOL);
fscanf(PRMFILE,"%d",&NumLayers);

printf("&Qzero=%lf",Qzero);
printf("&delta=%lf",delta);
printf("T=%lf",T);
printf("MAXITER=%ld",MAXITER);
printf("ERRTOL=%lf",ERRTOL);

printf("\nNumber of layers=%d\n",NumLayers);
for (int i=0; i<NumLayers; i++) {
  fscanf(PRMFILE,"%d",&LayerSize[i]);
  printf("Number of neurons in layer %d = %d\n",i,LayerSize[i]);
  }
OutLayerIndx = NumLayers-1;                         // accommodate 0 org'd arrays
fclose(PRMFILE);

NumWeights=0;
for (int lyr=OutLayerIndx; lyr>0; lyr--)            //Visit each layer
  NumWeights += LayerSize[lyr] *  LayerSize[lyr-1];
printf("\n%d\n",NumWeights);
}

/*****************************************************************************
* FUNCTION SetRandomWeights                                                 *
*                                                                           *
*   Initialize weights between 0 and 1                                      *
*****************************************************************************/
```

```
void ALOPEX::SetRandomWeights(){
int i,j,k;
double zWT;
  srand(6);
  for (i=1; i<NumLayers; i++) {
    for (k=0; k<LayerSize[i]; k++) {
      for (j=0; j<=LayerSize[i-1]; j++) {              // One extra for bias neuron
        zWT=(double)rand();
        W[i][j][k] =zWT/32767.0;
        }
      }
    }
}


/*****************************************************************************
* FUNCTION GetInputs                                                         *
*                                                                            *
*   Loads input layer neurons with current pattern                           *
*****************************************************************************/

void ALOPEX::GetInputs(){
int i,j,k;
EpochFlag=FALSE;
CurrIter++;
CurrPat++;                                              // Update the current pattern
if (CurrPat>=NumPatterns){
  EpochFlag=TRUE;
  CurrPat=0;
  Epoch++;
  }

for (i=0; i<LayerSize[0]; i++) {
  Neuron[0][i]=InPattern[CurrPat][i];                   // Show it to the neurons
  }
}


/*****************************************************************************
* FUNCTION RunNet                                                            *
*                                                                            *
*   Calculates outputs for all network neurons                               *
*****************************************************************************/

void ALOPEX::RunNet(){
int lyr;    // layer to calculate
int dNeuron; // dest layer neuron
int sNeuron; // src layer neuron
double SumNet;
double out;

for (lyr=1; lyr<NumLayers; lyr++) {
  Neuron[lyr-1][LayerSize[lyr-1]]=1.0; //force bias neuron output to 1.0
  for (dNeuron=0; dNeuron<LayerSize[lyr]; dNeuron++) {
    SumNet=0.0;
```

```
    for (sNeuron=0; sNeuron <= LayerSize[lyr-1]; sNeuron++) {          //add 1 for bias
      SumNet += Neuron[lyr-1][sNeuron] * W[lyr][sNeuron][dNeuron];
       }
    out=Sigmoid(SumNet,Qzero);
    Neuron[lyr][dNeuron] = out;
     }
   }
}


/******************************************************************************
* Function RecordErrors ( )                                                   *
*   Save error results by neuron & pattern for later use                      *
******************************************************************************/

void ALOPEX::RecordErrors(){
   int j;
for (j=0; j<LayerSize[OutLayerIndx]; j++) {
  ERROR[CurrPat][j] = ErrFunc(Desired[CurrPat][j], Neuron[OutLayerIndx][j]);
   } /* endfor */
}

/******************************************************************************
* IsConverged                                                                 *
*   Calculates error for each neuron and each pattern                         *
******************************************************************************/


int ALOPEX::IsConverged(){
double dNETj;
double SumErr;                                        //Cumulative error
int i, j;
int LocalConvFlg=TRUE;

SumErr=CalcSumErr();
if (SumErr>ERRTOL) {
   return FALSE;
 } else {
   return TRUE;
   } /* endif */
}


/******************************************************************************
* Function ArchivePut()                                                       *
*   Create a record of training results                                       *
******************************************************************************/

void ALOPEX::ArchivePut(){
int i, j;

AvgErr =CalcSumErr()/NumPatterns;

if (ArchOn==AVERAGE) {
  if ((ARCGRAN*(Epoch/ARCGRAN)) == Epoch) {        // only save ARCGRANth epochs
```

```
    fprintf(ARCHIVE,"%ld %lf\n",Epoch,AvgErr );
    printf("\nEpoch:%d ",Epoch);
    printf("Avg=%lf\n",AvgErr);
     } /* endif */
  }
}

/*****************************************************************************
* Function ErrFunc                                                           *
*    Calculates the error for a given output neuron & target value           *
*****************************************************************************/

double ALOPEX::ErrFunc(double Target, double Actual) {
  double E;
if (Target>=.99) {
   E=log(1/Actual);
   }
  else {
   E=log(1/(1-Actual));
   } /* endif */
 return E;
 }

/*****************************************************************************
* Function CalcSumErr()                                                      *
*    Calculates total error over all (ouput neurons) and patterns            *
*****************************************************************************/

double ALOPEX::CalcSumErr(){
  int j,Pat;
  double E,desire;
 E=0.0;
 for (Pat=0;Pat<NumPatterns ; Pat++) {
  for (j=0; j<LayerSize[OutLayerIndx]; j++) {
     E += ERROR[Pat][j];
     }
   } /* endfor */
 return E;
 }

/*****************************************************************************
* Function AdaptWeights()                                                    *
*    Calculates new weights based on the ALOPEX algorithm                    *
*****************************************************************************/

void ALOPEX::AdaptWeights(){
double E,R,P, DELTAc;
double DELTAWEIGHT, DELTAERROR;
int lyr, i, j,k;

E=CalcSumErr();
DELTAERROR=E-Ep1;
```

```
Ep1=E;                                              //Prev iter errors
for (lyr=OutLayerIndx; lyr>0; lyr—) {               // Visit each layer
  for (j=0; j<LayerSize[lyr]; j++) {                // Visit each neuron
    for (k=0; k<=LayerSize[lyr-1];k++ ) {           // Visit all Weights (incl. bias)
      DELTAWEIGHT= W[lyr][j][k]-Wprev1[lyr][j][k];
      DELTAc= DELTAWEIGHT*DELTAERROR;   //Calculate the correlation
      SumDELTAc+=fabs(DELTAc);          //Running sum of correlation over all weights
      Wprev1[lyr][j][k]=W[lyr][j][k];
      if (Epoch > 1) {                              //Wait till system is primed
        P = 1/(1 + exp(ETA*DELTAc/T));              // Probability fn once it is
        }
      else
        P= 0.5;                                     //50% till then
      R=(double)rand();
      R=R/32767.0;
      if (R>P) {
        W[lyr][j][k]=W[lyr][j][k] - delta;
        }
      else {
        W[lyr][j][k]=W[lyr][j][k] + delta;
        } /* endif */
      } /* endfor */
    } /* endfor */
  } /* endfor */

//Establish a new temp for the alopex probability fn base on avg correlation
if ((AnnealCount>=10)&&(SumDELTAc>0.0)) {
  T=SumDELTAc/ AnnealCount;
  AnnealCount=0;
  SumDELTAc=0.0;
  } else {
  AnnealCount++;
  } /* endif */
}


/*******************************************************************************
* FUNCTION SaveWeights                                                         *
*                                                                              *
*   Output weights to default file DFLT.WGT                                    *
*******************************************************************************/
void  ALOPEX::SaveWeights(char *WgtName) {
int lyr,s,d;
double zWT;
FILE *WEIGHTFILE;

  WEIGHTFILE = fopen(WgtName,"w");
  if (WEIGHTFILE==NULL){
    printf("Unable to open weight file for output:%s\n",WgtName);
    exit(0);
    }
  printf("SAVING CALCULATED WEIGHTS:\n\n");
  fprintf(WEIGHTFILE,"0.00\n");                         // Threshold always 0
  fprintf(WEIGHTFILE,"%lf\n",Qzero);                    // Temperature
  fprintf(WEIGHTFILE,"%d\n",NumLayers);                 // Number of layers
```

```
  for (lyr=0; lyr<NumLayers; lyr++) {                  // Save topology
    fprintf(WEIGHTFILE,"%d\n",LayerSize[lyr]);          // Number of neurons/layer
    }
  for (lyr=1; lyr<NumLayers; lyr++) {                  // Start at 1st hidden
    for (d=0; d<LayerSize[lyr]; d++) {
      for (s=0; s<=LayerSize[lyr-1]; s++) {            // One extra for bias
         zWT=W[lyr][s][d];
         fprintf(WEIGHTFILE,"%lf\n",zWT);
         }
      }
    }
  fclose(WEIGHTFILE);
}


/*****************************************************************************
* FUNCTION Sigmoid                                                           *
*                                                                            *
*                                                                            *
*   Output non-linear squashing function                                     *
*****************************************************************************/

double ALOPEX::Sigmoid(double Net, double Tempr){
  double x;
x = Net/Tempr;
if ( x >100.0)
  return 1.0;
if ( x < -100.0)
  return 0.0;
return 1.0/(1.0 + exp(-x));
}


/*****************************************************************************
* FUNCTION train                                                             *
*   Trains the network                                                       *
*    Input: Training data file name and Parameter file name                  *
*    Output:non-linear squashing function                                    *
*****************************************************************************/

int   ALOPEX::train(char *TrnFname, char *ParmFname){
if (ArchOn) ARCHIVE=fopen("archive.lst","w");
GetParms(ParmFname);
GetTrnSet(TrnFname);
SetRandomWeights();
int Converged=0;
while ((!Converged) && (CurrIter < MAXITER) ) {
  GetInputs();
  RunNet();
  RecordErrors();
  if (CurrPat==0)                            // Test for full epoch
    ArchivePut();
   Converged=IsConverged();
    AdaptWeights();
```

```
  }
if (ArchOn) fclose(ARCHIVE);
return Converged;
}

//------------------------------------------------------------------------------


/*******************************************************************************
* FUNCTION ShowResults                                                         *
*                                                                              *
*   Output Summary specific to the parity-3 problem                            *
*******************************************************************************/

void ShowResults(int ConvergeFlg, long int CurrEpoch, double temp){
printf("\n---------------------------------------------------------------\n");
if (ConvergeFlg) {
  printf("SUCCESS: Convergance has occured at iteration %ld\n",CurrEpoch);
  }
 else {
  printf("FAILURE: Convergance has NOT occured!!\n");
  } // endif
printf("Temperature = %lf\n",temp);
printf("\n---------------------------------------------------------------\n");
}

ALOPEX  Alopex;

/*******************************************************************************
* MAIN                                                                         *
*******************************************************************************/


int main(int argc, char *argv[])
{
int Converged ;

if (argc>3) {
  Converged = Alopex.train(argv[1],argv[2]);
  }
 else {
  printf("USAGE: ALOPEX TRAINING_FILE PARMS_FILE WEIGHT_FILE\n");
  exit(0);
  }

// Show how we did
ShowResults(Converged, Alopex.QueryEpoch(), Alopex.QueryQzero());
if (Converged)
  Alopex.SaveWeights(argv[3]);  // Save the weights for later use

return 0;
}
```

REFERENCES AND BIBLIOGRAPHY

Ackley, D. H., Hinton, G. E., and Sejnowski, T. J., "Learning algorithm for Boltzman machines," *Cognit. Sci.*, vol. 9, pp. 147–169, 1985.

Baker, T. and McCartor, H., "A comparison of neural network classifiers for optical character recognition," *Proc. SPIE, vol. 1661, pp. 191–202, 1991.*

Bebbis, G. N., Georgiopoulis, M., Papadourakis, G. M., and Heilman, G. L., "Increasing classification accuracy using multiple neural network schemes," *Proc. SPIE Applications of Neural Networks III*, vol. 1709, pp. 221–231, 1992.

Cybenko, G., "Approximation by superpositions of a sigmoidal function," *Math. Control, Signals, Syst.*, vol. 2, pp. 303–314, 1989.

Funahashi, K., "On the approximate realization of continuous mappings by neural networks," *Neural Networks*, vol. 2, no. 3, pp. 183–192, 1989.

Hammerstrom, D., "Working with neural networks," *IEEE Spectrum*, July 1993, pp. 46–53, 1993.

Harth, E., "Visual perception: a dynamic theory," *Biol. Cybern.*, vol. 22, pp. 169–180, 1976.

Harth, E. and Tzanakou, E., "ALOPEX: a stochastic method for determining visual receptive fields," *Vision Res.*, vol. 14, pp. 1475–1482, 1974.

Harth, E., Unnikrishnan, K. P., and Pandya, A. S., "The inversion of sensory processing by feedback pathways: a model of visual cognitive functions," *Science*, vol. 237, pp. 187–189, 1987.

Harth, E. and Pandya, A. S., "Dynamics of the ALOPEX Process: Applications to the Optimization Problem," in *Biomathematics and Related Computational Problems*, L. M. Ricciardi (Ed.), Reidel Publ., Amsterdam, pp. 459–471, 1988.

Hartman, K., Keeler, J. D., and Kowalski, J. M., "Layered neural networks with Gaussian hidden units as universal approximations," *Neural Computation*, vol. 2, pp. 210–215, 1990.

Haykin, S., *Neural Networks, A Comprehensive Foundation*, IEEE Society Press, Macmillan College Publishing, New York, 1994.

Herman, G. T., Odhner, D., and Yeung, K. T. D., "Optimization for pattern classification using biased random search techniques," *Ann. Operations Res.*, (in press).

Hinton, G. E., "Connectionist learning procedures," *Artif. Intell.*, vol. 40, pp. 184–234, 1989.

Hornik, K., Stinchcombe, M., and White, H., "Multilayer feedforward networks are universal approximators," *Neural Networks*, vol. 2, pp. 359–366, 1989.

Jacobs, R. A., "Increased rates of convergence through learning rate adaptation," *Neural Networks*, vol. 1, pp. 295–307, 1988.

Jean, J. S. N. and Wang, J., "Weight smoothing to improve network generalization," *IEEE Trans. Neural Networks*, vol. 5, no. 5, 1994.

Kneer, S., Personnaz, L., and Dreyfus, G., "Handwritten digit recognition by neural networks with single-layer training," *IEEE Trans. Neural Networks*, vol. 3, no. 6, pp. 962–968, 1992.

Kolen, J. F. and Pollack, J. B., "Backpropagation is sensitive to initial conditions," *Complex Syst.*, vol. 4, pp. 269–280, 1990.

Kung, S. Y., *Digital Neural Networks*, Prentice-Hall, Englewood Cliffs, NJ, 1993.

Lee, Y., Oh, S. H., Song, H. K., and Kim, M. W., "Design rules of multilayer perceptron," *Proc. SPIE, Science of Artificial Neural Networks*, vol. 1710, pp. 329–339, 1992.

Masters, T., *Practical Neural Network Recipes in C++*, Academic Press, Boston, San Diego, and New York, 1993.

Pandya, A. S. and Szabo, R., "A fast learning algorithm for neural network applications," *Conf. Proc. 1991 IEEE Int. Conf. Syst. Man Cybern.*, vol. 3, pp. 1569–1573, 1991.

Pandya, A. S. and Venugopal, P., "A stochastic parallel algorithm for supervised learning in neural networks," *IEEE Trans. Inf. Syst.*, E77-D, no. 4, pp. 376–384, 1994.

Rosenfeld, A., "Picture processing: 1986," *Comput. Vision, Graphics, Image Process.*, vol. 38, pp. 147–225, 1987.

Rumelhart, D. E., Hinton, G. E., and Williams, R. J., "Learning Internal Representations by Error Propagation," in *Parallel Distributed Processing, Vol. 1: Foundations*, D. E. Rumelhart and J. L. McClelland (Eds.), MIT Press, Cambridge, MA, pp. 318–362, 1986.

Soucek, B. *Fast Learning and Invariant Object Recognition*, John Wiley & Sons, New York, 1992.

Thrun, S. B., Bala, J., et al., *The MONK's Problems: A Performance Comparison of Different Learning Algorithms,* Carnegie-Mellon University, Pittsburgh, CMU-CS-91–197, 1991.

Tsay, S., Hong, P., and Chieu, B., "Handwritten digit recognition via OCON neural network by selective pruning," *IEEE Proc. 11th Int. Conf. Pattern Recognition,* pp. 656–659, 1992.

Tzanakou, E., Michalak, R., and Harth, E., "The ALOPEX process visual receptive fields by response feedback," *Biol. Cybern.,* vol. 35, pp. 161–174, 1979.

Venugopal, K. P. and Pandya, A. S., "ALOPEX algorithm for training multilayer neural networks," *Proc. IJCNN, Singapore,* pp. 182–190, 1991.

Wang, J. and Jean, J., "Resolving multifont confusion with neural networks," *Pattern Recognition,* vol. 26, no. 1, pp. 175–187, 1993.

Zurada, J. M., Zigoris, D. M., Arohime, P. B., and Desai, M., "Classification of printed characters using multi-layer feedforward neural networks," *IEEE Proc. 34th Midwest Symp. Circuits Syst.,* vol. 2, pp. 191–202, 1991.

5 Some Other Types of Neural Networks

5.1 GENERAL

In this chapter we will describe two important variations on the feed-forward network. Both of these use neurons that are different from the familiar perceptron with a sigmoidal activation function.

5.2 RADIAL BASIS FUNCTION NETWORKS

Design of a neural network for pattern classification may be viewed as a curve-fitting problem in a hyperspace, where learning weights amounts to finding a hypersurface that provides a "best fit" to the training data. Generalization in this case involves use of this hypersurface to interpolate the test data. Cover [1965] showed that a complex pattern classification problem cast in high-dimensional space nonlinearity is more likely to be linearly separable than a low-dimensional space. The method of radial-basis functions (RBF) is a technique for interpolation in a high-dimensional space. Broomhead and Lowe [1988] were among the first users of the RBF technique to provide an alternative to learning in neural networks.

A radial-basis function network consists of a hidden layer of high enough dimension which provides a nonlinear transformation from the input space. The output layer in these networks provides a linear transformation from the hidden-unit space to the output space. Thus RBF networks in the basic form have only a single hidden layer compared to multilayer perceptrons (MLPs) which may have one or more hidden layers.

5.2.1 Network Architecture

Figure 5.1 shows the transformations imposed on the input vector by each layer of the RBF network. The hidden neuron performs the activation function on the input (not shown) and the classifier output is simply a weighted linear summation of the activation functions. Note that, unlike neurons in MLP which share a common neuron model regardless of their location in a layer, hidden layer neurons in an RBF network are quite different than output layer neurons.

Use of RBFs for transformation in hidden layer neurons allows higher dimension for the hidden-unit space. An activation function called RBF responds to a field of view around a fixed location "c" such that $f(x)$ is largest for $x = c$ and as $|x - c|$ increases $f(x) \to 0$ (see Figure 5.1). The Gaussian potential function is an example of such a function.

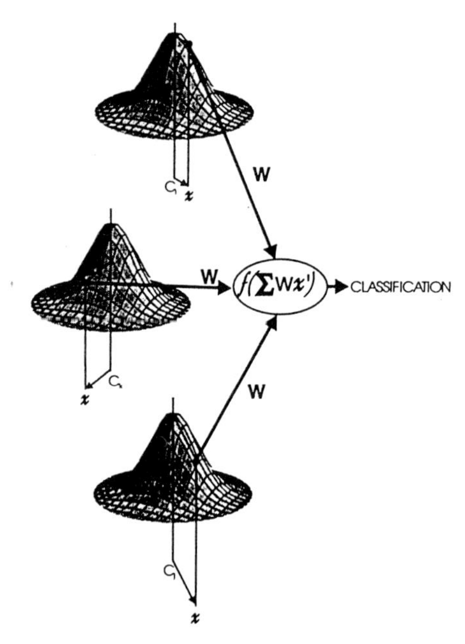

FIGURE 5.1 RBF network, transformation of input space. A response of hidden layer neurons is a function of how close x is to c_i.

When using radial basis function neurons, a category of patterns can be regarded as a Gaussian distribution of points in pattern space. The RBF neuron fires when its input is sufficiently close to activate the Gaussian. Each hidden layer neuron in a multilayer perceptron (MLP) evaluates a weighted sum of inputs. Instead, hidden layer neurons in RBFs encode the inputs by computing a measure of how close they are to a receptive field (e.g., distance between the input vector and the centroid of that neuron). The RBF classifiers thus belong to the group of receptive field classifiers. Their output nodes create complex decision regions by utilizing the overlapping localized regions found by the activation functions of the hidden layer neurons. Center and width are the two important parameters associated with the activation function.

For a set of inputs x_j and weights w_{ij}, the output of a radial basis function neuron will be

$$\text{out} = g\left(\sum_{j}^{n}\left(x_j - w_{ij}\right)^2\right) \tag{1}$$

where $g(\ )$ is the Gaussian function:

$$g\left(r^2\right) = ce^{-r^2/\sigma^2} \tag{2}$$

The distance function is taken as the Euclidean distance since it is the most popular one. The Gaussian function in this case serves as the activation function. For a detailed discussion on whether a particular function can be used as an RBF, see Micchelli [1986].

Figure 5.2 contrasts the MLP and RBF networks. In particular, notice the heterogeneous neurons in the RBF network. In contrast to MLPs which construct global approximations to nonlinear input-output mappings, RBFs construct local approximations using exponentially decaying localized nonlinearities (equation 2). Thus RBFs have greater local clustering power than the conventional perceptron (e.g., MLP).

Wong [1991] shows that because of the nature of the Gaussian, RBFs may have difficulty in learning the high frequency part of a mapping. This may be due to the fact that in order to represent an input-output mapping to a desired degree of smoothness, the number of RBFs required to span the input space adequately may have to be very large. Wong [1991] demonstrates the use of back propagation to train RBF networks. Kung [1993] indicates that the effectiveness of back propagation in this context is diminished.

5.2.2 RBF Training

Training in an RBF network involves finding the centers, widths, and weights connecting the hidden neurons to the output neurons. Clustering algorithms, such as the K-nearest-neighbor algorithm, can be used to address this problem. The corresponding center of the RBF neuron representing a cluster can be taken as the cluster mean. For a discussion of clustering algorithms see Duda and Hart [1973] and Fukunaga [1990].

Here we describe an iterative clustering algorithm, originally proposed by Musavi et al. [1991] as a possible candidate. This algorithm is particularly advantageous in that it minimizes the number of hidden layer neurons. The clustering algorithm is defined as follows:

1. Initialize by assigning each training point to a cluster.
2. Randomly label each of the clusters ($L = 1, \ldots, C$).
3. Select the first cluster ($L = 1$).
4. Locate any cluster in the same class as the selected cluster.
5. Merge the two clusters and compute the new mean.
6. Find the distance, d_{opp}, between the new mean value and the mean of the nearest cluster of the opposite class.
7. Compute the distance from the new mean to the farthest point belonging to its own cluster. This distance is defined as the radius, R, of the cluster.
8. Let α be defined as the clustering parameter:
   a. If $d_{opp} > \alpha R$, then accept the move defined in step 5. (The current value of L is associated with the newly created cluster. C is incremented to $C = C + 1$.)
   b. Otherwise, reject the merge. Recover the two original clusters and go to step 5 with L incremented to $L = L + 1$ and with C unaltered.
9. Repeat step 4 through step 8 until $L = C$.

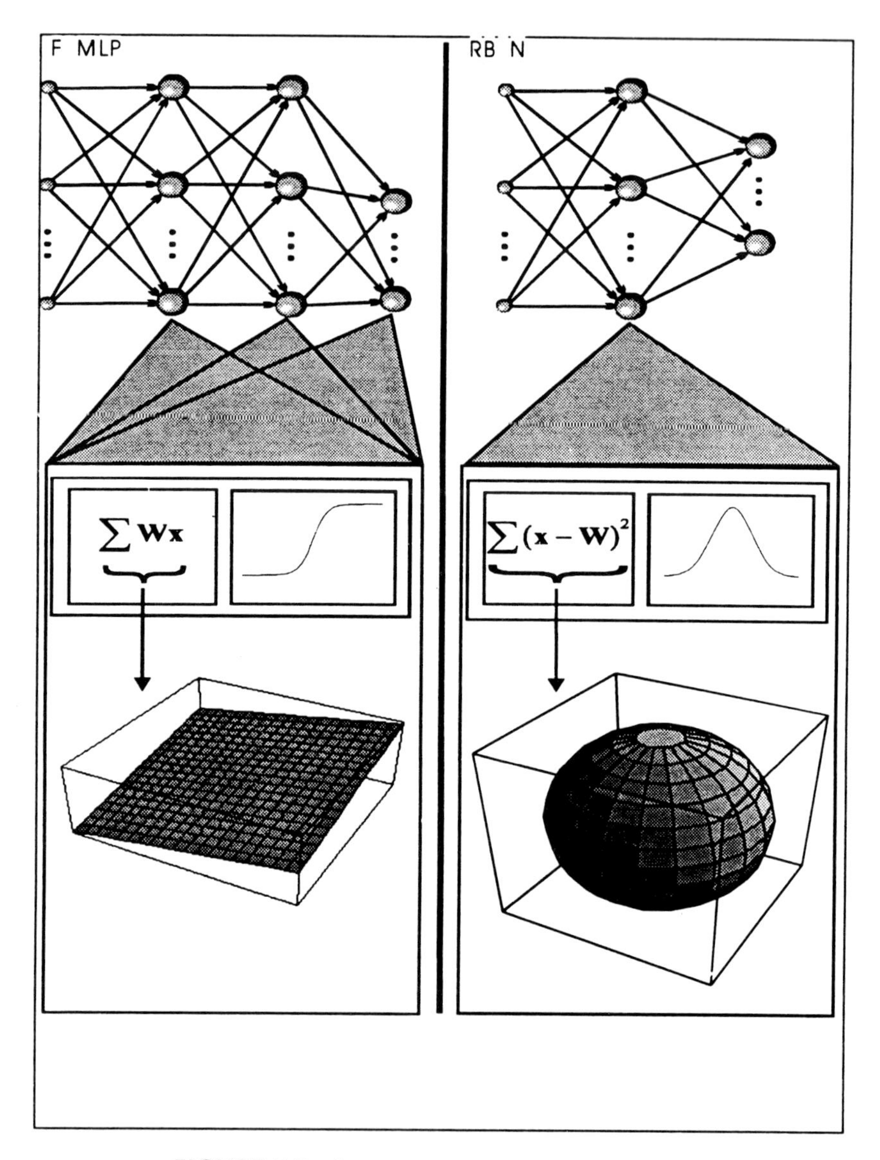

FIGURE 5.2 Structures of RBF and MLP networks.

In this algorithm the clustering parameter, α, is a constant which controls the extent to which clustering will take place. High values of α leads to lesser node reduction but will increase accuracy over the training set due to the overlapping reduction.

Note that this algorithm addresses only two class problems. Multiclass problems can be solved by using multiple single class networks. For a given class there may be several clusters depending on the membership function. The training results in designating one hidden neuron per cluster. Musavi et al. [1991] show an example where two data sets consisting of 400 points generated by Gaussian random vectors were separated into two classes. Figure 5.3 shows the pattern space with patterns belonging to both classes. The clustering algorithm discussed earlier was used to train the network which formulated 86 clusters. Thus the center and radius parameters for the RBF network neurons were determined. The center of each cluster is also shown in Figure 5.3. A simple delta rule was then used to train the weights to the output neuron which was trained to give output equal to zero for class one and output equal to one for class two. Figure 5.4 shows the output surface for the trained network for the samples shown in Figure 5.3.

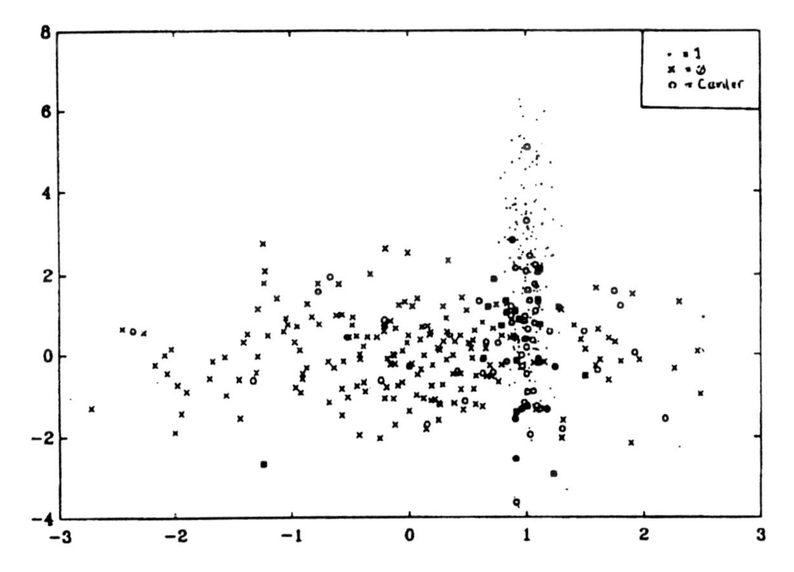

FIGURE 5.3 Test samples and centers of clusters. (From Musavi, M. T. et al., ACMANNA — 91, *Anal. Neural Network Appl.,* 1991. With permission.)

5.2.3 Applications of RBF Networks

RBF networks, also referred to as local networks, are capable of fast learning and reduced sensitivity to the order of presentation of training patterns. Moody and Darken [1989] demonstrated that fast learning can be achieved by using RBF networks, while Poggio and Giorsi [1990] have applied regularization theory for improving generalization capabilities of RBF networks. For those interested in pursuing the RBF networks, Haykin [1994] provides an in-depth discussion.

Renals [1989] applied RBF networks for pattern classification in speech domain. RBFs have been applied to the problem of handwritten digit recognition by Lee [1991]. In this experiment no front-end feature extraction was utilized. The network inputs were a raw bit map. The results yield a 5.15% error rate for the conventional perceptron and a 4.77% error rate for radial-basis functions, too small a difference

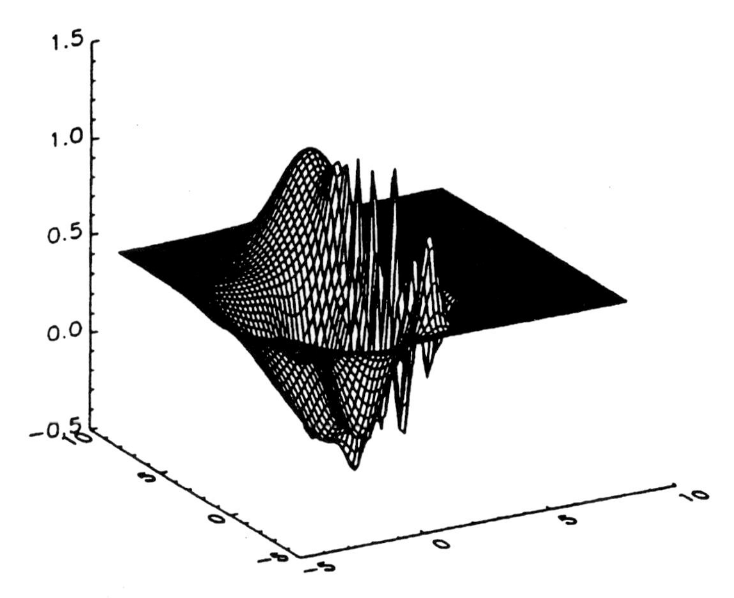

FIGURE 5.4 Output surface for samples from Figure 5.3. (From Musavi, M. T. et al., ACMANNA — 91, *Anal. Neural Network Appl.,* 1991. With permission.)

to draw conclusions. It is, however, enough to demonstrate that RBFs can be brought to bear on the problem.

RBF networks have been used for a wide range of applications such as image processing [Poggio and Edelman, 1990; Saha et al., 1991]; medical diagnosis [Lowe and Webb, 1990]; time series analysis [He and Lapedes, 1991; Broomhead and Lowe, 1988; Kadirkamanathan et al., 1991]; and speech recognition [Ng and Lippman, 1991; Niranjan and Fallside, 1990].

5.3 HIGHER ORDER NEURAL NETWORKS

Later, in chapter 6, we will discuss feature extraction. In the process of that discussion it will be seen that various forms of invariance may be achieved by a two-stage process in which an invariant feature set is first created and then presented to a recognition engine. An alternative is to build the desired invariance into the structure of the network itself. This is done by the higher order neural networks in this section.

5.3.1 INTRODUCTION

The objective in object recognition is that an input image must be recognized regardless of its position, size, and angular orientation, as shown in Figure 5.5. Multilayer perceptrons (MLPs), the neocognitron (discussed in Chapter 11), and higher order neural networks (HONNs) [Giles and Maxwell, 1987] are three of the more successful neural network architectures applied to achieve this objective. The most important advantage of the HONNs is that invariance to geometric transformations

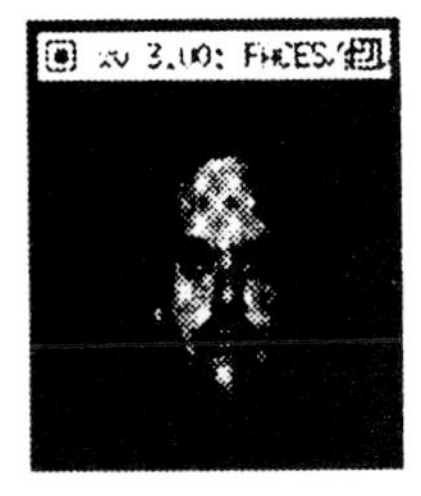

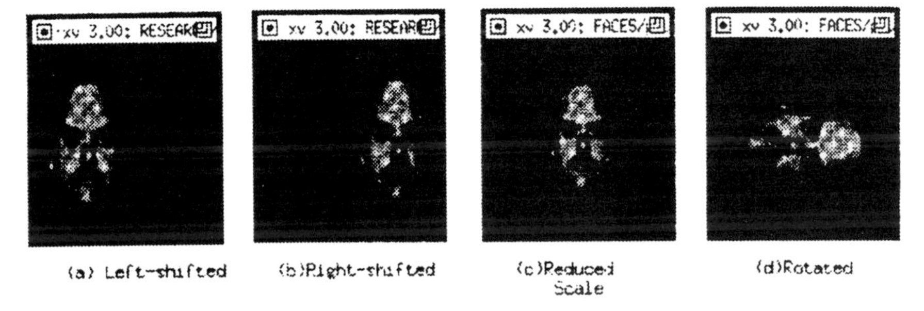

FIGURE 5.5 Sample shifted, scaled, and rotated image.

can be incorporated into the network and need not be learned through iterative weight updates.

For MLPs to learn to perform such recognition, the training set must include a large subset of transformed views of objects [Rumelhart, 1989; Troxel et al., 1988]. Typically, they require a large number of training passes to generalize the concepts behind the geometric transformations. The neocognitron has a hierarchical structure of simple and complex cells which allow it to perform object recognition independent of its position in the input field, a slight change in size, or a slight deformation [Fukushima, 1992]. The limitation is that the number of cells in the model increases almost linearly with the number of objects it is required to learn to distinguish.

In contrast to these two approaches, in the case of HONNs, the desired invariances are built directly into the architecture of the network by exploiting the known relationships in the input pattern. Thus the network is "pretrained" and does not need to learn invariance to geometric transformations, and HONNs need to be trained on just one view of each object rather than on numerous transformed views.

HONNs can be designed to be invariant to two-dimensional coordinate transformation of images by adjusting their weights properly. A second-order version of such a network could be made insensitive to translation and scale distortions. A third-order neural network can be used to perform translation, scale, and rotation invariant object recognition with a significant reduction in training time over other neural net paradigms such as MLPs.

5.3.2 ARCHITECTURE

In a general HONN, the output, y_i, of a node, i, is given as follows:

$$O = f\left(\sum_{i=0}^{N-1} w_i x_i + \sum_{j=0}^{N-1} \sum_{k=0}^{N-1} w_{jk} x_j x_k + \sum_{j=0}^{N-1} \sum_{k=0}^{N-1} \sum_{l=0}^{N-1} w_{jkl} x_j x_k x_l + \ldots \right) \tag{3}$$

where f is a nonlinear transformation function; x_j x_k x_l are inputs from other nodes; and w_{ij}, w_{ijk}, w_{ijkl} determine the weight that each input is assigned in summation (see Figures 5.6 (a) and 5.7).

HONNs can be trained by back propagation where the additional high order terms are regarded as additional inputs.

Figures 5.6 (a) and (b) show examples of simple first- and second-order networks. The shaded squares between the input and output layers in the second-order network figure (see Figure 5.7) represent product units where the products of the incoming signals are computed and passed on.

In a higher order network of order N, inputs are taken as products, up to N at a time (see equation 3). In a strictly N^{th} order network inputs are taken N at a time and the lower order terms will be missing. The output, y_i, for a strictly third-order neural network (see Figure 5.7) is given as follows:

$$y_i(x) = S\left[\sum_{j=1}^{N} \sum_{k=1}^{N} \sum_{l=1}^{N} W_{ijk} x_j x_k x_l \right] \tag{4}$$

Notice that first-order or second-order terms do not appear in this formulation. Also note, that the inputs are first combined in triplets and then the weighted sum of these products is used to determine the output. S is a nonlinear transformation while all the other variables have representations that are the same as indicated in equation 3.

The third-order network can be trained using a simple delta rule and the hyperbolic tangent function as the activation function. The tangent function is preferred due to its bipolarity, a property which has proved to be advantageous in reducing the training times of various networks.

5.3.3 INVARIANCE TO GEOMETRIC TRANSFORMATIONS

In order to build invariance to the three geometric transformations, the neural network needs to learn the relationship between input pixels. The weights of HONN are constrained such that all combinations of three pixels (triples) which define similar triangles are connected to the output with the same weight. The neural net structure used can be a single-layer, feed-forward net with one output neuron per category. In the above method, all similar triangles are mapped to the same weight. In this case, "similar" means "possessing the same set and sequence of internal

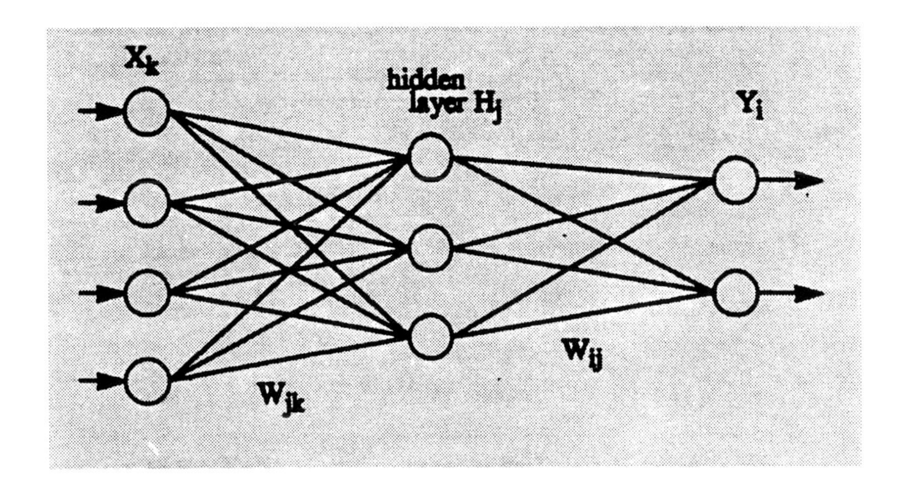

(a)

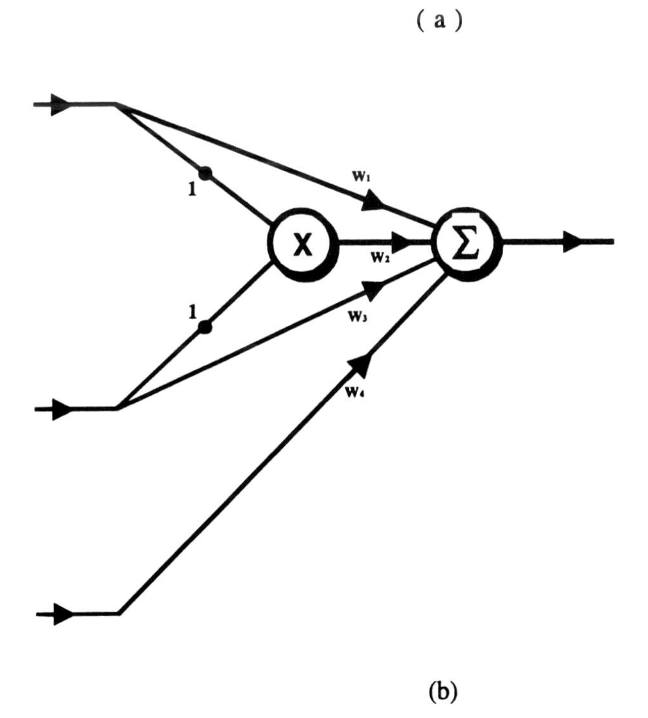

(b)

FIGURE 5.6 The multilayer perceptron in (a) is a first-order network, where each weight modifies a single input. For a second-order network in (b) the first two inputs are multiplied together (at the cross) and then multiplied by a weight before summing. The other connections are similar to the MLP.

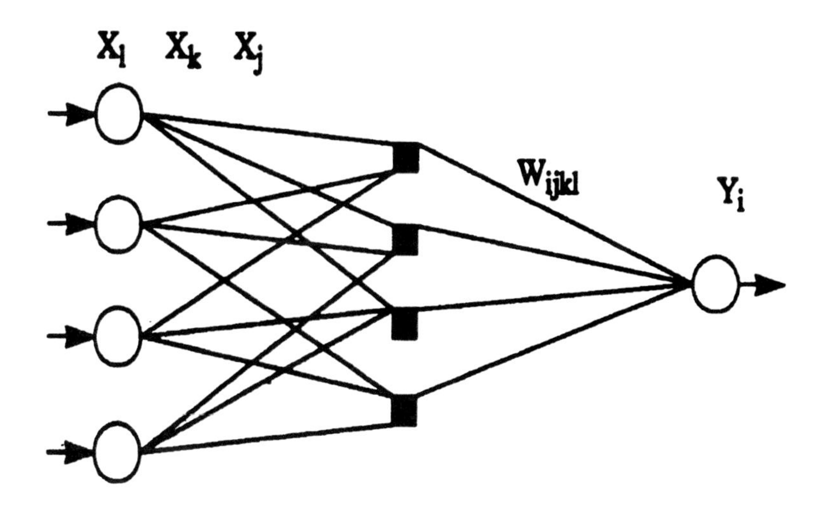

FIGURE 5.7 Strictly third-order networks involve multiplication of three inputs before multiplying by a weight. Then all four product terms (one from each square) are summed.

angles when the triangles are traversed in the same direction". Figure 5.8 contains examples of similar and dissimilar triangles (triplets).

The triangles are always traversed in the same direction when computing the angles because the order of the internal angles is an important variable, as it allows us to distinguish between a triangle and its laterally inverted counterpart. In this case, the two triangles are indeed different because it would not be possible (using any combination of translation, scale change, or rotation) to map one triangle onto its laterally inverted counterpart. The internal angles for every possible triple in the image have to be calculated and stored.

One difficulty inherent in HONNs is that the number of weights grows very rapidly with the order of the net and with the input dimensionality. In an $N \times N$ image resolution there are N^2-choose-3 ways to choose a combination of three pixels. Thus a 10×10 pixel input field requires 100-choose-3 = 161,700 interconnections. Increasing the resolution to 128×128 pixels increases the number of possible triplets to 7.3×10^{11}. Such a large number of interconnections is far too large to store on most machines. In most of the practical implementations, ways had to be found to limit this effect (see, for example, Spirkovaska and Reid [1993]; Lisboa and Perantonis [1991]).

5.3.4 An Example

Consider that the task is to design a strictly third-order network for differentiating between two simple geometric figures (a rectangle and a triangle) as shown in Figure 5.9.

The first step is to define an appropriate plane bisector passing through one of the vertices of the triangle (Figure 5.10). The internal angles are calculated by using

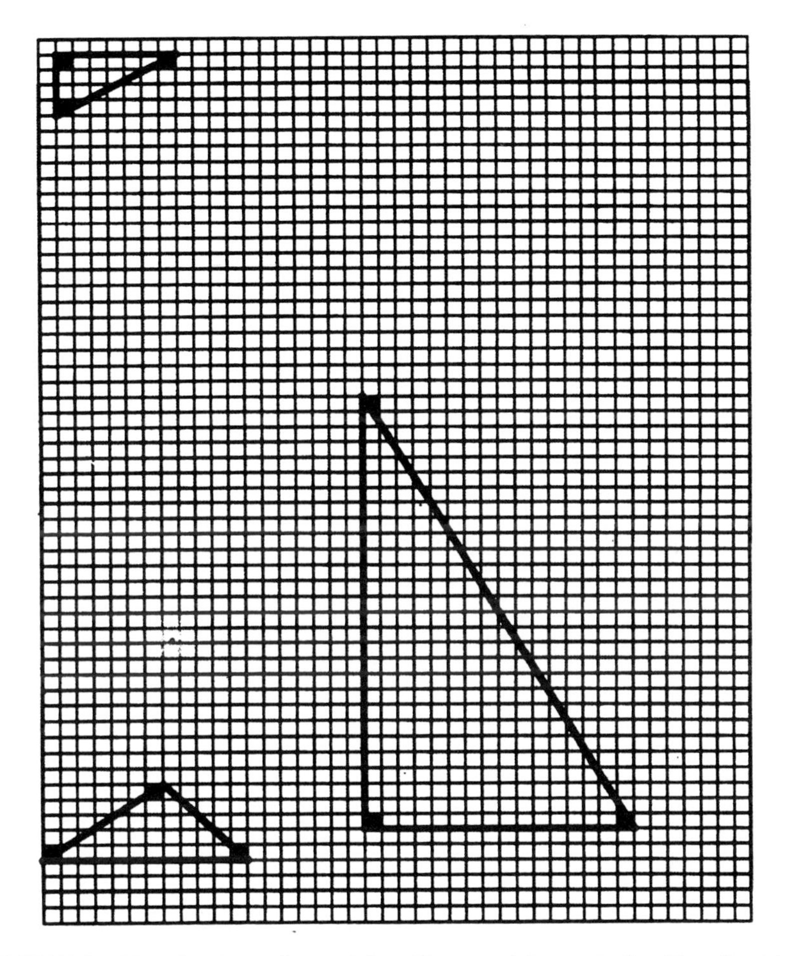

FIGURE 5.8 The triangles at the top left and bottom right are similar. The other triangle is dissimilar to both of these.

this vertex and the bisector as reference points. The figure shows the four possible cases/combinations of vertex position and bisector position. After the three angles are found, the centroid of the triangle is computed and used to determine the vertex sequence required for clockwise traversal (Figure 5.11). In practice, any internal angle will suffice as a reference, but it is just as easy to use the centroid as a reference.

In practice, the angles calculated can be quantized into the nearest groups of 4 degrees to allow for possible noise disturbance in the image data. Experiments have been performed to test the abilities of the third-order net in differentiating between two simple geometric monochrome figures (a rectangle and a triangle) with varying amounts of noise (see Figures 5.12 and 5.13). The network was trained to produce an output of +1 for a rectangle, and an output of –1 for a triangle. Some of the images used for testing the trained network had pixels added (i.e., "snow"). Others

FIGURE 5.9 The simple shapes used were a rectangle and a triangle. The numerical portion (extension) of the shape name represents the angle (in degrees) through which the figure has been rotated (clockwise) in order to produce the image.

had pixels missing. It was found that in practice, snowy images were more prone to being misclassified than those that had pixels missing.

Figure 5.14 shows the results of the simulation for a pair of simple geometric shapes — "box2" and "v". As can be seen, all the triangles used to test the network classified correctly (to –1). However, two of the boxes were misclassified — "box" and "box.270".

5.3.5 Practical Applications

Duren and Peikari [1991] propose a second-order neural network for translation and orientation invariance. Once again, the nature of the neuron is different from the perceptron. More specifically for a second-order network (see Figure 5.6 [b]) the output of a neuron becomes:

$$O = f\left(\sum_{i=0}^{N-1} w_i x_i + \sum_{j=0}^{N-1}\sum_{k=0}^{N-1} w_{jk} x_j x_k\right) \tag{5}$$

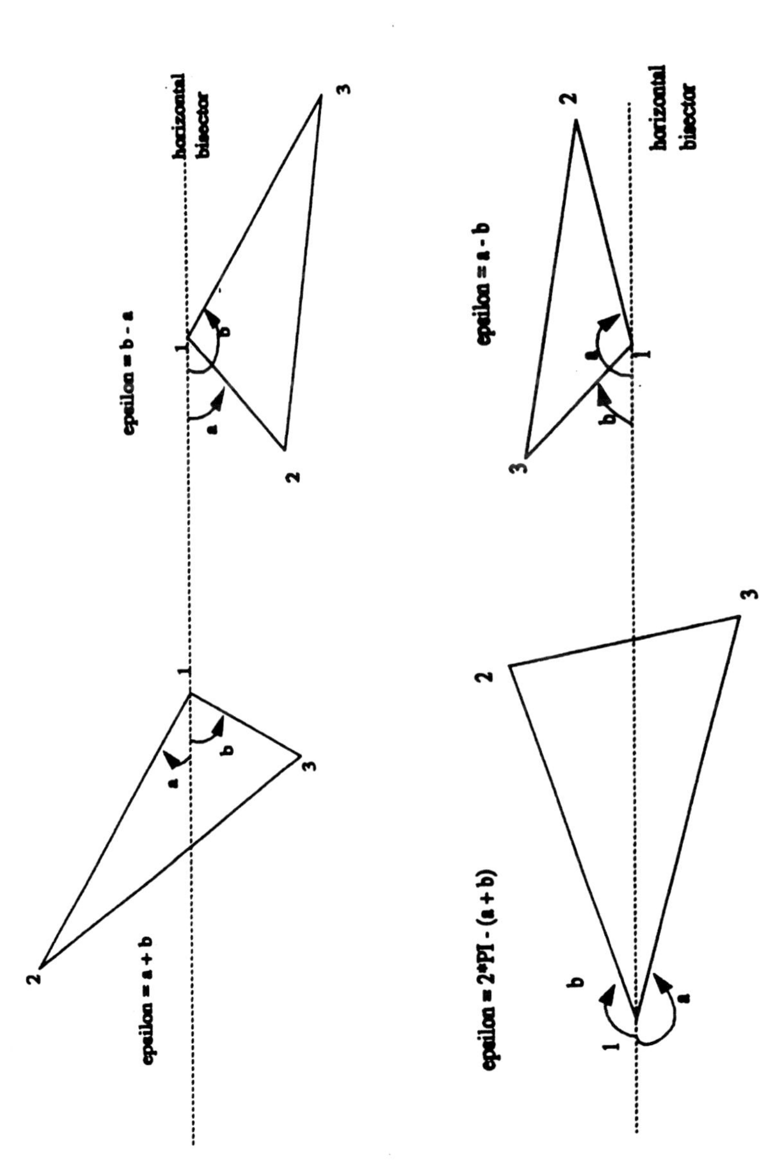

FIGURE 5.10 Combinations of vertex position and bisector position.

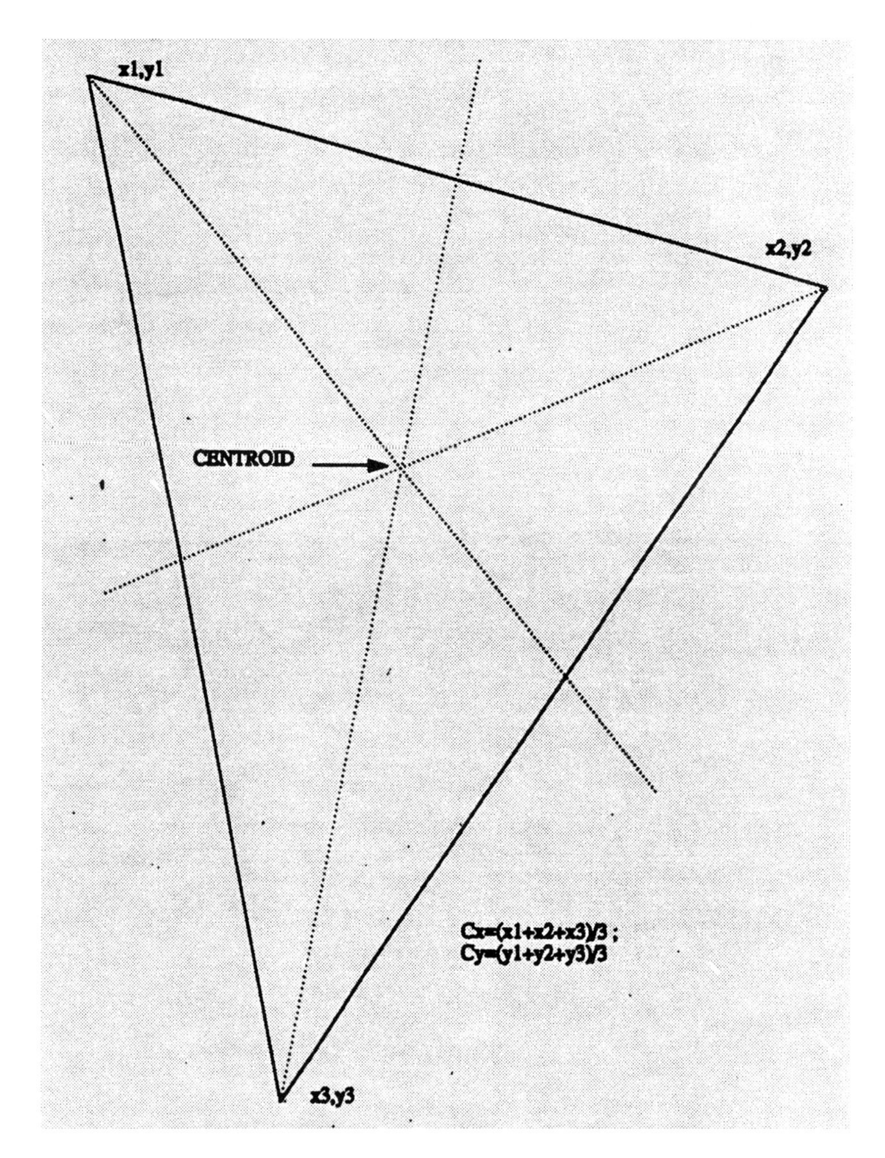

FIGURE 5.11 Centroid of a triangle.

Translation invariance was achieved by constraint the weights to be a function of the distance between the inputs. The reported success rate is 94.8% for an experiment involving handwritten digit recognition.

Lisboa and Perantonis [1991] used third-order neural networks in the context of handwritten digits; thus in this case processing proceeds with triplets of pixels. The desired invariance was for rotation and scale. The patterns used were 20 × 20 pixel bit maps of handwritten digits. Results for rotation and scaling were both impressive.

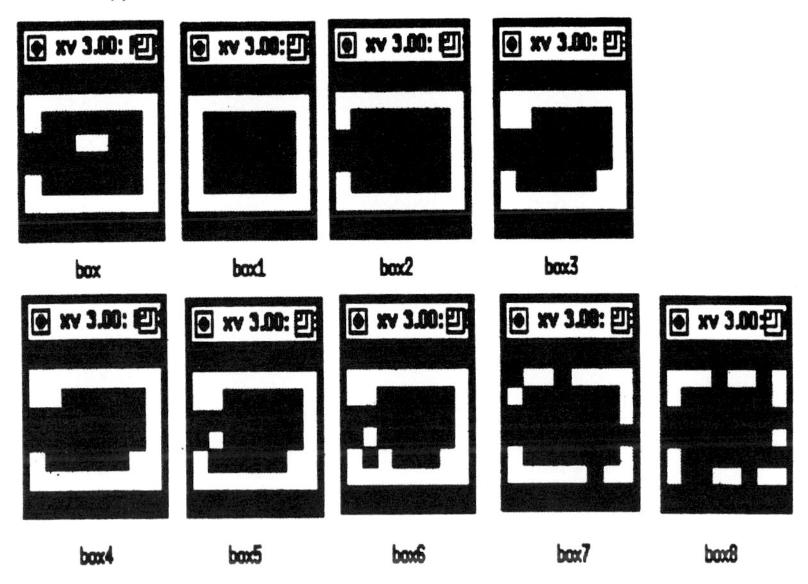

FIGURE 5.12 Noisy rectangular images.

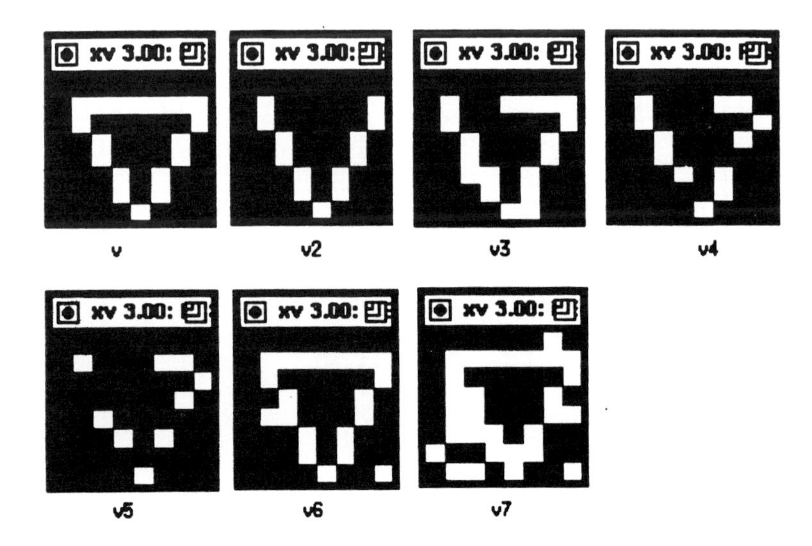

FIGURE 5.13 Noisy triangular images.

angle resolution : 4 degrees
weight initialization : RANDOM
activation function : HYPERBOLIC TANGENT
target outputs : +1 , -1
error threshold : 0 .0000001
learning rate : 0 . 1
training image 1 : box2 (o/p : +1)
training image 2 : v (o/p : -1)

Image Name	Network Output	Image Name	Network Output
box	-1	v	-1
box.270	-1	v.left	-1
box1	1	v.90	-1
box1.90	1	v.180	-1
box2	1	v.270	-1
box2.90	1	v2	-1
box2.180	1	v2.90	-1
box3	1	v2.180	-1
box3.90	1	v2.270	-1
box3.180	1	v3	-1
box3.270	1	v3.90	-1
box4	1	v3.180	-1
box4.90	1	v3.270	-1
box4.180	1	v4	-1
box5	1	v4.90	-1
box5.90	1	v4.180	-1
box5.180	1	v5	-1
box6	-1	v5.90	-1
box6.90	-1	v5.180	-1
box6.180	1	v6	-1
box7	1	v6.90	-1
box7.90	1	v6.180	-1
box7.180	1	v7	-1
box8	1	v7.90	-1
box8.90	1	v7.180	-1
box8.180	1		

FIGURE 5.14 Simulation results for simple geometric shapes.

REFERENCES AND BIBLIOGRAPHY

Broomhead, D. S. and Lowe, D., "Multivariable functional interpolation and adaptive networks," *Complex Syst.*, vol. 2, pp. 321–355, 1988.

Cover, T. M., "Geometrical and statistical properties of systems of linear inequalities with applications in pattern recognition," *IEEE Trans. Electron. Comput.*, EC-14, pp. 326-334, 1965.

Duda, R. O. and Hart, P. E., *Pattern Classification and Scene Analysis,"* John Wiley & Sons, New York, 1973.

Duren, R. and Peikari, B., "A comparison of second order neural networks to transform based methods for translation and orientation invariant object recognition," Neural Networks for Signal Processing, Proc. of the IEEE Workshop, pp. 236–245, 1991.

Fukushima, K., "Character recognition with neural networks," *Neurocomputing*, vol. 4, pp. 221-233, 1992.

Giles, G. L. and Maxwell, T., "Learning, invariances and generalizations in high-order neural networks," *Appl. Opt.*, vol. 26, pp. 2972-2978, 1987.

Haykin, S., *Neural Networks, A Comprehensive Foundation,* MacMillan College Publishing, New York, 1994.

He, X. and Lapedes, A., "Nonlinear modeling and prediction by successive approximation using radial basis functions," Technical report LA-UR-91–1375, Los Alamos National Laboratory, NM, 1991.

Kadirkamanathan, V., Niranjan, M., and Fallside, F., "Neural networks and radial basis functions in classifying static speech patterns," *Comput. Speech Language,* vol. 4, pp. 275–289, 1990.

Kadirkamanathan, V., Niranjan, M., and Fallside, F., "Sequential adaptation radial basis function neural networks." *Adv. Neural Inf. Process. Syst.*, vol. 3, pp. 721–727, 1991.

Kung, S. Y., *Digital Neural Networks,* Prentice-Hall, Englewood Cliffs, NJ, 1993.

Lee, Y., "Handwritten digit recognition using K nearest-neighbor, radial basis function, and backpropagation neural networks," *Neural Computation,* no. 3, pp. 440–449, 1991.

Lippman, R. P., "Pattern classification using neural networks," *IEEE Commun. Mag.*, vol. 27, no. 11, pp. 47–64, 1992.

Lisboa, P. J. G. and Perantonis, S. J., "Invariant pattern recognition using third-order networks and zernike moments," *IEEE Int. Joint Conf. Neural Networks (IJCNN) 91,* pp. 1421–1425, 1991.

Lowe, B. and Webb, A. R., "Exploiting prior knowledge in network optimization: an illustration from medical prognosis," *Network,* vol. 1, pp. 291–323, 1990.

Minsky, M. L. and Papert, S., *Perceptrons,* MIT Press, Cambridge, MA, 1969.

Moody, J. E. and Darken, C. J., "Fast learning in networks of locally-tuned processing units," *Neural Comput.*, Vol. 1, pp. 281-294, 1991.

Musavi, M. T., Faris, K. B., Chan, K. H., and Ahmed, W., "On the implementation of RBF technique in neural networks," ACM ANNA — 91, *Anal. Neural Network Appl.*, pp. 110–115, 1991.

Nirangan, M. and Fallside, F., "Neural networks and radial basis functions in classifying static speech patterns," *Comput. Speech Lang.*, vol. 4, pp. 275-289, 1990.

Ng, K. and Lippman, R. P., "Practical characteristics of neural networks and conventional pattern classifiers, *Adv. Neural Inf. Process. Syst.*, vol. 3, pp. 970–976, 1991.

Poggio, T. and Giorgi, F., "Regularization algorithm for learning that are equivalent to multilayer networks, *Science,* Vol. 247, pp. 978- 982, 1990.

Poggio, T. and Edelman, S., "A network that learns to recognize three-dimensional objects," *Nature,* vol. 343, pp. 263–266, 1990.

Renals, S., "Radial basis function network for speech pattern classification," *Electron. Lett.*, vol. 25, pp. 437-439, 1989.

Rosenblatt, F., Principles of Neurodynamics, Spartan Books, New York, 1962.

Rumelhart, D. E., *Neural Networks,* vol. 2, pp. 348–352, 1989.

Saha, A., Christian, J., Tang, B. S., and Wu, C. L., "Oriented non-radial basis functions for image coding and analysis," in ...analysis," in *Advances in Neural Information Processing Systems, Vol. 3,* R. P. Lippmann, J. E. Moody, and D. S. Touretzeky (Eds.), Morgan Kaufmann Press, San Mateo, CA, pp. 728–734, 1991.

Spirkovska, L. and Reid, M.B., "Coarse-coded higher order neural networks for PSRI optic recognition," *IEEE Trans. Neural Networks,* vol. 4, no. 2, pp. 276-283, 1993.

Troxel, S. E., Rogers, S. K., and Kabrisky, M., "The use of neural networks in PSRI recognition," in Proc. Joint Int. Conf. Neural Networks, San Diego, CA, July 24–27, pp. 593–600, 1988.

Wong, Y., "How radical basis functions work," Proc. Int. Joint Conf. Neural Networks, Vol. 2, Seattle, WA, pp. 133-138, 1991.

6 Feature Extraction I: Geometric Features and Transformations

6.1 GENERAL

In one sense many neural networks perform a kind of automatic feature extraction. For example, the hidden layer nodes in a feed-forward network trained by back propagation can be thought of as extracting features which will ultimately be resolved into a classification at the output layer. if there are multiple hidden layers, the hidden layer neurons in each successive layer extract features of increasing complexity and/or abstraction. Nevertheless, we do not choose the features that the network will train to and these features may or may not have desirable properties. They are features in a fairly abstract sense and are unlikely to resemble anything a human would recognize as a feature such as line segments or loops or arcs. Therefore we now wish to examine the use of explicit feature selection and feature extraction to accomplish particular purposes. The first such motivation leading to feature extraction is our desire to achieve a reduction in dimensionality.

A fully connected network with enough discriminative power to perform recognition of a real problem may have too many parameters to generalize well when raw bit maps provide the input. A judicious selection of features to reduce the dimensionality of the input vector while hopefully preserving its salient features is an approach that will improve generalization. In so doing, the size of the required training set is also reduced as discussed in chapter 4. This chapter will discuss several approaches to extract and utilize such features. We begin by examining the extraction of intuitive geometric features. We then illustrate their use in a feed-forward multilayer perceptron (FFMLP) network. An alternative approach uses features in a feature map. This involves constraining the first few layers of the network to be local thereby ensuring the detection of local features at these neurons [Le Cun et al., 1990]. Feature maps are discussed in section 6.3.

Another benefit that can be derived from a judicious choice of features is that some desired forms of invariance may be achieved. By invariance it is meant that the input may be allowed to vary in some particular way without affecting the recognition result. It is frequently desirable to produce a representation that is invariant with respect to rotation, translation, and scale changes. We will illustrate the use of such a feature choice while employing transformation techniques to achieve a reduction in dimensionality as well.

Clearly, the choice and quality of features will have a profound effect on the characteristics of a recognition system. In the context of the character recognition problem, different sets of features can cause different characters to become ambiguous in the recognizer [Wang and Jean, 1993]. Later, in chapter 12, we will return to this topic when we discuss multiple recognizer systems and confusion matrices.

For the above and other reasons (including the impact that feature selection will have on the neural network size and performance) feature extraction is an important part of pattern recognition. We will examine the selection and extraction of features to be used in a recognizer network in some depth.

6.2 GEOMETRIC FEATURES (LOOPS, INTERSECTIONS, AND ENDPOINTS)

One approach to feature extraction would be the use of features which are intuitive in the sense that they are directly perceptible to humans. Loops, intersections, and endpoints all fall into this category. This fairly small collection of geometric features gave surprisingly good recognition results in the context of handwritten character recognition [Darwiche et al., 1992]. Here endpoints, intersections, and loops were the only features used in a character recognition problem. The spatial resolution was intentionally low and information about the connectivity between intersections and endpoints was not used.

6.2.1 INTERSECTIONS AND ENDPOINTS

The following methods for the detection of intersection points and endpoints all assume that we operate on a skeletonized bit map. Recall that the skeletonization process was addressed in section 3.5. In the discussion that follows we refer to a pixel that is "on" as a 1-pixel and a pixel that is "off" as a 0-pixel.

Endpoints are found by examining all individual 1-pixels in the skeletonized bit map image. As a consequence of skeletonization an endpoint will have one and only one of its 8 contiguous neighbors as a 1-pixel. We may therefore sum the 8 neighbors and recognize an endpoint when the sum is one.

Intersections are found in a similar fashion. Each of the 1-pixels in the image is examined, and the number n of contiguous 1-pixel to the focus pixel is counted (i.e., each of the 8 neighboring pixels is examined). If the count n of neighboring pixels exceeds 2 ($n> = 3$), then the focus pixel is considered to be an intersection.

We will now present a code example which illustrates the detection of intersections and endpoints. Listing 6.1 provides the class definitions for our feature extraction class, cThinner, as follows:

Listing 6.1

```
struct tgrid {
  int x;
  int y;
};

struct tLoop {
  int x;            // Center of mass X coord
  int y;            // Center of mass Y coord
  int m;            // Total Mass (Useful to disqualify ornamentation)
};
class cThinner {
private:
 int   xMax, yMax;                     // Size of the bitmap
 int   PelxMax, PelyMax;
 int   PelxMin, PelyMin;
 int   M[32][32];                      // Space for the whole bitmap
 int   Stage1[32][32];                 // orig image archive
 int   W[32][32];                      // Work Space save area
   .
   .
   .
 tMap  Map[7];
 int   isEndPt(int, int);
 int   isIntersect(int, int);
 int   isIntersect2(int, int);
 int   EndPtCnt;                       // number of end points
 tgrid EndPoint[324];                  // endpoint coords
 int   IntPtCnt;                       // number of intersects
 tgrid IntersectPoint[324];            // intersect coords
 int   LoopCnt;
 tLoop LoopLocation[324];
 int ValidZeroChild(int, int, int, int);
 void Frame2Ones();
 void NQZeroes(coordq *, tgrid);
 void KillOneLoop(int&, int&, int&);   // deal w/ 1 loop

public:
 cThinner(); //Constructor
 void  SetGridSize(int Mx, int My){xMax=Mx; yMax=My;}
 void  Store(int x,int y,int Pel){Stage1[x][y]=M[x][y]=Pel;}
 int   QueryPel(int x,int y){return M[x][y];}
 void  Thin2();
   .
   .
   .
 void  LocateEndPts();
 void  LocateIntersects();
 int   QueryEndCount(){return EndPtCnt;}
 int   QueryIntCount(){return IntPtCnt;}
 tgrid QueryEndPoint(int n);
 tgrid QueryIntPoint(int n);
 void  Grid2Work();                    //save pixel map in work buffer
```

```
  void   Work2Grid();                    //save pixel map in work buffer
  int    isLoop();                       //1= there are loops 0=no loops
  void   LocateLoops();  //find all independent loops. (work buf is left w/ Endpoints excised)
  int    QueryLoopCnt(){return LoopCnt;}
  tLoop  QueryLoop(int k){return LoopLocation[k];} //returns info re Nth loop
  int    FindNeighbor(int&, int &);
  void   Kill1Path(int);
  void   KillEndPaths();
  void   IsolatedPelcleanup();
  int    isAnyPelOn();
  int    isAnyPelOff(int&, int&);
};
```

The methods of class cThinner that are most relevant to the detection of endpoints and intersections can be found in listing 6.2.

Listing 6.2

```
void cThinner::LocateEndPts() {
  int x,y,i,j;
EndPtCnt=0;
for (y=0; y<yMax; y++) {
  for (x=0; x<xMax; x++) {
    if ((M[x][y]==1) && isEndPt(x,y)) {
      EndPoint[EndPtCnt].x=x;           // we have an endpoint so record it.
      EndPoint[EndPtCnt].y=y;
      EndPtCnt++;
      } /* endif */
    } /* endfor */
  } /* endfor */
}

void cThinner::LocateIntersects() {
  int x,y,i,j;
IntPtCnt=0;
for (y=0; y<yMax; y++) {
  for (x=0; x<xMax; x++) {
    if ((M[x][y]==1) && isIntersect(x,y)) {
      // we have an intersection point so do something with it.
      IntersectPoint[IntPtCnt].x=x;
      IntersectPoint[IntPtCnt].y=y;
      IntPtCnt++;
      } /* endif */
    } /* endfor */
  } /* endfor */
}

tgrid  cThinner::QueryEndPoint(int n){
  tgrid RetVal;
```

```
RetVal.x= EndPoint[n].x;
RetVal.y= EndPoint[n].y;
return RetVal;
}
tgrid  cThinner::QueryIntPoint(int n){
  tgrid RetVal;
RetVal.x= IntersectPoint[n].x;
RetVal.y= IntersectPoint[n].y;
return RetVal;
}
```

6.2.2 Loops

There are many methods of loop detection. For example, in Darwiche [1992], loop detection was based on the presence of circular or elliptical shapes within the character. At this point, we describe two methods which may be used for loop detection. Each has its own advantages. Our first method builds a tree of the character and therefore also establishes the connectivity of the character. The second method builds a membership list of pixels within the loop thus allowing us to calculate the "center of mass" of the loop, another relevant character metric.

Two issues which may arise in connection with feature loops are (1) broken or incomplete loops and (2) ornamental loops. In the first instance the loop is not fully closed. Such a broken or incomplete loop could easily be missed. An example of this situation is shown in Figure 6.1 (a). We discuss this issue later in our discussion of loops. In the second instance the loop is small and part of a flourish which is not salient to the recognition. An example of this situation is shown in Figure 6.1 (b).

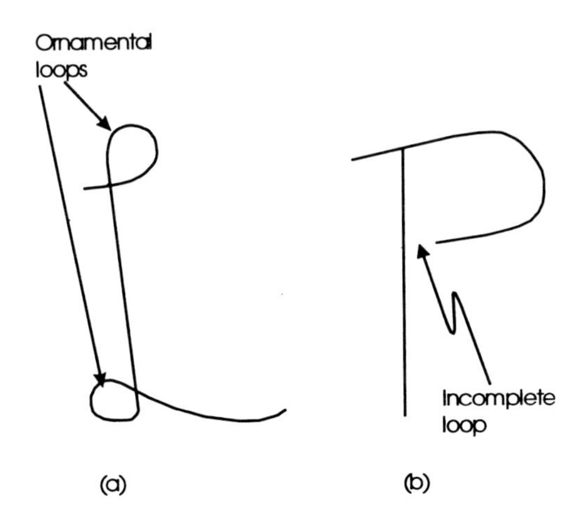

FIGURE 6.1 (a) Ornamental, and (b) incomplete loops.

Notice that once our method of loop detection has created a membership list, we can calculate the total mass for each loop as the sum of all member pixels. That is, each pixel is considered to have a unit mass. A threshold, T, may then be established such that, if total mass within the loop is below the threshold, the loop is considered to be ornamental.

A reasonable approach for loops that have been determined to be ornamental might be to 1-fill such loops (i.e., set all pixels interior to the loop to 1-pixels) and to then re-thin the character.

Loops can be extracted from the bit map of an image and described in terms of the labeled skeletal contours or paths between connected endpoints and intersection points. In the process, a tree of the character connectivity is built, the nodes of which are the endpoints and the intersection points. The edges are the labeled pixel paths connecting nodes. For the sake of this discussion it is assumed that the character is merged and thus all pixels in the character are connected. In this case if the number of interconnections, I, and the number of endpoints, E, are both zero, then the character consists of a single loop. Otherwise, we visit each of the interconnection points in turn. All loops associated with a given interconnection point I_k can be established by building a tree with I_k as its root. The tree is built in the following manner. The focus node of the root is I_k. For each node build a set of child nodes consisting of all possible paths from the parent focus node. Such paths are those reachable by following a trail until the first intersection point or endpoint is reached with edges already traversed and nodes already visited (except for the root node) specifically excluded. The focus of a child is the destination node of the pixel trail. As each child is created the edge is labeled. Terminal nodes consisting of the root node are loops. By following a terminal node that is a loop back to the root, the loop path is known. After this process is complete, duplicates must be eliminated as well as loops fully subsumed by two or more other loops. Where two loops L_1, L_2 taken together cover all the pixel paths in a third loop L_3, then L_3 should be eliminated if an arbitrary interior point in L_1 and an arbitrary point in L_2 are both also interior points of L_3.

Figures 6.2 and 6.3 show the trees generated as described above for the characters "R" and "8", respectively.

It should be strongly noted that in the process of traversing the pixel paths, other potentially useful possible action could be taken. Such an action might, for example, be marking some number of intermediate points at intervals along the edges for a richer description of character shape.

We now briefly digress to discuss the subject of incomplete loops as promised. Suppose we establish a minimum threshold radius around each endpoint in the system. If the circle defined by that radius contains a pixel or pixels disconnected from the point in focus, then we will create a pixel bridge to the closest of the disconnected pixels. (If the location of cusps is available, the same should be done for these.) Although it is easy enough to find examples of reasonable handwriting that contain incomplete loops at points other than endpoints or cusps, this procedure would render the above technique for loop detection effective for many of the common incomplete loops.

Loops may also be detected by a process of building membership lists. The process begins by amputating all edges terminating in endpoints from the character. If no 1-pixels remain, there are no loops and the process is complete. Otherwise, at

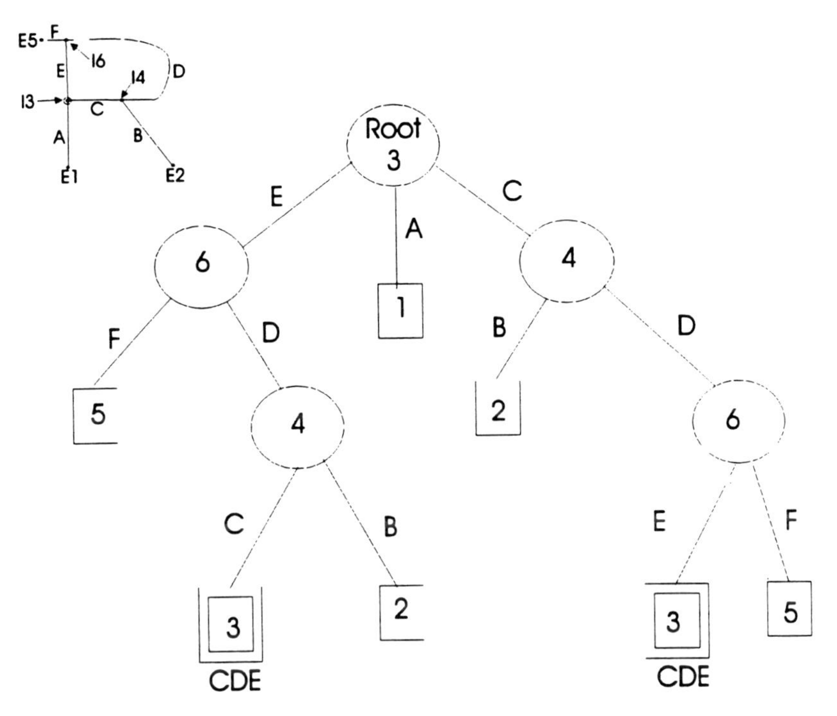

FIGURE 6.2 Loop tree for the letter "R".

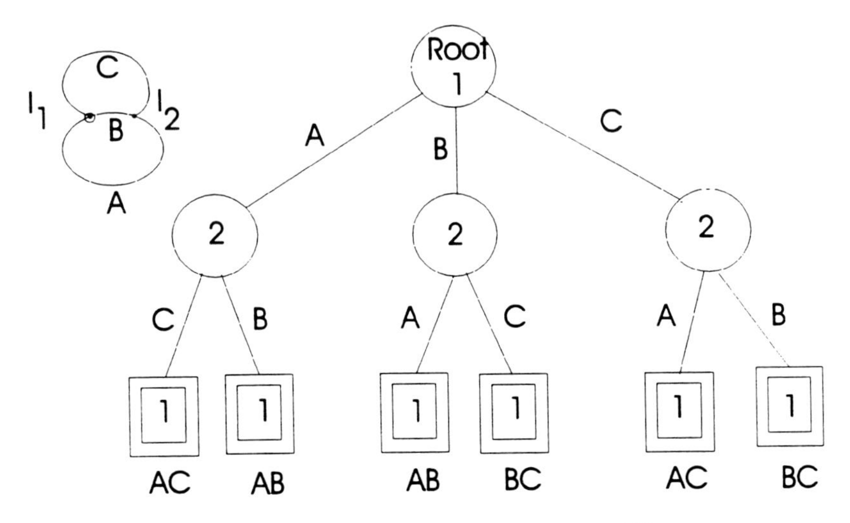

FIGURE 6.3 Loop tree for the number "8".

least one loop is present. We proceed by building a list, Q_0, of all 0-pixels (excluding those that are contiguous with the bit map border). Each pixel in Q_0 is within some loop. We may remove an arbitrary pixel from Q_0, and build a collection of all its 0-pixel neighbors that do not cross a boundary formed by 1-pixels. This collection is the membership for a single loop, L. All of the L member pixels are removed from Q_0. The process continues until Q_0 is empty and all loops within the character have been detected. Methods of the cThinner class (see listing 6.1 for class definition) which implement this technique of loop detection are given in listing 6.3. The classes "coordq" and "coordqel" collectively provide support for maintaining the queues in the algorithm and are also provided in listing 6.3.

Now we will explore the geometric placement of a loop on the bit map. One way to do this is by calculating its center of mass. The center of mass of N masses can be calculated in two dimensions as follows:

$$x_{cm} = \frac{\sum_{i=1}^{N} m_i x_i}{\sum_{i=1}^{N} m_i} \qquad y_{cm} = \frac{\sum_{i=1}^{N} m_i y_i}{\sum_{i=1}^{N} m_i} \tag{1}$$

where N is the number of pixels in the loop and x_i and y_i are the coordinates the mass m_i.

In our case each mass, m_i, is a pixel (that is a loop member) which we will take as having a unit mass. Therefore to calculate the center of mass for the loop, equation 1 will then become:

$$x_{cm} = \frac{1}{N}\sum_{i=1}^{N} x_i \qquad y_{cm} = \frac{1}{N}\sum_{i=1}^{N} y_i \tag{2}$$

which is simply the average.

Listing 6.3

```
//-----------------------------------------------------------------
class coordqel {
private:
  tgrid   coord;
  coordqel *next;
public:
  coordqel(){next=NULL;}
  void setnext(coordqel *p){next=p;}
  void setxy(int x,int y){coord.x=x; coord.y=y;}
  void query(int& x, int& y){x=coord.x; y=coord.y;}
  coordqel *querynext(){return next;}
};
```

```
//------------------------------------------------------------------------
class coordq {
private:
 coordqel *head;
 int     count;
public:
 coordq(){head=NULL;}
 void enQTop(int x, int y);
 void enQTop(tgrid pel){enQTop(pel.x, pel.y);}
 int deQTop(int& x, int& y);
 int Qdelete(coordqel *p);
 int find(int x, int y);
 int DatAvail();
};

int coordq::DatAvail(){
if (head==NULL)
  return 0;
 else
  return 1;
}

int coordq::find(int x, int y){
  int x1,y1;
  coordqel *tmp;
tmp=head;
while (tmp!=NULL) {
 tmp->query(x1,y1);
 if ((x==x1) && (y==y1))
    return 1;
  else
    tmp=tmp->querynext();
  } /* endwhile */
 return 0;
 }

 int coordq::Qdelete(coordqel *p){
   coordqel *cur;
   coordqel *prev;
   cur=head; prev=NULL;
   while ((cur!=NULL) && (cur!=p)) {
     prev=cur;
     cur=cur->querynext();
     } /* endwhile */
   if (cur!=NULL) {
     if(prev!=NULL)
      prev->setnext(cur->querynext());
      else
       head=cur->querynext();
     delete cur;
     return 1;
     }
    else
     return 0;                    // bad pointer, Q empty or just not found. delete failed.
}
```

```
void coordq::enQTop(int x, int y){
  coordqel *tmp;
  tmp=new coordqel;
  tmp->setxy(x,y);
  tmp->setnext(head);
  head=tmp;
}

int coordq::deQTop(int& x, int& y){
  coordqel *tmp;
  if(head!=NULL) {
    head->query(x,y);
     tmp=head;
    head=head->querynext();
     delete tmp;
     return 1;
     }
   else
     return 0;
}

//-----------------------------------------------------------------------

//*********************************************************
// Copy fns between pixel map & temporary work space.   *
//*********************************************************
void cThinner::Grid2Work(){
int x,y;
for (x=0; x<xMax; x++) {
  for (y=0; y<yMax; y++) {
    W[x][y]=M[x][y];
    } /* endfor */
  } /* endfor */
}

void cThinner::Work2Grid(){
int x,y;
for (x=0; x<xMax; x++) {
  for (y=0; y<yMax; y++) {
    M[x][y]=W[x][y];
    } /* endfor */
  } /* endfor */
}

//*********************************************************
//Deletes pel iff all surrounding pells are zero
//*********************************************************

void cThinner::IsolatedPelcleanup() {
int x,y;
for (x=0; x<xMax; x++) {
  for (y=0; y<yMax; y++) {
    if (M[x][y]==1){                        // faster than a for loop
     if ( (M[x-1][y-1]==0) && (M[x-1][y]==0) && (M[x-1][y+1]==0) && (M[x][y-1]==0)
       && (M[x][y+1]==0) && (M[x+1][y-1]==0) && (M[x+1][y]==0) && (M[x+1][y+1]==0) )
         M[x][y]=0;
```

```
      }
    } /* endfor */
  } /* endfor */
}

//**********************************************************
// Returns 1 if there is a nearest neighbor 1 pel to x,y *
//        0 otherwise                                    *
// also modifies x,y to be the neighbors coord           *
//**********************************************************
int cThinner::FindNeighbor(int& x, int& y){
 int i,j;
for (i=0; i<3; i++) {
  for (j=0; j<3; j++) {
    if (M[x-1+i][y-1+j]==1) {
      x=x-1+i; y=y-1+j;
      return 1;
      } /* endif */
    } /* endfor */
  } /* endfor */
return 0;
}

int cThinner::isAnyPelOn(){
int x,y;
for (x=0; x<xMax; x++) {
  for (y=0; y<yMax; y++) {
    if (M[x][y]==1) return 1;
    } /* endfor */
  } /* endfor */
return 0;
}

int cThinner::isAnyPelOff(int& x, int& y){
for (x=0; x<xMax; x++) {
  for (y=0; y<yMax; y++) {
    if (M[x][y]==0) return 1;
    } /* endfor */
  } /* endfor */
return 0;
}

//**********************************************************
// Starting at a given endpoint eliminate all pixels until *
// either:                                                 *
//      A) Another endpoint is encountered (inclusive)*
//   or                                                    *
//      B) An Intersect is encountered  (exclusive)  *
//**********************************************************
void cThinner::Kill1Path(int EPindx){
  int x,y;
  int Done;
```

```
x = EndPoint[EPindx].x;
y = EndPoint[EPindx].y;
Done=0;
while (!Done) {
  M[x][y] = 0;                          // kill the current pel
  if (!FindNeighbor(x,y)) Done =1;      //No neighbor found...we're done
  if (isEndPt(x,y)) {                   //check whether x,y is an endpoint
    M[x][y]=0;                          //endpoint<->endpoint delete is inclusive
    Done=1;
    }
  if (isIntersect2(x,y)) Done=1;
  } /* endwhile */
}

void cThinner::KillEndPaths(){
  int i;
for (i=0; i<EndPtCnt; i++) {
  if(M[EndPoint[i].x][EndPoint[i].y])
    Kill1Path(i);
  } /* endfor */
}

int cThinner::isEndPt(int x, int y) {
  int sum,i,j;
sum=-1;                                 //account for centrer pixel being on
for (j=y-1; j<=y+1; j++) {
  for (i=x-1; i<=x+1; i++) {
    if((i>=0) && (j>=0) &&(i<xMax) && (j<yMax))
      sum+=M[i][j];                     // count the neighbors
    } /* endfor */
  } /* endfor */
if (sum==1)                             // Exactly one neighbor ON
  return 1;                             // defines an endpoint.
return 0;
}

int cThinner::isIntersect(int x, int y) {
  int sum,i,j;
sum=-1;                                 //account for center pixel being on
for (j=y-1; j<=y+1; j++) {
  for (i=x-1; i<=x+1; i++) {
    if((i>=0) && (j>=0) && (i<xMax) && (j<yMax))
      sum+=M[i][j];                     // count the neighbors
    } /* endfor */
  } /* endfor */
if (sum>=3) {                           // three or more neighbors 'ON'
  return 1;                             // defines an intersection.
  }/* endif */
return 0;
}

void cThinner::LocateLoops() {
  int x,y,m;
LoopCnt=0;
Grid2Work();                            //Save copy of orig
while (QueryEndCount()>0){
```

```
  LocateEndPts();                          //Calc Endpoints
  LocateIntersects();                      //  and intersects
  KillEndPaths();                          //Remove the pixel-path
  Thin2();                                 //Re-Thin
  }
IsolatedPelcleanup();                      // Remove pels w/ all zero neighbor
if (!isAnyPelOn()) {
  LoopCnt=0;                               //No loops detected
  }
 else {
  Frame2Ones();
  if (QueryIntCount()==0) {
     LoopCnt=1;                            //There is exactly 1 loop
     isAnyPelOff(x,y);
     // Now locate the loops
     KillOneLoop(x,y,m);                   // kill the pel and all its friends
     LoopLocation[0].x=x;
     LoopLocation[0].y=y;
     LoopLocation[0].m=m;
      }
    else {
    //Here there are multiple loops, so establish how many & locate them
    LoopCnt=0;
     while (isAnyPelOff(x,y)) {
       KillOneLoop(x,y,m);                 // kill the pel and all its friends
       LoopLocation[LoopCnt].x=x;
       LoopLocation[LoopCnt].y=y;
       LoopLocation[LoopCnt].m=m;
        LoopCnt++;
        } /* endwhile */
     } /* endif */
  } /* endif */
Work2Grid();                               //Restore orig grid
LocateEndPts();                            //Played with these so...
LocateIntersects();                        // ...they need restoration too
}

int cThinner::isIntersect2(int x, int y){
int found,i;
found=0;
for (i=0; i<IntPtCnt; i++) {
  if ( (IntersectPoint[i].x==x) && (IntersectPoint[i].y==y))
     found=1;
  } /* endfor */
return found;
}

void cThinner::Frame2Ones(){
  coordq f21;
  tgrid  fpoint;
  int x,y;
fpoint.x =0;
fpoint.y =0;
NQZeroes(&f21,fpoint);
while (f21.DatAvail()) {
```

```
  f21.deQTop(x, y);
  M[x][y]=1;
  } /* endwhile */
}

int cThinner::ValidZeroChild(int cx, int cy, int px, int py){
if ((cx<0) || (cy<0) || (cx>17) || (cy>25)) // stay on grid
        return 0;
if (M[cx][cy]!=0)                           // Pel must be zero
        return 0;
if ((cx==px) || cy==py)                     // no x-ings possible for
        return 1;                           // directly hor/vert neighbor
if ((M[cx][py]==1) && (M[px][cy]==1))       // Check diagonal neighors
        return 0;
return 1;
}

void cThinner::NQZeroes(coordq *targQ, tgrid seedpel){
  coordq open;
  int x,y,i,j;
if (M[seedpel.x][seedpel.y]==0) open.enQTop(seedpel);
      else return ;                         //exit if seed nonzero
while (open.DatAvail()) {
  open.deQTop(x, y);
  targQ->enQTop(x,y);
  // Now examine each NN of pel at (x,y) ...
  // ...if its valid & not already present we'll  place it on open.
  for (j=-1; j<=1; j++) {
    for (i=-1; i<=1; i++) {
      if (!((i==0) && (j==0)) ) {           //each NN visited here
        if(ValidZeroChild(x-i,y-j,x,y)){
          if ((open.find(x-i,y-j)==0) && (targQ->find(x-i,y-j)==0)){
            open.enQTop(x-i,y-j);
            }/* endif */
          }/* endif */
        } /* endif */
    } /* endfor */
  } /* endfor */
  } /* endwhile */
}

void cThinner::KillOneLoop(int& x, int& y, int& m) {
   coordq targQ;
   tgrid  seedpel;
   int sx,sy;
m=0;
seedpel.x=x;
seedpel.y=y;
sx=sy=0;
NQZeroes(&targQ, seedpel);                //all interior loop-pels to targq
while (targQ.DatAvail()) {                //while they last
  targQ.deQTop(x, y);                     //examine them 1 by 1
  sx+=x;                                  // to determine loop
  sy+=y;                                  //  coords  and
```

```
  m++;                                  //  mass
  M[x][y]=1;                            // and delete
  } /* endwhile */
if (m>0){
  x=sx/m;
  y=sy/m;}
}
```

The use of the geometric feature detection methods in a calling program is illustrated in listing 6.4. Note that character thinning must be invoked first before the methods to detect features can be used. Character thinning was discussed in chapter 3. Subsequent to the location of features the number and position of points belonging to each feature type can be queried and displayed.

Listing 6.4

```
cThinner     Thinner;

int main(int argc, char *argv[]) {
  int       mx,my,i,j,d,chIndx, Lcnt;
  tgrid fpt;
  int fcnt;
  tLoop Lpt;
   .
   .

Thinner.Thin2();                        // First Thin the character
   .
   .
Thinner.LocateEndPts();                 // Build a list of all End Points
Thinner.LocateIntersects();             // Build a list of all intersections
Thinner.LocateLoops();                  // Build a list of all loops

fcnt= Thinner.QueryEndCount();          //Display any endpoints
printf("\n  End Points:");
for (i=0; i<fcnt; i++) {
  fpt=  Thinner.QueryEndPoint(i);
  printf("[%d,%d]", fpt.x,fpt.y);
    } /* endfor */
fcnt= Thinner.QueryIntCount();          //Display any intersections
printf("\n  Intersects:");
for (i=0; i<fcnt; i++) {
  fpt =  Thinner.QueryIntPoint(i);
  printf("[%d,%d]", fpt.x,fpt.y);
  } /* endfor */
Lcnt=Thinner.QueryLoopCnt();            //Display any loops
if (Lcnt==0) {
  printf("\nNo loops detected.");
} else {
  printf("\n  Loops:    ");
  for (i=0; i<Lcnt; i++) {
    Lpt =  Thinner.QueryLoop(i);
    printf("[m=%d,(%d,%d)]", Lpt.m,Lpt.x,Lpt.y);
     } /* endfor */
  printf("\n");
} /* endif */
return 0;
}
```

While we have chosen to provide details on the detection of endpoints, intersection points, and loops, it should be mentioned that these features represent only a fraction of those which can be found and used. To provide a more complete representation of the shape, it is useful to distinguish between lines and arcs. Additionally points of inflection and cusps are significant and interesting points that will be useful in representing the shape of a digital curve. For more detail on the detection of points of inflection and cusps see Pikaz and Dinstein [1994].

6.3 FEATURE MAPS

In addition there is another way of looking at geometric features. Feature maps take advantage of the idea that if a localized feature is useful at one point in the image, it is likely to be useful in other areas as well. If a distorted character contains salient features that may be somewhat displaced from those of a typical character, the feature map will be useful in their detection.

The feature map then consists of a plane of neurons whose weights are constrained to be equal. Thus within a given feature map each neuron performs the same operation, which is equivalent to saying it detects the same feature. However, each neuron receives its input from different parts of the image. The area or portion of the image from which a given neuron receives information is called the neuron receptive field. Both the constrained weights within a feature map and the receptive fields of the feature map neurons are illustrated in Figure 6.4. As shown in the figure, multiple feature maps, one per feature to be detected, are common.

Note also from Figure 6.5 that the receptive field of neurons in a feature map may be overlapping or contiguous. Feature map outputs are passed on to higher stages of the network. Frequently the first layer will consist of overlapping receptive fields and the next layer will have contiguous fields, the purpose of which is to limit the effect of translation of the feature. This arrangement bears a striking resemblance to the S-layer of the neocognitron network described later in chapter 11. Multiple feature maps are necessary to capture different features in the same image. Feature map layers can be applied to subsequent hidden layers to extract higher order features, in a manner similar to the neocognitron. Such higher order features do not require positional information to be as precise as it must be for simpler features. Thus a layer following the feature map layer and resembling the neocognitron C-layer may be employed. (See chapter 11, Figure 11.1.)

6.4 A NETWORK EXAMPLE USING GEOMETRIC FEATURES

Having discussed at length the extraction of loops, intersections, and endpoints it seems appropriate to show their subsequent application to a neural net recognizer as well.

Figure 6.6 illustrates one possible network which makes use of several of the geometric features discussed so far. The intent is more to create a vehicle to explore

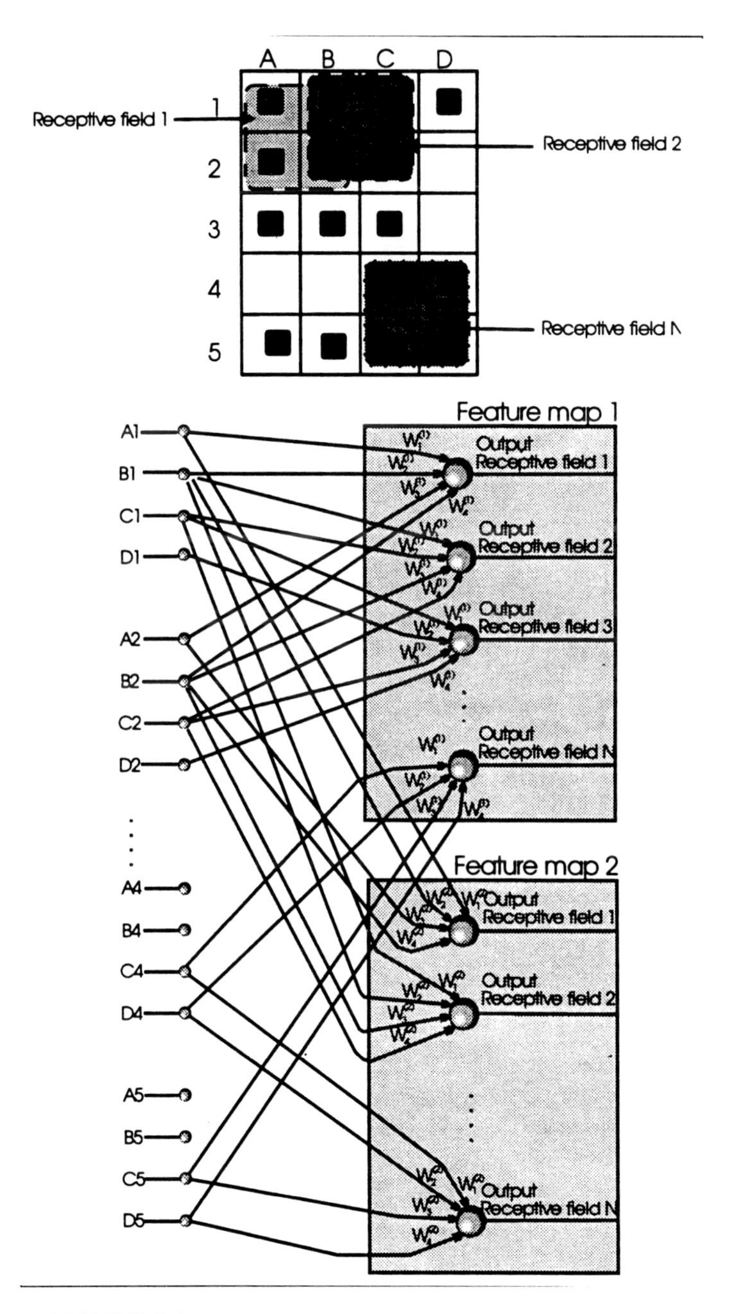

FIGURE 6.4 Receptive fields and corresponding feature maps.

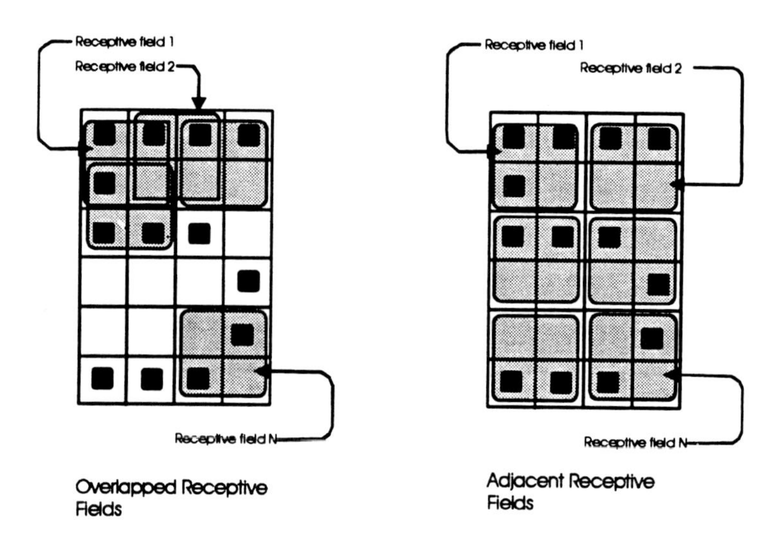

FIGURE 6.5 Overlapped and adjacent receptive fields.

the exploitation of geometric features than to propose an optimal solution. The architecture is similar to that in Darwiche et al. [1992] so it is reasonable to expect that it will possess similar desirable properties (see section 6.2).

Feature points as shown (see Figure 6.6) for the network consist of n quadruples (x, y, I, E). Coordinates x and y are the normalized position on the pixel grid. That is, for an M × M pixel grid, a pixel in the ith row and j^{th} column would have $x = j/M$ and $y = i/M$. I and E are binary valued and designate the feature point as an endpoint or an intersection, respectively. When neither is asserted, the referenced point would be an intermediate point along the traversal of a pixel path. A number of such intermediate points would be necessary to define the character without ambiguity. The inclusion of such intermediate points in some sense can be thought of as providing connectivity information. Clearly, terms for additional features could easily be added. For example, the inclusion of the loop feature (see section 6.2.2) leads to the 5-tuple (x, y, I, E, L).

Not all characters will have the same number of feature points. The separation of the network into n 4-tuples was motivated by the knowledge that different characters (in fact different instances of the same character) can and will have varying numbers of feature points. For characters having fewer than n feature points the inputs to the surplus input layer neurons are set to zero. The feature points should be presented to the network in a consistent, repeatable manner. One simple and perfectly adequate choice would be to sort points first by x then y to determine the order of presentation.

Two salient points require mention here. First is the choice to represent the position of the feature point as an analog value. It is expected that this will aid in interpolative generalization (at least in comparison to a raw bit map input). While the geometric nature of the representation is not intrinsically invariant with respect

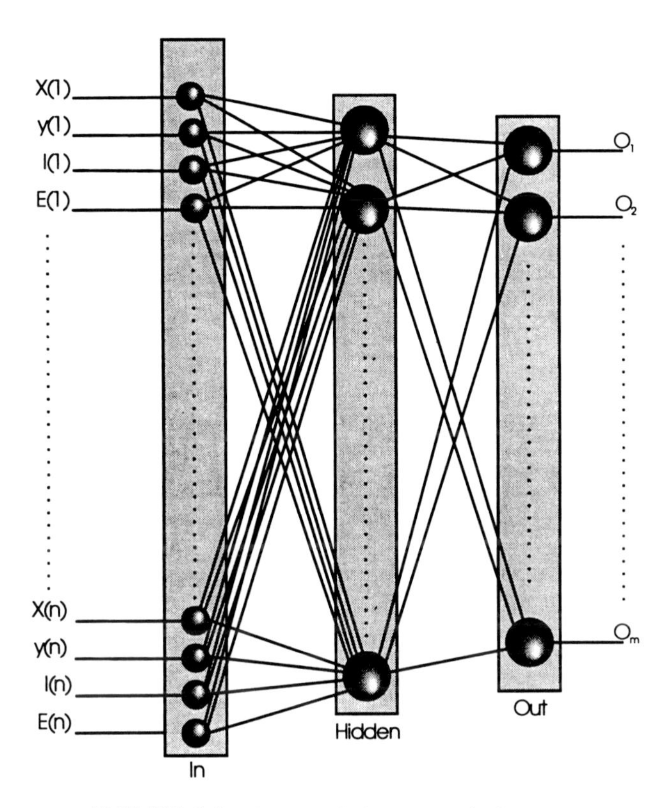

FIGURE 6.6 A network for geometric features.

to transformations such as translation rotation and scaling, within some perturbation limit the abovementioned generalization will render this network tolerant, especially with the addition of appropriate training set vectors.

A second salient point is that although the network shown above (Figure 6.6) is represented as an ACON network, no particular importance should be attached to this aspect. The important aspect of the network is its input side. An OCON version for this architecture can be easily constructed if desired.

6.5 FEATURE EXTRACTION USING TRANSFORMATIONS

In the preceding sections we examined various geometric features and some methods by which they are calculated. In the sections that follow we present two of the major transformations which have been applied to various pattern recognition applications. In connection with the discussion of these transformations we will discuss features which will preserve various forms of invariance. In particular, the transformations we will discuss are the Gabor transformation and Fourier descriptors (FD).

6.6 FOURIER DESCRIPTORS

Fourier descriptors (FD) provide one method which can be used to describe the boundaries of nonoverlapping objects. This method possesses the added advantage of producing a representation that is invariant with respect to rotation, translation, and scale changes. For a detailed discussion of the properties of Fourier descriptors see Boas [1966]. Following the discussion in Venugopal et al. [1992], consider the closed curve, γ shown in Figure 6.7.

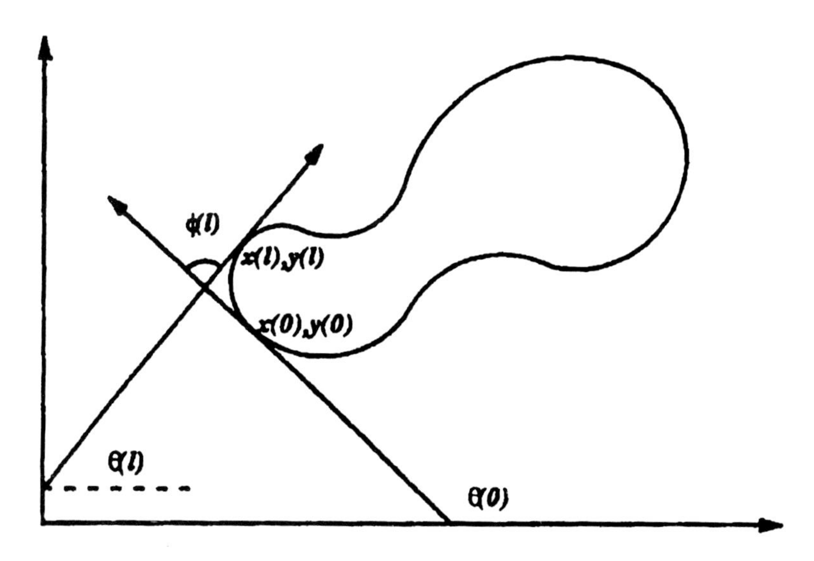

FIGURE 6.7 Definition of the shape function for a closed curve γ.

In the figure, $x(0)$, $y(0)$ represent an arbitrary starting point along the curve. $\theta(l)$ is the tangential direction at distance l along the curve from the starting point. If the total length of the curve is L then the arc length l is $0 \le l \le L$. The function $\varphi(l) = \theta(l) - \theta(0)$ provides the relative angle at arc length l from the starting point.

Normalizing with respect to length yields $\varphi^*(t) = \varphi(Lt/2\pi) + t$ where $0 \le t \le 2\pi$. Notice that by representing the shape in terms of the normalized shape function, φ^*, the desired invariance with respect to rotation, scale, and translation is achieved.

We may now expand $\varphi^*(\mathbf{t})$ as a Fourier series. This yields:

$$\varphi^*(t) = \mu_0 + \sum_{k=1}^{\infty} a_k \cos kt + b_k \sin kt \tag{3}$$

where in the continuous case:

$$\mu_0 = \frac{1}{2\pi} \int_0^{2\pi} \varphi^*(t)\, dt \tag{4}$$

$$a_n = \frac{1}{\pi}\int_0^{2\pi} \varphi^*(t)\cos(nt)\,dt \tag{5}$$

$$b_n = \frac{1}{\pi}\int_0^{2\pi} \varphi^*(t)\sin(nt)\,dt \tag{6}$$

Converting to polar coordinates yields:

$$\varphi^*(t) = a_0 + \sum_{k=1}^{\infty} A_k \cos(kt - \lambda_k) \tag{7}$$

$$A_k = \sqrt{a_k^2 + b_k^2} \tag{8}$$

$$\lambda_k = \tan^{-1}\left(\frac{b_k}{a_k}\right) \tag{9}$$

The set $\{A_k, \lambda_k, k = 1, \ldots, \infty\}$ are the Fourier descriptors for the curve.

Finally for the special case of a polygonal curve (see Figure 6.8) with m vertices $\mathbf{V}_0 \ldots \mathbf{V}_{m-1}$ the Fourier descriptors can be calculated as:

$$a_k = -\frac{1}{k\pi}\sum_{n=1}^{m} \Delta\varphi_n \sin\frac{2\pi l_n}{L} \tag{10}$$

$$b_k = \frac{1}{k\pi}\sum_{n=1}^{m} \Delta\varphi_n \cos\frac{2\pi l_n}{L} \tag{11}$$

$$l_n = \sum_{i=1}^{n} \Delta l_i \tag{12}$$

where l_i is the length of the edge $(\mathbf{V}_{i-1}, \mathbf{V}_i)$ and $\Delta\varphi_n$ is the angular direction at vertex $\mathbf{V}_n$.

This situation is of special interest for character recognition where it is natural to represent character maps in just such a piecewise fashion. In general, an effort is made to represent the character by a minimal set of line segments that preserves the shape.

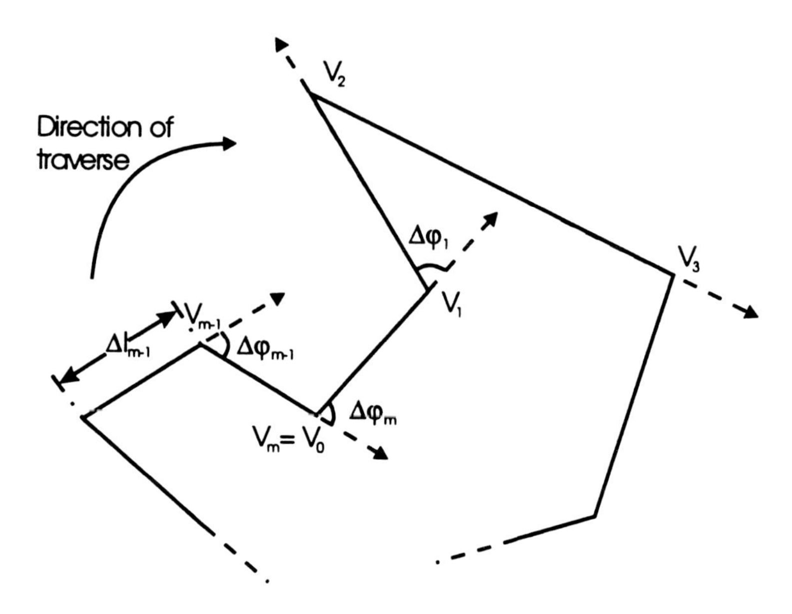

FIGURE 6.8 Representation of a shape as a piecewise polygonal curve.

Notice that the desired invariance characteristics have already been introduced with the definition of the feature φ^*. The motivation behind calculating the Fourier descriptors then becomes that of achieving a dimensionality reduction of the input vector.

In Venugopal et al. [1992] the first 15 Fourier descriptors were used as features leading to a 30-component input vector in a feed-forward network. The network, trained by ALOPEX (discussed in chapter 4), achieved a 98% accuracy over the domain of handwritten digits. The error function used was

$$e_k = \log\left(\frac{1}{\text{out}_k}\right) \text{ when desired is 1} \tag{13}$$

$$e_k = \log\left(\frac{1}{1-\text{out}_k}\right) \text{ when desired is 0} \tag{14}$$

$$E = \sum_k e_k \tag{15}$$

where e_k is the error at the kth output neuron and E is the total error.

6.7 GABOR TRANSFORMATIONS AND WAVELETS

The Gabor transformation is a biologically inspired transformation which may be used to transform raw input data (into a potentially more useful form) that encodes the pattern for further classification. Biological visual systems, in particular the simple cells of the visual cortex, employ localized frequency signatures similar to those obtained from Gabor operators [Porat and Zeevi, 1988]. That is, to follow the biological example, we are seeking a means to represent an image in terms of a set of functions which will contain information in both the spatial domain and frequency domain.

A bit map image contains only spatial information. Kosko [1992] points out that the bit map may also be thought of as a transformation where the basis functions are the delta function. The Fourier transform representation which expresses the image in terms of sines and cosines provides only frequency domain information. It gives the spatial periodicity of the image belonging to a given frequency but loses all measure of locality. Putting aside for the moment the biological justifications, it seems intuitively obvious that features containing both kinds of information would be superior to those limited to only one while excluding the other. It remains to be seen how this may be best accomplished.

To begin, the set of basis functions we are looking for are the product of a spatial term and a frequency term as follows:

$$f_k = g(x)\exp(ikx) \tag{16}$$

The precision to which a function (that is, the product of widths in time and frequency domains, respectively) can be specified is limited by an uncertainty principle. It turns out that the minimum uncertainty is achieved when the spatial term, $g(x)$, is a Gaussian window, leading to the definition of a one-dimensional Gabor function:

$$\gamma_k = a_{k,\sigma}\exp\left(-\frac{x^2}{2\sigma^2}\right)\exp(ikx) \tag{17}$$

where a_{ks} is a normalization constant. The following definition for Morlet wavelets, based on the Gabor functions is

$$\psi_k = b_{k,\sigma}\exp\left(-\frac{-k^2x^2}{2\sigma^2}\right)\exp(ikx) \tag{18}$$

A sample Morlet wavelet is shown in Figure 6.9 below.

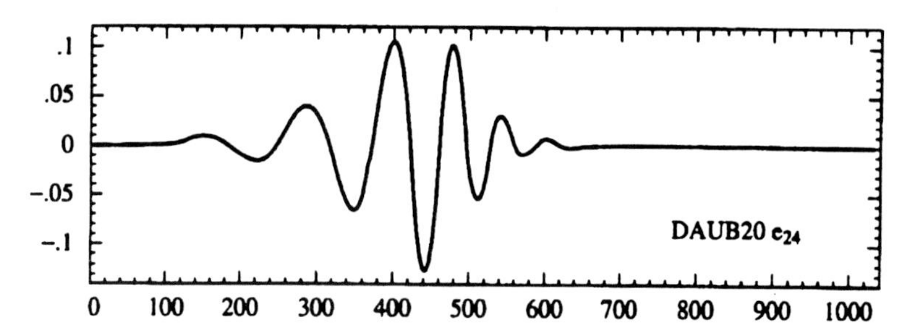

FIGURE 6.9 A sample Morlet wavelet. (From Press, W. H. et al., *Numerical Recipes in C : The Art of Scientific Computing,* Cambridge University Press, 1992. With permission.)

For the two-dimensional case the Morlet filter becomes:

$$\psi_{\mathbf{k}} = n_{k,\sigma} \exp\left(-\frac{-\mathbf{k}^2\mathbf{x}^2}{2\sigma^2}\right)\exp(i\mathbf{kx}) \tag{19}$$

which can be reconstituted as:

$$\psi\left(n, y, \sigma, k_x, k_y\right) = \exp\left(-\frac{-k_x^2 k_y^2}{2\sigma^2}\left(x^2 + y^2\right)\right)\exp\left\{i\left(k_x\mathbf{x} + k_y y\right)\right\} \tag{20}$$

where (x, y) are the positional variables in the spatial domain, (k_x, k_y) are the wave numbers and σ is the Gaussian window parameter (i.e., the standard deviation of the Gaussian envelope). From this it is evident that the wavelet has an orientation angle, θ, and a frequency, ω_0 as follows:

$$\theta = \tan^{-1}\left(\frac{k_y}{k_x}\right) \tag{21}$$

$$\omega_0 = \sqrt{k_x^2 + k_y^2} \tag{22}$$

The Morlet transformations can be expressed as the convolution of the image with $I(\mathbf{x})$ with the Morlet set of basis functions as follows:

$$G\left(\mathbf{k}, \mathbf{x}_0\right) = \int \psi\left(\mathbf{x}_0 - \mathbf{x}\right) I(\mathbf{x})\, d^2x \tag{23}$$

where $\mathbf{k}$ and $\mathbf{x}_0$ range over the image and the frequency planes.

In effect the convolution represents the sum of the product of each picture element in a window that is the size of the filter with a corresponding filter values; and $\mathbf{x}_0$ serves to locate the filter window in the image. The Gabor transformation would of course be similar, simply substituting γ for ψ.

To fully specify the wavelet based filter set values of orientation angle, θ, frequency, ω_0 and the Gaussian spread, σ, must all be chosen. Typically, while only a single value of σ is used, a range of values must be chosen for θ and ω_0 in order to achieve a good representation of the image. Since σ determines how sharply the spatial component falls off as we move away from the center of the window, its choice is strongly influenced by the window size.

One such set of parameters, used in a study [Lu et al. 1991] in which textures were clustered based on their wavelet representation, was $\sigma = \pi$ with 4 scales $\omega_0 = \pi/8, \pi/4, \pi/2, \pi$ and 6 orientations $\theta = 0, 30, 60, 90, 120, 150$ degrees for a total of 24 features within each filter window. In this study the full 24 element feature set was obtained for a window centered at each pixel. Clearly, this gets rather expensive very quickly. For a 64×64 window there would be 4096 vectors, each of dimension 24. The choice of window size and the degree to which windows will overlap is therefore an important choice that must be made. If we return to the biological model, an overlap for windows of 80% is indicated. Perhaps this represents a figure of merit from which to start. In any case, it is certainly a great deal more economical than the choice that centers a window at every pixel. The clustering in this study [Lu et al. 1991] was accomplished using a self-organizing feature map.

The Gabor wavelet transformation has been applied successfully to the problem of address block location on envelopes [Jain and Bhattacharjee 1992]. In this case the following even symmetric version of the Gabor filter was used:

$$G(x,y) = \exp\left\{-\frac{1}{2}\left[\frac{x^2}{\sigma_x^2} + \frac{y^2}{\sigma_y^2}\right]\right\}\cos(2\pi u_0 x + \phi) \tag{24}$$

Note that the Gaussian window parameter now depends on orientation and that the exponential is replaced by the cosine. Four orientation angles were used while the number of radial frequencies was $\log_2$ of the number of columns in the image.

So far Gabor transforms have been most frequently applied to the segmentation of textures within images. However, they have been applied to the character recognition problem. Garris et al. [1991] used Gabor features as input to a network trained by back propagation. Training and testing involved only a single OCR font with 100% accuracy. Wilson et al. [1990] used Gabor filtering with an adaptive resonance theory (ART) network. Their accuracy was 99% on machine print and 80% on handwritten characters. The applicability of wavelet transformations to image recognition, in general, and character recognition, in particular, is seductive. The major caveat is that the technique does appear to be computationally expensive.

It turns out that the Gabor transforms can be used to represent the shape of a character in a manner that is nearly identical to that used in the earlier description of Fourier descriptors. In fact this is exactly what Eichman et al. [1990] have done.

Recalling the shape function from the earlier discussion of Fourier descriptors:

$$\varphi(l) = \theta(l) - \theta(0) \tag{25}$$

$$\varphi'(t) = \varphi\left(\frac{Lt}{2\pi}\right) + t \tag{26}$$

A minor change in notation is that we now use φ' prime instead of φ*. This is to avoid confusion since in what follows the "*" will be used to designate the complex conjugate. In the discussion of FDs, the function φ was expanded into a Fourier series where the sines and cosines provided the basis functions. The major disadvantage of this representation is that the Fourier descriptors do not reflect local spectral information. Local distortions of the shape function will affect the entire set of Fourier descriptors. This deficit can be remedied by representing the shape function in terms of a set of Gabor elementary functions.

$$f_{mn} = w(t - mD)\, e^{inWt} \tag{27}$$

The function $\varphi'(t)$ can now be represented in terms of this set of elementary functions:

$$\varphi'(t) = \sum_{n=-\infty}^{\infty} \sum_{m=-\infty}^{\infty} a_{mn} f_{mn}(t) \tag{28}$$

As before we choose **w(t)** to be a Gaussian:

$$\mathbf{w}(t) = g(t) = \left(\frac{\sqrt{2}}{D}\right)^{1/2} e^{\left[-\pi\left(\frac{t}{D}\right)^2\right]} \tag{29}$$

Then the coefficients $\mathbf{a}_{mn}$ can be obtained by:

$$\mathbf{a}_{mn} = \int \varphi'(t)\, \gamma^*(t - mD)\, e^{-inWt}\, dt \tag{30}$$

It is necessary to use an auxiliary biorthogonal function, γ(t), because the Gabor elementary functions are not themselves orthogonal. γ(t) is given by:

$$\gamma(t)=\left(\frac{1}{\sqrt{2}D}\right)^{1/2}\left(\frac{K_0}{\pi}\right)^{-3/2} e^{\left[\pi\left(\frac{t}{D}\right)^2\right]} \sum_{n+1/2\geq t/D}(-1)^n e^{\left[-\pi(n+1/2)^2\right]} \tag{31}$$

$$t=0,1,\ldots,D-1$$

K_0 = 1.8540746 is a normalization factor.

Analogous to the earlier Fourier descriptors, the complex Gabor coefficients, $\mathbf{a}_{mn}$, now represent the shape of the image boundary and may be used as a feature set for a recognition engine.

It is pointed out that although the Gaussian is optimal in the sense described above, other spatial window functions are possible. One of these put forward in the reference is

$$w(t)=rect(t/D)=\begin{cases}1 & -D/2<t<D/2\\ 0 & \text{Elsewhere}\end{cases} \tag{32}$$

with a corresponding biorthogonal function:

$$\gamma(t)=\frac{1}{D}\operatorname{rect}\left(\frac{t}{D}\right) \tag{33}$$

Figures 6.10 through 6.13 from Eichman et al. [1990] provide a useful graphical representation of the stages of the process by which an image is first expressed in terms of its Gabor coefficients and then restored by the dual transformation. For this example the parameters used were M = 16 and $N = D$ = 32.

The original character, the letter "B", is shown below in Figure 6.10.

The spectrum of its associated shape function, $\varphi'(t)$, is shown in Figure 6.11.

A representation of the character in terms of its Gabor coefficients appears below in Figure 6.12.

Finally Figure 6.13 shows the character as it is reconstructed from the Gabor coefficients.

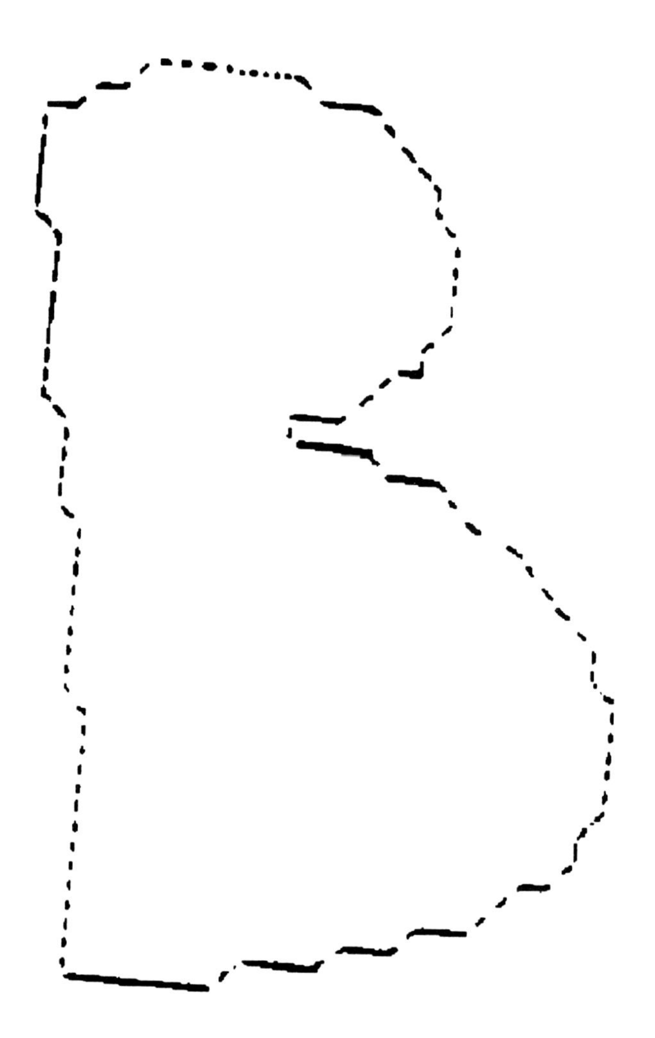

FIGURE 6.10 Original letter “B” image. (From Eichman, G. et al., *SPIE,* vol. 1297, pp. 86–94, 1990. With permission.)

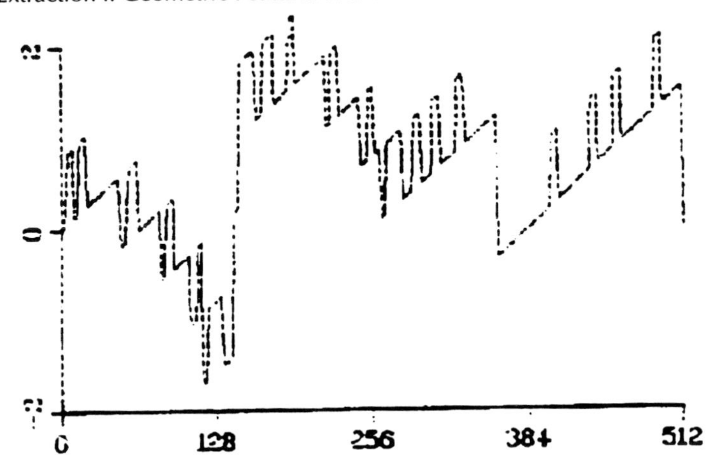

FIGURE 6.11 Spectrum of the shape function, φ'(t) for the letter "B". (From Eichman, G. et al., *SPIE,* vol. 1297, pp. 86–94, 1990. With permission.)

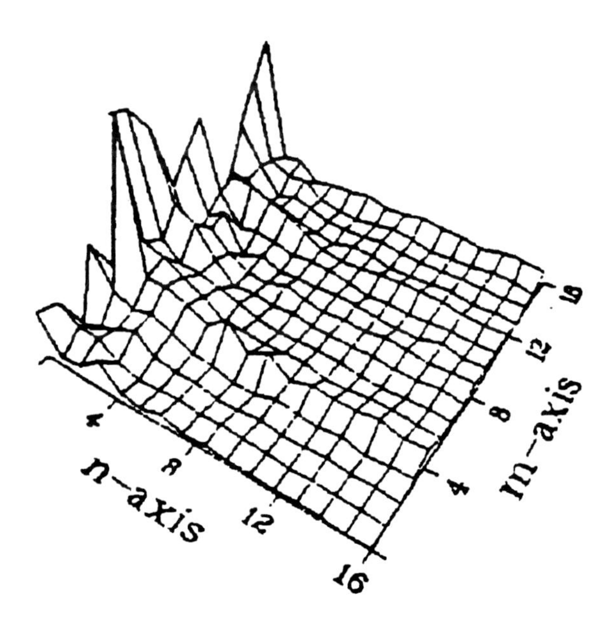

FIGURE 6.12 Representation of the letter "B" in terms of its Gabor coefficients. (From Eichman, G. et al., *SPIE,* vol. 1297, pp. 86–94, 1990. With permission.)

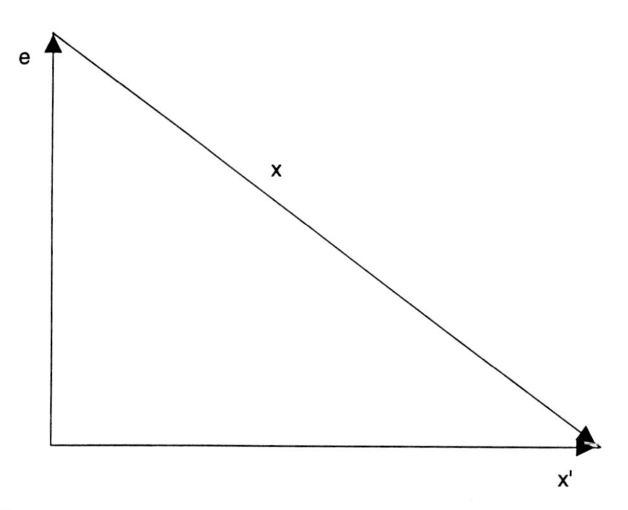

FIGURE 6.13 Letter "B" reconstructed from Gabor coefficients. (From Eichman, G. et al., *SPIE,* vol. 1297, pp. 86–94, 1990. With permission.)

REFERENCES AND BIBLIOGRAPHY

Boas, J. L., *Mathematical Methods in the Physical Sciences,* John Wiley & Sons, New York, 1966.

Darwiche, E., Pandya, A. S., and Mandalia, A. D., "Automated optical recognition of degraded handwritten characters," *SPIE,* vol. 1661, pp. 203–214, 1992.

Eichman, G., Lu, C., Jankowski, M., and Tolimieri, R., "Shape representation by Gabor expansion," Hybrid Image and Signal Processing II, *SPIE,* vol. 1297, pp. 86–94, 1990.

Garris, M. D., Wilkinson, R. A., and Wilson, C. L., "Analysis of a biologically motivated neural network for character recognition," *ACM ANNA,* pp. 160–175, 1991.

Jain, A. K. and Bhattacharjee, S. K., "Address block location on envelopes using Gabor filters," *Pattern Recognition,* vol. 25, no. 12, pp. 1459–1477, 1992.

Kosko, B., *Neural Networks for Signal Processing,* Prentice-Hall, Englewood Cliffs, NJ, 1992.

Le Cun, Y. L., Boser, B., Denker, J. S., Henderson, D., Hubbard, W., and Jackel, L. D., "Handwritten Digit Recognition with a Backpropagation Network," *Advances in Neural Information Systems,* Vol. 2, Morgan Kaufman, San Mateo, CA, pp. 396–404, 1990.

Lu, S., Hernandez, J. E., and Clark, G. E., "Texture segmentation by clustering Gabor feature vectors," *IEEE Joint Conf. Neural Networks,* vol. 1, pp. 683–687, 1991.

Pikaz, A. and Dinstein, I., "Using simple decomposition for smoothing and feature point detection of noisy digital curves," *IEEE PAMI,* vol. 16, no. 8, pp. 808–813, 1994.

Porat, M. and Zeevi, Y. Y., "The Generalized Gabor scheme of image representation in biological and machine vision," *IEEE PAMI,* vol. 10, no. 4, pp. 452–468, 1988.

Press, W. H., Tenkoiseley, S. A., Vetterling, W. T., and Flannery, B. P., *Numerical Recipes in C: The Art of Scientific Computing,* Cambridge University Press, England, 1992.

Venugopal, K. P., Pandya, A. S., and Sudhakar, R., "Invariant recognition of 2-D objects using Alopex," *Proc. SPIE,* vol. 1709, pp. 182–190, 1992.

Wang, J. and Jean, J., "Resolving multifont character confusion with neural networks," *Pattern Recognition,* vol. 5, no. 1, pp. 175–187, 1993.

Wilson, C. L., Wilkinson, R. A., and Garris, M. D., "Self organizing neural network character recognition on a massively parallel computer," *IEEE IJCNN,* vol. 2, pp. 325–329, 1990.

7 Feature Extraction II: Principal Component Analysis

7.1 DIMENSIONALITY REDUCTION

The analysis of high-dimensional data sets is often a complex task. The pattern recognition systems often suffer from the curse of high dimensionality. The process of feature extraction prior to a stage in which objects are classified (see Figure 1.1) may be regarded as a reduction in dimensionality utilizing a transformation of variables which causes some (Watanabe, 1985) variables to be more meaningful than others. These more important variables may be thought of as features. If those variables of lesser importance can be ignored then the desired reduction in dimensionality is achieved. Moreover, it is also likely that such statistically less important components of the transformed variables may arise from noise (or extraneous flourishes not relevant to the intrinsic nature of the pattern). Then their exclusion may be useful for that reason.

The process of feature extraction transforms the data space (pattern space) into a feature space that is of much lower dimension compared to the original data space, yet it retains most of the intrinsic information content of the data. The dimensionality can be reduced using a number of methods, such as principal component analysis, factor analysis, feature clustering, etc. Factor analysis achieves data reduction by forming linear combinations of features, such that the lower dimensional representation accounts for the correlations among the features. Feature clustering approach merges features which are highly correlated in order to eliminate redundant information [Duda and Hart, 1973].

Principal component analysis (PCA) provides a means by which to achieve such a transformation where the feature space accounts for as much of the total variation as possible. Principal component analysis is a well-known technique in multivariate analysis [Preisendorfer, 1988; Jolliffe 1986] where the importance of variables is evaluated statistically. In addition, by providing a quantitative measure of the error introduced by omitting any given components of transformed vector, the number and identity of such expendable components will be known and the dimensionality of the input random vector may be reduced with minimum loss.

Suppose a n-dimensional pattern vector, $\mathbf{X}$ is to be mapped on to the feature vector $\mathbf{Y}$ in an m-dimensional space, where $m < n$. Let E be the mean square error equal to the sum of the variances of the elements eliminated from **X** in order to truncate it and obtain **Y**. Principal component analysis finds the invertible linear transformation T such that truncation of **X** to:

$$\mathbf{Y} = T \bullet \mathbf{X} \tag{1}$$

is optimum in the mean square error sense.

This transformation based on statistical properties of vector representations is commonly referred to as eigenvector, principal component analysis, Hotelling transform, or Karhunen-Loeve transformation. In 1901, Pearson first introduced this transformation in a biological context to recast linear regression analysis into a new form. Hotelling [1933] used it in context of psychometry to transform discrete variables into uncorrelated coefficients. Kramer and Mathews [1956] rediscovered the Hotelling transformation years later.

In 1947, it appeared independently in the setting of probability theory [Karhunen, 1947] for transforming continuous data. Loeve [1948, 1963] subsequently generalized it. Hence, it was coined the Karhunen-Loeve (K-L) transform. Koschman [1954] showed that the K-L transformation minimizes the mean square truncation error. For a good treatment of principal component analysis, readers may refer to a multivariate analysis textbook such as Preisendorfer [1988] or Jolliffe [1986].

For readers with a signal processing background, Kung [1993] provides a good review of principal component neural networks. He presents the problem of principal component analysis in the context of Wiener filtering which is a popular technique for the least-squares-error restoration of original signal. In this regard principal eigenvectors represent directions where the signal has maximum energy.

Figure 7.1 shows a two-dimensional random vector in a Cartesian coordinate system and the principal components for that vector.

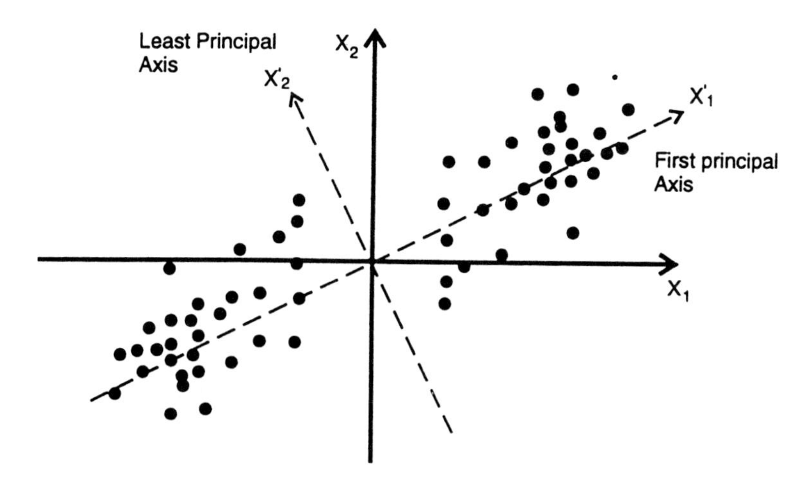

FIGURE 7.1 A pattern space with principal component axes. (Adapted from Kung [1993])

Note that in Figure 7.1 neither the x_1-axis nor the x_2-axis effectively discriminates points in the distribution from one another, while the first principal component axis serves this purpose rather well. It can also be observed that the least principal component axis contributes minimally.

7.2 PRINCIPAL COMPONENTS

This section begins by describing principal components, and the Karhunen-Loeve transform for obtaining them is described in the next section. Finally, an example is described where the principal components are extracted as features for samples in two different classes.

Consider a population of n-dimensional pattern vectors, $\mathbf{X}$, given by:

$$\mathbf{X} = \begin{pmatrix} x_1 \\ x_2 \\ \vdots \\ x_{n-1} \\ x_n \end{pmatrix} \tag{2}$$

In general, given the probability distribution function, $\{\mathbf{X}\}$, the mean vector of the population is given by:

$$\mu_x = E[\mathbf{X}] \tag{3}$$

where $E[\mathbf{X}]$ is the expected value of pattern vector, $\mathbf{X}$.

Therefore, the mean vector can be approximated from discrete sample, in the absence of the probability density function, by:

$$\mu_x = \frac{1}{M}\sum_{k=1}^{M} x_k \tag{4}$$

The covariance matrix of the pattern vector population in general is defined as:

$$\Sigma_x = E\left[(x_k - \mu_x)(x_k - \mu_x)^T\right] \tag{5}$$

which is real, symmetric, and on the order of $n \times n$. The element σ_{ij} of Σ_x is the covariance between elements x_i and x_j of the $\mathbf{X}$ vector. If x_i and x_j are uncorrelated, the covariance is zero, i.e.

$$\sigma_{ij} = \sigma_{ji} = 0 \tag{6}$$

Thus, the covariance matrix will be

$$\Sigma_x = \frac{1}{M-1}\sum_{k=1}^{M}(x_k - \mu_x)(x_k - \mu_x)^T \tag{7}$$

In general, if the probability distribution function {**X**}, is known *a priori*, the autocorrelation matrix is given by:

$$R_x = E\left[\mathbf{X}\mathbf{X}^T\right] \tag{8}$$

For this autocorrelation matrix, R_x, $\{\Phi_i\}$ are the normalized eigenvectors, i.e.:

$$\left\|\Phi_i\right\| = \sqrt{\Phi_i^T \Phi_i} = 1 \tag{9}$$

and the corresponding eigenvalues are $\{\lambda_i\}$.

The pattern vector **X** can now be projected onto the one-dimensional subspace spanned by the eigenvector Φ_i in order to get its eigencomponent. As shown in Figure 7.1 the first principal component (i.e., the largest) of the pattern vector **X** can be given by:

$$\mathbf{X}_i = \Phi_i^T \mathbf{X} \tag{10}$$

Likewise, the rest of the *m* principal components can be obtained through projections. Note that the last component (see Figure 7.1) will have the least value.

In practice *a priori* knowledge of the autocorrelation matrix for a population of pattern vectors is seldom available. However, for a large number of sample input vectors {**X**(t), for t = 1, ..., M} an estimate of the autocorrelation matrix, R_x, can be made by averaging the sample vectors:

$$\tilde{R}_x \frac{1}{M}\sum_{t=1}^{M}\mathbf{x}(t)\mathbf{x}^T(t) \tag{11}$$

Also it may be assumed that:

$$R_x = \lim_{M \to \infty} \tilde{R}_x \tag{12}$$

so that, for M sufficiently large, the estimate will be in some sense satisfactory.

7.2.1 PCA EXAMPLE

Consider, four pattern vectors given by:

$$x_1 = \begin{pmatrix} 1 \\ 0 \\ 1 \end{pmatrix}, \quad x_2 = \begin{pmatrix} 2 \\ 3 \\ 1 \end{pmatrix}, \quad x_3 = \begin{pmatrix} 0 \\ 1 \\ 1 \end{pmatrix}, \quad \text{and} \quad x_4 = \begin{pmatrix} 1 \\ 4 \\ 1 \end{pmatrix} \tag{13}$$

the mean vector can be obtained by using equation 4:

$$\mu_x = \frac{1}{4}\begin{pmatrix} 1+2+0+1 \\ 0+3+1+4 \\ 1+1+1+1 \end{pmatrix} = \begin{pmatrix} 1 \\ 2 \\ 1 \end{pmatrix} \tag{14}$$

Using equation 7 the covariance matrix in this case can be given as:

$$\Sigma_x = \frac{1}{4}\left\{\begin{pmatrix} 1-1 \\ 0-2 \\ 1-1 \end{pmatrix}(0 \ -2 \ 0) + \begin{pmatrix} 1 \\ 1 \\ 0 \end{pmatrix}(1 \ 1 \ 0) + \begin{pmatrix} -1 \\ -1 \\ 0 \end{pmatrix}(-1 \ -1 \ 0) + \begin{pmatrix} 0 \\ 2 \\ 0 \end{pmatrix}(0 \ 2 \ 0)\right\} \tag{15}$$

$$= \frac{1}{4}\begin{pmatrix} 0+1+1+0 & 0+1+1+0 & 0+0+0+0 \\ 0+1+1+0 & 4+1+1+4 & 0+0+0+0 \\ 0+0+0+0 & 0+0+0+0 & 0+0+0+0 \end{pmatrix}$$

$$= \begin{pmatrix} 0.5 & 0.5 & 0 \\ 0.5 & 2.5 & 0 \\ 0 & 0 & 0 \end{pmatrix}$$

7.3 KARHUNEN-LOEVE (K-L) TRANSFORMATION

Principal components are linear combinations of random variables having special properties with respect to variance. The first principal component is the normalized linear combination with maximum variance. The second has the next largest variance and so forth. In effect the principal components are ranked by their ability to distinguish among classes. The K-L transformation is an orthonormal transformation of an n-dimensional vector **X** to an n-dimensional vector **Y** which achieves this property. It turns out that the principal components will be the characteristic vectors of the covariance matrix.

A linear combination of principal vectors can provide a feature vector. The associated eigenvalues reflect the importance of each eigenvector. The important issue is that given a particular situation, how many principal components are sufficient? In some cases a threshold value is used and principal components with associated eigenvalues less than the threshold are dropped. Sometimes the number of components is fixed *a priori*, as in the case of situations requiring visualization of feature space limitations that are imposed due to two- or three-dimensional space requirements.

To find the principal components, it remains to obtain the eigenvectors Φ_i and the eigenvalues λ_i of the covariance matrix Σ_x.

$$\Sigma_x \Phi_i = \lambda_i \Phi_i \tag{16}$$

With this accomplished, the n-dimensional random input vector $\mathbf{X}$ is expressed in terms of its principal components as:

$$\mathbf{Y} = \Phi^T \mathbf{X}$$

with components (17)

$$\mathbf{y}_i = \Phi_i^T x_i$$

Equation 17 is also referred to as Hotelling transformation.

In the context of pattern recognition, each component y_i can be viewed as a feature of pattern vector $\mathbf{X}$. The features are mutually uncorrelated, so that the covariance matrix of $\mathbf{Y}$ is diagonal. Thus,

$$\Sigma_y = \Phi^T \mathbf{X} \Phi = \begin{pmatrix} \lambda_1 & 0 & 0 & \cdots & 0 \\ 0 & \lambda_2 & 0 & \cdots & 0 \\ \vdots & \vdots & & \ddots & \vdots \\ 0 & 0 & 0 & \cdots & \lambda_n \end{pmatrix} \tag{18}$$

Further, it can be shown that the mean square error in representing $\mathbf{X}$ by using only the first $m < n$ principal components ($\lambda_1 > \lambda_2 > \lambda_3 > \ldots > \lambda_{n-1} > \lambda_n > 0$) is

$$\overline{\varepsilon^2}(m) \sum_{i=m+1}^{n} \lambda_i \tag{19}$$

The approximation of pattern vector $\mathbf{X}$ by this truncation is $\mathbf{X}'$ given by:

$$\mathbf{X}' = \sum_{i=1}^{m} y_i \phi_i \tag{20}$$

and the error vector, **e**, is given by:

$$\mathbf{e} = \mathbf{X} - \mathbf{X}' \tag{21}$$

The mean square error is given by:

$$\|\mathbf{e}\| = \|\mathbf{X} - \mathbf{X}'\| = \left(\sum_i (x_i - x_i')^2 \right)^{1/2} \tag{22}$$

Figure 7.2 illustrates the relationship between the error vector, **e**, and the original pattern, **X**.

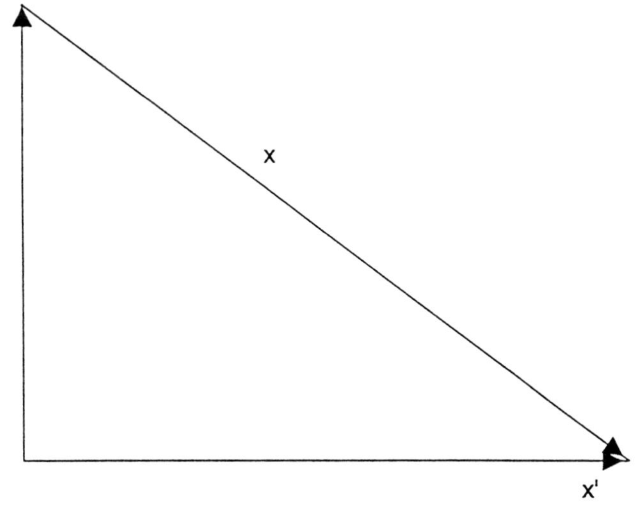

FIGURE 7.2 The original pattern vector and its relationship with the approximated pattern vector.

7.3.1 K-L Transformation Example

Consider two classes each with four patterns given by:

$$C_1 = \left\{ \begin{pmatrix} -2 \\ -2 \end{pmatrix}, \begin{pmatrix} -4 \\ -4 \end{pmatrix}, \begin{pmatrix} 4 \\ 4 \end{pmatrix}, \begin{pmatrix} 2 \\ 2 \end{pmatrix} \right\}$$

$$C_2 = \left\{ \begin{pmatrix} -2 \\ 2 \end{pmatrix}, \begin{pmatrix} -4 \\ -4 \end{pmatrix}, \begin{pmatrix} 4 \\ 4 \end{pmatrix}, \begin{pmatrix} 2 \\ -2 \end{pmatrix} \right\} \tag{23}$$

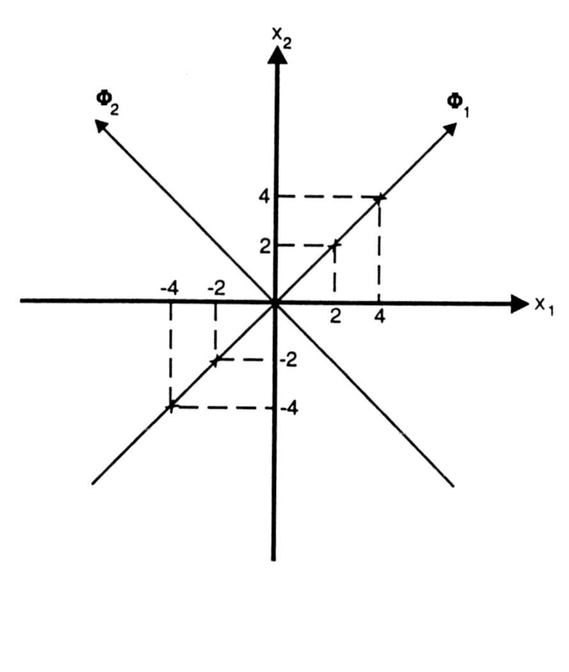

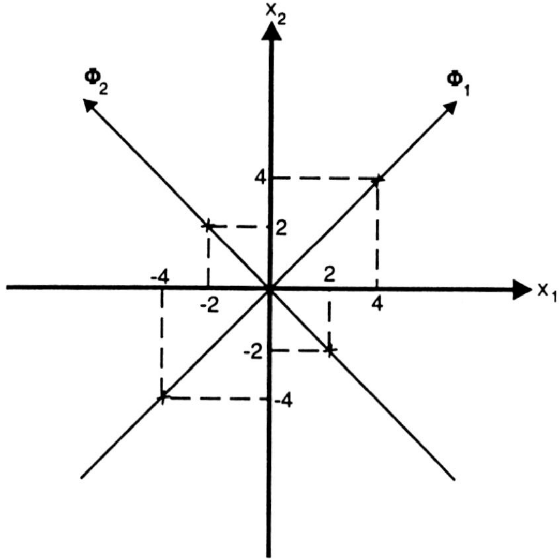

FIGURE 7.3 Patterns plotted along with the eigenvectors.

The patterns can be plotted in a two-dimensional pattern space as shown in Figure 7.3.

The mean vector can be obtained using equation 4. Thus:

$$\mu_1 = 0 \quad \text{and} \quad \mu_2 = 0 \tag{24}$$

The covariance matrix for each class can be obtained using equation 7:

$$\Sigma_a = \frac{1}{4}\sum_{i=1}^{4} x_i x_i^T \tag{25}$$

$$= \frac{1}{4}\left\{\begin{pmatrix}-4\\-4\end{pmatrix}\begin{pmatrix}-4 & -4\end{pmatrix} + \begin{pmatrix}-2\\-2\end{pmatrix}\begin{pmatrix}-2 & -2\end{pmatrix} + \begin{pmatrix}4\\4\end{pmatrix}\begin{pmatrix}4 & 4\end{pmatrix} + \begin{pmatrix}2\\2\end{pmatrix}\begin{pmatrix}2 & 2\end{pmatrix}\right\}$$

$$= \frac{1}{4}\begin{pmatrix}40 & 40\\40 & 40\end{pmatrix} = \begin{pmatrix}10 & 10\\10 & 10\end{pmatrix} \quad \text{for class } C_1$$

$$\Sigma_b = \frac{1}{4}\sum_{i=1}^{4} x_i x_i^T \tag{26}$$

$$= \frac{1}{4}\left\{\begin{pmatrix}-2\\2\end{pmatrix}\begin{pmatrix}-2 & 2\end{pmatrix} + \begin{pmatrix}4\\4\end{pmatrix}\begin{pmatrix}4 & 4\end{pmatrix} + \begin{pmatrix}-4\\-4\end{pmatrix}\begin{pmatrix}-4 & -4\end{pmatrix} + \begin{pmatrix}2\\-2\end{pmatrix}\begin{pmatrix}2 & -2\end{pmatrix}\right\}$$

$$= \frac{1}{4}\begin{pmatrix}40 & 24\\24 & 40\end{pmatrix} = \begin{pmatrix}10 & 6\\6 & 10\end{pmatrix} \quad \text{for class } C_2$$

Now using the K-L transformation we can evaluate the eigenvalues λ_i's and the eigenvectors of Σ for class C_1:

$$\lambda_{1a} = 20, \quad \lambda_{2a} = 0$$

$$\varphi_{1a} = \begin{pmatrix}\frac{1}{\sqrt{2}}\\ \frac{1}{\sqrt{2}}\end{pmatrix}, \quad \varphi_{2a} = \begin{pmatrix}\frac{-1}{\sqrt{2}}\\ \frac{1}{\sqrt{2}}\end{pmatrix} \tag{27}$$

For class C_2:

$$\lambda_{1b} = 16, \quad \lambda_{2b} = 0$$

$$\varphi_{1b} = \begin{pmatrix}\frac{1}{\sqrt{2}}\\ \frac{1}{\sqrt{2}}\end{pmatrix}, \quad \varphi_{2b} = \begin{pmatrix}\frac{-1}{\sqrt{2}}\\ \frac{1}{\sqrt{2}}\end{pmatrix} \tag{28}$$

These eigenvectors can now be plotted in the pattern space as 45 degree lines (see Figure 7.3).

For dimensionality reduction let us consider the effect of eliminating the less significant principal components for class, C_1; since $\lambda_{2a} = 0$, we can see using equation 22 that the mean square error will also be zero. Also it is clear from Figure 7.3 that all four patterns of class 1 can be expressed using only Φ_1. For patterns in class 2, elimination of Φ_2 will result in an error of 4, since $\lambda_{2b} = 4$. As shown in Figure 7.3, points $(-4, 4)^T$ and $(4, 4)^T$ can be expressed using only Φ_1 but $(-2, 2)^T$ and $(2, -2)^T$ will result in an error.

The mean square error in this case will be

$$\|e\| = \left(0^2 + 0^2 + \left(16^2\right)^{1/2} + \left(16^2\right)^{1/2}\right)/4 = 4 \tag{29}$$

Figure 7.4 illustrates the applications of principal component analysis to pattern vectors (data points) plotted in a two-dimensional pattern space. The natural coordinates x and y are represented by the horizontal and vertical axes, respectively, in the diagram. Once the principal components are obtained (as the example in Section 7.3.1) the rotated axes 1 and 2 can be plotted. Then the density plots can be formed by projecting the cloud representing the patterns (data points) onto each of the two axes 1 and 2 as shown in Figure 7.4. The important point in this simple example is that the projection of the data set onto axis 1 captures the salient feature of the data, namely, the fact that it is bimodal (i.e., the structure consists of two clusters), which is completely obscured by projections onto the original x and y axes.

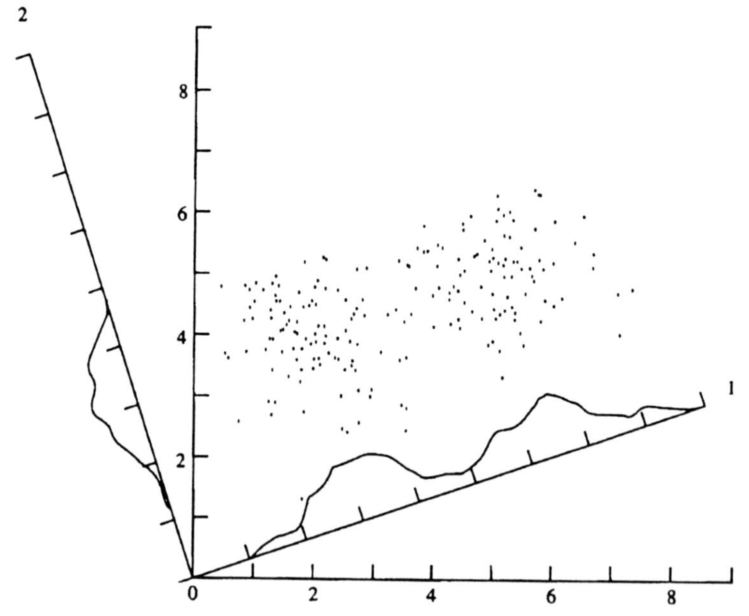

FIGURE 7.4 Principal component analysis reveals the bimodal nature of the cloud of data points. (From Linsker, R., *Computer,* vol. 21, pp. 105–117, 1988. With permission of IEEE.)

7.4 PRINCIPAL COMPONENT NEURAL NETWORKS

Principal components and the Karhunen-Loeve transform have been used quite successfully for feature extraction, image/data compression, and pattern classification. Numerical methods exist for the computation of eigenvectors and eigenvalues (for example, see Press et al., [1992]). However, one drawback appears to be that these calculations can be computationally very intensive. In this section we look briefly at a neural network implementation which can be made to perform that task.

In these approaches the neural networks are used for extracting the most representative low-dimensional subspace from a high-dimensional pattern vector space. Figure 7.5 shows a multilayer perceptron network which can be used to extract most critical and representative features of the data. The number of neurons in the input and output layer is determined based on the dimensionality of pattern vector **X** and desired output vector **Y**, respectively. The idea here is to use a relatively few hidden neurons in order to achieve data reduction.

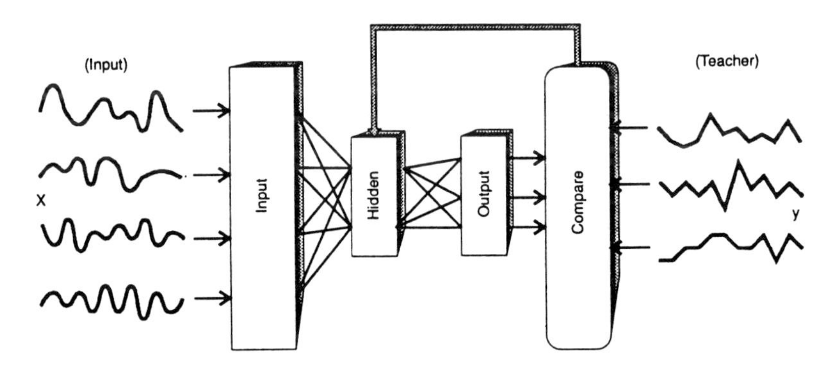

FIGURE 7.5 An associative neural network for extracting relevant features. (Adapted from Kung [1993].)

This configuration is for supervised training since the **y** vector acts as the teacher which produces the necessary signal for error correction in the weights. Learning algorithms such as back propagation or ALOPEX can be used to train the weights.

The network shown in Figure 7.5 can be used as an autoassociative network also. In this case, the network is self-supervised in the sense that the outputs, y are to be trained to mimic the inputs x. During training the comparison stage can provide appropriate feedback to accomplish this because the desired output is equal to the input. It can be shown that if the network consists exclusively of linear nodes, then the network weights can, with appropriate training, be made to converge to the principal components discussed above [Kung, 1993].

The concept of this autoassociative network remains intriguing even in isolation from the rigorous concept of principal component analysis. Suppose that such a network were to be used for feature extraction using a nonlinear squashing function (such as the sigmoid function) and a hidden layer significantly smaller than the input/output layers. Under these conditions the network can be expected to find efficient ways of encoding the information contained in the input data set. Intuitively, one expects a greater discriminative power to result from the nonlinear neurons. (We might also consider such variations on the theme as radial-basis function neurons.) What remains is a front-end self-supervised feature extraction stage whose features would consist of the outputs from the hidden neurons. That is, after training, the output layer is discarded and the hidden layer is fed directly into the recognizer network.

Oja [1982] describes the use of a single linear neuron model as a maximum eigenfilter. A single linear neuron can evolve into a filter for the first principal component of the pattern distribution by adapting its synaptic strengths using a Hebbian-type learning rule. Haykin [1994] shows that the synaptic strength vector, **W**, of a self-organized linear neuron operating under the modified Hebbian rule converges with a probability 1 to a vector of unit Euclidean length. Moreover, this vector lies in the maximal eigenvector direction of the correlation matrix of the pattern vector.

Kung and Diamantaras [1990] proposed a neural network learning algorithm called adaptive principal component extraction (APEX) for the recursive computation of the principal components. The algorithm functions in an iterative fashion and computes the $(j + 1)^{th}$ principal component, given the first j principal components. For a detailed description of this algorithm, the reader may refer to Kung [1993] or Haykin [1994].

7.5 APPLICATIONS

The modified Hebbian rule of Oja [1982] can be used to train a single linear neuron as a maximum eigenfilter. This rule can be generalized to train a feed-forward network composed of a single layer of linear neurons. The learning algorithm coined as generalized Hebbian algorithm (GHA) is discussed in detail by Haykin [1994]. Sanger [1989] described how to train such a network to perform principal component analysis of arbitrary size of pattern vector.

In this application the GHA is used in the context of the image compression problem. The image of the three children, shown in Figure 7.6 (a) was used for training the weights of the network.

In order to present the information to the neural network, the image was digitized to form a 256×256 pixel map with 256 gray levels. The linear feed-forward network had a single layer of eight neurons and each received 64 inputs. These inputs were 8×8 nonoverlapping blocks of the image obtained by scanning it from left to right and top to bottom. The network weights were trained using the GHA and 64 weights associated with each neuron formed the 8×8 masks shown in Figure 7.7. The white pixels in Figure 7.7 represent positive weights (excitatory synaptic connections), whereas the black pixels represent the negative weights (inhibitory synapses). The gray pixels indicate zero weights (i.e., no connection). The final image compression for the image in Figure 7.6 was obtained by using the following procedure (Sanger, 1989):

1. Eight coefficients for image compression were obtained by presenting each 8 × 8 block of image to the network and calculating the resulting output at each of the eight neurons.
2. The logarithm of variance of each coefficient over the entire image was used to approximately determine the number of bits used to represent that coefficient.
3. Thus, the coefficients were uniformly quantized such that the first two masks generated 5 bits each, the third mask generated 3 bits, and the remaining five masks generated 2 bits each.
4. As a result each 8 × 8 block of pixels was compressed to 23 bits, resulting in a data compression of 0.36 bits per pixel.

Figure 7.6 (b) shows the reconstructed children's image obtained by using the quantized coefficients. First, all the masks were weighted by their quantized coefficients, and then they were added to reconstitute each block of the image.

More interesting is the application of the same masks (Figure 7.7) to another image with statistics probably similar to those of the children's image (Figure 7.6 (a)). Figure 7.8 (a) shows the image of a dog used for this purpose.

In this case, the coefficients were uniformly quantized such that:

- The first two masks generated 7 bits each.
- The third mask generated 5 bits.
- The forth mask generated 4 bits.
- The remaining four masks generated 3 bits each.

As a result each 8 × 8 block of pixels was compressed to 35 bits, resulting in a data compression of 0.55 bits per pixel.

The reconstructed image of the dog, obtained by using the quantized coefficients derived from the masks of Figure 7.7 is shown in Figure 7.8 (b).

The Karhunen-Loeve transform has been applied to the problem of handwritten character recognition on a large scale [Grother, 1992]. The study encompassed 76,753 training set characters from 944 different writers. The system achieved 96.1% recognition on a test set of 15,000 characters. The implementation involves a multilevel neural net with the first layer performing principal component feature extraction. Since the eigenvectors can be calculated using a neural approach (as discussed in section 7.4), they can be regarded as a trained weight layer. The approach taken here was to first calculate the principal components conventionally by the Karhunen-Loeve transformation and then to use the eigenvectors directly as weights in this principal component feature extraction stage. By doing so, the computationally intensive job of finding the eigencomponents is limited to the training stage and as a result it does not affect operational performance. This stage was followed by training a feed-forward perception network using a conjugate gradient algorithm. Note that while the first feature extraction stage is linear, the subsequent classification stage is not. The classification network used a single hidden layer.

(a)

(b)

FIGURE 7.6 (a) The image to be compressed using the neural network. (b) The reconstructed image using the compression coefficients determined using the neural network. (From Sanger, T. D., *Neural Networks,* vol. 12, pp. 459–473, 1989. With permission from Elsevier Science Ltd.)

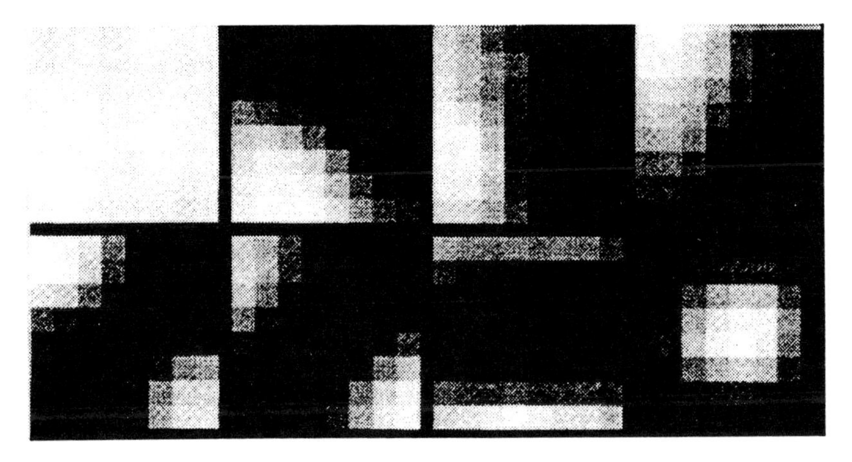

FIGURE 7.7 The eight 8 × 8 masks obtained by training the weights of the single-layer network. The pixel intensities represent the weight values as discussed in the text. (From Sanger, T. D., *Neural Networks,* vol. 12, pp. 459–473, 1989. With permission from Elsevier Science Ltd.)

Two topologies were used:

1. A 32 input × 32 hidden × 10 output — Using the first 32 principal components and yielding an accuracy of 93.7%
2. A 48 input × 48 hidden × 10 output — Using the first 48 principal components and yielding an accuracy of 94.5%

Very often it is assumed that the Karhunen-Loeve transform will be applied to the raw (bit map) input data. This need not necessarily be the case. The Karhunen-Loeve transformation could in fact be used to achieve similar purposes when applied to a previously calculated feature set. The question is why should we do so when the potential exists for the loss of statistically relevant information that is present in the input but not in the chosen feature set? One answer is found in situations where some form of invariance is explicitly sought. By designing the feature set to be invariant with respect to the desired property, that property is preserved. Subsequent application of principal component analysis could then provide a statistically optimal reduction of the input vector dimension without loss of the desired invariance.

(a)

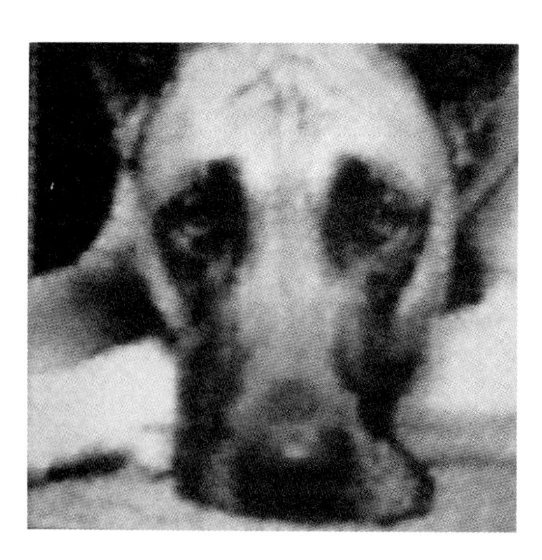

(b)

FIGURE 7.8 (a) An image of a dog which is compressed using the masks trained by the image of three children. (b) The reconstructed image. (From Sanger, T. D., *Neural Networks,* vol. 12, pp. 459–473, 1989. With permission from Elsevier Science Ltd.)

REFERENCES AND BIBLIOGRAPHY

Anderson, T. W., *An Introduction to Multivariate Statistical Analysis,* John Wiley & Sons, New York, 1984.

Duda, R. O. and Hart, P. E., *Pattern Classification and Scene Analysis,* John Wiley & Sons, New York, 1973.

Fu, L., *Neural Networks in Computer Intelligence,* McGraw-Hill, New York, 1994.

Fukunaga, K., *Introduction to Statistical Pattern Recognition,* Academic Press, New York and London, 1972; 2nd ed., 1985.

Gonzalez, R. C. and Woods, R. E., *Digital Image Processing,* Addison-Wesley, Reading, MA, 1992.

Grother, P. J., "Karhunen-Loeve Feature Extraction for Neural Handwritten Character Recognition," NISTR 4824, U.S. Dept. of Commerce, pp. 1–12, 1992.

Haykin, S., *Neural Networks: A Comprehensive Foundation,* Macmillan College Publishing, New York, 1994.

Hotelling, H., "Analysis of a complex of statistical variables into principal components," *J. Educ. Psychol.,* vol. 24, pp. 417–441, 498–520, 1933.

Jolliffe, I. T., *Principal Component Analysis,* Springer-Verlag, New York, 1986.

Karhunen, K., "Uber lineare methoden in der Wahrscheinlichkeitsrechung," Annales Academiae Scientiarum Fennicae, Series A1: Mathematica-Physica, vol. 37, pp. 3–79, 1947 (Translation: Rep. T-131, RAND Corp., Santa Monica, CA, 1960).

Koschman, A., "On the filtering of nonstationary time series," *Proc. 1954 Natl. Electron. Conf.,* p. 126, 1954.

Kramer, H. P. and Mathews, M. V., "A linear coding for transmitting a set of correlated signals," *IRE Trans. Inf. Theory,* vol. IT-2, pp. 41–46, 1956.

Kung, S. Y., *Digital Neural Networks,* Prentice-Hall, Englewood Cliffs, NJ, 1993.

Kung, S. Y. and Diamantaras, C. I., "A neural network learning algorithm for adaptive principal component extraction (APEX)," *Proc. Int. Conf. Acoustics, Speech, and Signal Processing,* vol. 2, pp. 861–864, 1990.

Linsker, R., "Self-organization in a perceptual network," *Computer,* vol. 21, pp. 105–117, 1988.

Loeve, M., "Fonctions Aleatoires de Second Ordre," in P. Levy (Ed.), *Processus Stochastiques et Mouvement Brownien,* Hermann, Paris, 1948.

Loeve, M., *Probability Theory,* 3rd ed., Van Nostrand, New York, 1963.

Oja, E., "A simplified neuron model as a principal component analyzer," *J. Math. Biol.,* vol. 15, pp. 267–273, 1982.

Preisendorfer, R. W., *Principal Component Analysis in Meteorology and Oceanography,* Elsevier, New York, 1988.

Press, W. H., Teukoisky, S. A., Vetterling, W. T., and Flannery, B. P., *Numerical Recipes in C: The Art of Scientific Computing,* Cambridge University Press, Cambridge, 1992.

Sanger, T. D., "Optimal unsupervised learning in a single-layer feedforward neural network," *Neural Networks,* vol. 12, pp. 459–473, 1989.

Schalkoff, P. J., *Pattern Recognition Statistical Structural and Neural Approaches,* John Wiley & Sons, New York, 1992.

Watanabe, S., *Pattern Recognition: Human and Mechanical,* John Wiley & Sons, New York, 1985.

8 Kohonen Networks and Learning Vector Quantization

8.1 GENERAL

A common attribute of biological systems is the ability to extract and act upon regularities within their natural environment without benefit of a teacher. Likewise, this is the goal of self-organizing neural networks: to discover significant regularities within the input data without external supervision. In chapters 3, 6, and 7 the discussion centered upon mechanisms by which data from the external environment can be in some sense predigested for the benefit of a recognition engine waiting downstream. Chapters 6 and 7, in particular, dealt with the extraction of features to be used as input to supervised recognizer networks. By contrast, the task of feature detection viewed in a context of self-organizing networks falls to the network itself. Chapters 4 and 5 explored several of the various available networks based upon supervised learning.

This chapter centers around self-organizing systems which can, as biological systems do, discover the structure, patterns, or features directly from their environment. It is the Kohonen self-organizing feature map (SOFM), in particular, that partakes strongly of the biological inspiration to be found in the retinal cortex. Additionally, the SOFM, in part due to its biological roots possesses the unique property that cluster centers or features aggregate geometrically within the network output layer. That is, features that are similar by virtue of possessing a small Euclidean distance between them in feature space will stimulate responses in SOFM output neurons that are also geometrically close to each other.

We remark, however, that such self-organizing behavior is not limited to the domain/realm of neural networks. We begin our study in this chapter with a clustering algorithm drawn from statistical pattern recognition: the K-means algorithm. If this topic seems oddly placed, we observe that the K-means algorithm clearly does possess self-organizing characteristics; and the astute reader will notice similarities between this algorithm and the neural techniques discussed subsequently.

The networks we present in this chapter are distinguished by their use of competitive learning. Competitive learning can be described as a process in which output layer neurons compete among themselves to acquire the ability to fire in response to given input patterns. In the networks we describe this will be the neuron whose weight vector is closest in the input space to the current pattern vector. This is again at least reminiscent of K-means in which the cluster center with the shortest Euclidean norm distance to the input pattern vector acquires the pattern (and earns the right to respond to that pattern).

Some form of lateral feedback is another identifying trait common among the network presented here. Neurons utilize lateral feedback to alter network behavior in limited regions or neighborhoods surrounding output layer neurons. Lateral feedback may be established by means of a set of connections and associated weights among output layer neurons. The weights will typically decrease as the radius from the source to the destination neuron increases. If the weights are negative, the result will tend to inhibit the firing of other output layer neurons. Such lateral inhibition can be seen as contributing to the winner-take-all strategy/paradigm described above.

The Kohonen SOFM employs lateral feedback in a slightly different cause: to induce topological organization among the neurons of the output layer. Output layer neurons are organized in a grid or lattice structure of typically one or two dimensions. Higher dimensionality is possible but rare. The lateral feedback is responsible for motivating like features to form at geometrically close positions on the output layer lattice.

The chapter concludes with the presentation of the learning vector quantization network. This variation of the Kohonen model is interesting in that it merges self-organizing and teacher-based techniques. Competitive learning does still take place but the manner in which this occurs is supervised by a teacher. By this it is meant that competitive learning occurs locally, within groups or classes that specified by a training input.

8.2 THE K-MEANS ALGORITHM

Pattern vectors of n-dimensions may be considered as representing points within an n-dimensional Euclidean space. One of the most obvious means by which we may establish a measure of similarity between or among such pattern vectors is by means of their proximity to one another. Figure 8.1 illustrates this idea. The K-means algorithm is one of many clustering techniques that partakes of this notion of clustering by minimum distance. Simply stated, vectors which identify points that are geometrically close together may be taken in some sense as belonging together: guilt by association. Before presenting detail regarding the operation of the K-means algorithm, a more precise notion of the distance metric is needed. The Euclidean norm of a vector, $\mathbf{x} = [x_1, x_2, \ldots, x_n]^T$, is defined as follows:

$$\|\mathbf{x}\| = \left[\sum_{i=1}^{n} x_i^2 \right]^{1/2} \tag{1}$$

Equation 1 provides the length of the vector, $\mathbf{x}$. Since we desire the distance or length between two vectors within the pattern space, we need only to apply equation 1 to the vector difference as follows:

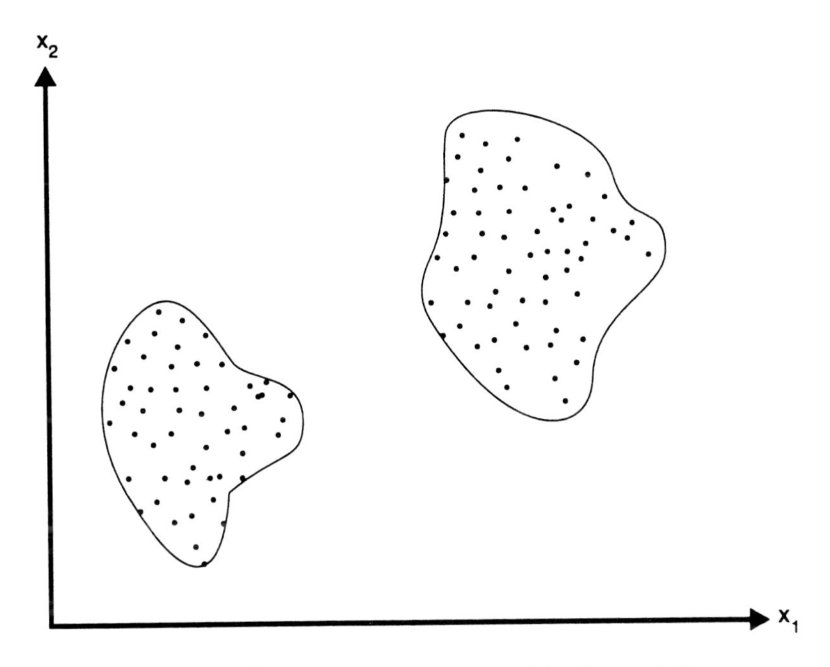

FIGURE 8.1 Pattern classes that are evident by proximity.

$$\|\mathbf{x}-\mathbf{z}\| = \left[\sum_{i=1}^{n}(x_i - z_i)^2\right]^{1/2} \tag{2}$$

where **x** and **z** are pattern vectors of order *n*.

Now that a measure of pattern similarity has been established, the task of establishing a procedure by which patterns are partitioned into cluster domains must be undertaken. That is, we require a procedure that will establish a set of clusters (with associated cluster centers) such that the distance between an input vector and the closest cluster center serves to classify the vector. The K-means algorithm represents one such method.

K-means makes the assumption that the number of cluster centers that will be required to adequately represent the sample space is known *a priori*. This assumption in itself somewhat limits the utility of the procedure. Other variations on minimum distance statistical clustering techniques lack this difficulty but may have other problems such as a sensitivity to the order in which input data are presented to the system. See Vector Quantization in chapter 10.3.

Before describing the K-means process, some nomenclature must be established. Let $\mathbf{x}^{(p)}$ represent the pth input space vector. The complete set of input vectors will then be $\{\mathbf{x}^{(1)}, \mathbf{x}^{(2)}, \ldots, \mathbf{x}^{(P)}\}$. The vector, **z**, represents the cluster center for each of the *K* clusters. That is, it points to the position in Euclidean space at which the cluster center is located. Since there are *K* clusters, there are also *K* cluster centers: $\mathbf{z}_1, \mathbf{z}_2, \ldots, \mathbf{z}_k$. Finally the notation, $S_j = \{\mathbf{x} | \mathbf{x}$ is closest to cluster $j\}$ will be used to represent the set of samples that belong to the jth cluster center. The K-means algorithm is implemented in the following steps.

Step 1. Initialize:

Choose the number of cluster, K. For each of these K clusters choose an initial cluster center:

$$\{\mathbf{z}_1(l), \mathbf{z}_2(l), \ldots, \mathbf{z}_k(l)\}$$

where $\mathbf{z}_j(l)$ represents the value of the cluster center at the lth iteration. The starting values can be arbitrary but are generally taken to be the value of the first K of the sample vectors.

Step 2. Distribute samples:

Distribute all sample vectors. By this is meant that each sample vector $\mathbf{x}^{(p)}$ is attached to one of the K clusters according to the following criteria:

$$\mathbf{x}^{(p)} \in S_j(l) \text{ if } \left\| \mathbf{x}^{(p)} - \mathbf{z}_j(l) \right\| < \left\| \mathbf{x}^{(p)} - \mathbf{z}_i(l) \right\| \tag{3}$$

$$\text{for all } i = 1, 2, \ldots, K, \quad i \neq j$$

$S_j(l)$ represents the population of cluster j at iteration l. Ties in equation 3 may be resolved arbitrarily.

Step 3. Calculate new cluster centers:

Using the new cluster membership sets established in step 2, recalculate the position of each cluster center such that the sum of the distances from each member vector to the new cluster center is minimized. Specifically we wish to minimize J_j where:

$$J_j = \sum_{\mathbf{x}^{(p)} \in S_j(1)} \left\| \mathbf{x}^{(p)} - \mathbf{z}_j(l+1) \right\|^2 \tag{4}$$

$$j = 1, 2, \ldots, K$$

The value of $\mathbf{z}_j(l + 1)$ which minimizes equation 4 is simply the mean taken over the samples of $S_j(I)$. Therefore the new cluster center is calculated using equation 5 as follows:

$$\mathbf{z}_j(l+1) = \frac{1}{N_j} \sum_{\mathbf{x}^{(p)} \in S_j(1)} \mathbf{x}^{(p)} \tag{5}$$

where N_j is the number of sample vectors attached to S_j during step 2.

Step 4. Check for convergence:
The condition for convergence is that no cluster center has changed its position during step 3. This condition can be expressed mathematically as follows:

$$\mathbf{z}_j(l+1) = \mathbf{z}_j(1) \quad j = 1,2,\ldots,K \tag{6}$$

If equation 6 is satisfied, then convergence has occurred. Otherwise iterate by going to step 2.

A number of factors may influence the behavior of the K-means algorithm. Among these are the number of cluster centers, the choice of initial cluster centers, and the geometric properties of the input data. Some experimentation with the choice of K and the initialization parameters may be required. Although no formal proof of convergence exists, the method can be expected to do well where the nature of the data is consistent with the assumption inherent in using the minimum distance as a similarity measure.

Listing 8.1 provides a full implementation of the K-means algorithm as presented here. Subsequently, in section 8.2.1 we will provide an example using the K-means algorithm.

Listing 8.1

```
/*****************************************************************************
*                                                                            *
* KMEANS                                                                     *
*                                                                            *
*****************************************************************************/

#include <stdio.h>
#include <stdlib.h>
#include <string.h>
#include <conio.h>
#include <math.h>

// FUNCTION PROTOTYPES


// DEFINES
#define     SUCCESS       1
#define     FAILURE       0
#define     TRUE          1
#define     FALSE         0
#define     MAXVECTDIM    20
#define     MAXPATTERN    20
#define     MAXCLUSTER    10
```

```
// ***** Defined structures & classes *****
struct aCluster {
  double     Center[MAXVECTDIM];
  int        Member[MAXPATTERN];          //Index of Vectors belonging to this cluster
  int        NumMembers;
};

struct aVector {
  double     Center[MAXVECTDIM];
  int        Size;
};

class System {
private:
  double     Pattern[MAXPATTERN][MAXVECTDIM+1];
  aCluster   Cluster[MAXCLUSTER];
  int        NumPatterns;                 // Number of patterns
  int        SizeVector;                  // Number of dimensions in vector
  int        NumClusters;                 // Number of clusters
  void       DistributeSamples();         // Step 2 of K-means algorithm
  int        CalcNewClustCenters();       // Step 3 of K-means algorithm
  double     EucNorm(int, int);           // Calc Euclidean norm vector
  int        FindClosestCluster(int);     //ret indx of clust closest to pattern
                                          //whose index is arg
public:
  system();
  int LoadPatterns(char *fname);          // Get pattern data to be clustered
  void InitClusters();                    // Step 1 of K-means algorithm
  void RunKMeans();                       // Overall control K-means process
  void ShowClusters();                    // Show results on screen
  void SaveClusters(char *fname);         // Save results to file
};

int System::LoadPatterns(char *fname){
  FILE *InFilePtr;
  int   i,j;
  double x;
if((InFilePtr = fopen(fname, "r")) == NULL)
   return FAILURE;
fscanf(InFilePtr, "%d", &NumPatterns);           // Read # of patterns
fscanf(InFilePtr, "%d", &SizeVector);            // Read dimension of vector
fscanf(InFilePtr, "%d", &NumClusters);           // Read # of clusters for K-Means
for (i=0; i<NumPatterns; i++) {                  // For each vector
  for (j=0; j<SizeVector; j++) {                 // create a pattern
    fscanf(InFilePtr,"%lg",&x);                  // consisting of all elements
    Pattern[i][j]=x;
   printf("Pattern[%d][%d]=%f\n",i,j,Pattern[i][j]);
    } /* endfor */
  } /* endfor */
printf("\n");
return SUCCESS;
}
```

```
//***************************************************************************
// InitClusters                                                             *
//  Arbitrarily assign a vector to each of the K clusters                   *
//  We choose the first K vectors to do this                                *
//***************************************************************************
void System::InitClusters(){
int i,j;
printf("Initial cluster centers:\n");
for (i=0; i<NumClusters; i++) {
  Cluster[i].Member[0]=i;
  for (j=0; j<SizeVector; j++) {
    Cluster[i].Center[j]=Pattern[i][j];
    printf("Cluster[%d].Center[%d]=%f\n",i,j,Cluster[i].Center[j]);
     } /* endfor */
  } /* endfor */
printf("\n");
}

void System::RunKMeans(){
 int converged;
converged=FALSE;
while (converged==FALSE) {
  DistributeSamples();
  converged=CalcNewClustCenters();
// converged=TRUE;
   } /* endwhile */
}

double System::EucNorm(int p, int c){          // Calc Euclidean norm of vector difference
double dist;                                   // between pattern vector, p, and cluster
int i;                                         // center, c.
dist=0;
for (i=0; i<SizeVector ;i++){
  dist += (Cluster[c].Center[i]-Pattern[p][i])*(Cluster[c].Center[i]-Pattern[p][i]);
  } /* endfor */
//dist = sqrt(dist);
return dist;
}

int System::FindClosestCluster(int pat){
  int i, ClustID;
  double MinDist, d;
MinDist =9.9e+99;
ClustID=-1;
for (i=0; i<NumClusters; i++) {
  d=EucNorm(pat,i);
// printf("Distance from pattern %d to cluster %d = %f \n",pat,i,d);
   if (d<MinDist) {
     MinDist=d;
     ClustID=i;
     } /* endif */
   } /* endfor */
if (ClustID<0) {
  printf("Aaargh");
```

```
  exit(0);
  } /* endif */
return ClustID;
}

void System::DistributeSamples(){
int i,pat,Clustid,MemberIndex;
//Clear membership list for all current clusters
for (i=0; i<NumClusters;i++){
  Cluster[i].NumMembers=0;
  }
for (pat=0; pat<NumPatterns; pat++) {
  //Find cluster center to which the pattern is closest
  Clustid= FindClosestCluster(pat);
  printf("patern %d assigned to cluster %d\n",pat,Clustid);
  //post this pattern to the cluster
  MemberIndex=Cluster[Clustid].NumMembers;
  Cluster[Clustid].Member[MemberIndex]=pat;
  Cluster[Clustid].NumMembers++;
  } /* endfor */
}

int System::CalcNewClustCenters(){
  int ConvFlag,VectID,i,j,k;
  double tmp[MAXVECTDIM];
ConvFlag=TRUE;
for (i=0; i<NumClusters; i++) {                //for each cluster
  for (j=0; j<SizeVector; j++) {               // clear workspace
    tmp[j]=0.0;
    } /* endfor */
  for (j=0; j<Cluster[i].NumMembers; j++) {    //traverse member vectors
    VectID=Cluster[i].Member[j];
    for (k=0; k<SizeVector; k++) {             //traverse elements of vector
      printf("Cluster[%d]  Pattern[%d][%d]=%f,
Member_ID=%d\n",i,VectID,k,Pattern[VectID][k],VectID);
//      x=Pattern[VectID][k];
//      tmp[k] = tmp[k]+x;                     // add (member) pattern elmnt into temp
      tmp[k] += Pattern[VectID][k];            // add (member) pattern elmnt into temp
      } /* endfor */
    } /* endfor */
  for (k=0; k<SizeVector; k++) {               //traverse elements of vector
    tmp[k]=tmp[k]/Cluster[i].NumMembers;
    if (tmp[k] != Cluster[i].Center[k])
      ConvFlag=FALSE;
    Cluster[i].Center[k]=tmp[k];
    } /* endfor */
  } /* endfor */
return ConvFlag;
}

void System::ShowClusters(){
  int cl;
for (cl=0; cl<NumClusters; cl++) {
  printf("\nCLUSTER %d ==>[%f,%f]\n", cl,Cluster[cl].Center[0],Cluster[cl].Center[1]);
  } /* endfor */
}

void System::SaveClusters(char *fname){
}
```

```
main(int argc, char *argv[]) {
  System kmeans;
if (argc<3) {
  printf("USAGE: KMEANS PATTERN_FILE(input) CLUSTER_FILE(output)\n");
  exit(0);
  }
if (kmeans.LoadPatterns(argv[1])==FAILURE ){
  printf("UNABLE TO READ PATTERN_FILE:%s\n",argv[1]);
  exit(0);
  }
kmeans.InitClusters();
kmeans.RunKMeans();
kmeans.ShowClusters();
}
```

8.2.1 K-Means Example

We will now illustrate the application of the K-means algorithm. In this example K-means will be utilized to cluster the points illustrated in Figure 8.2.

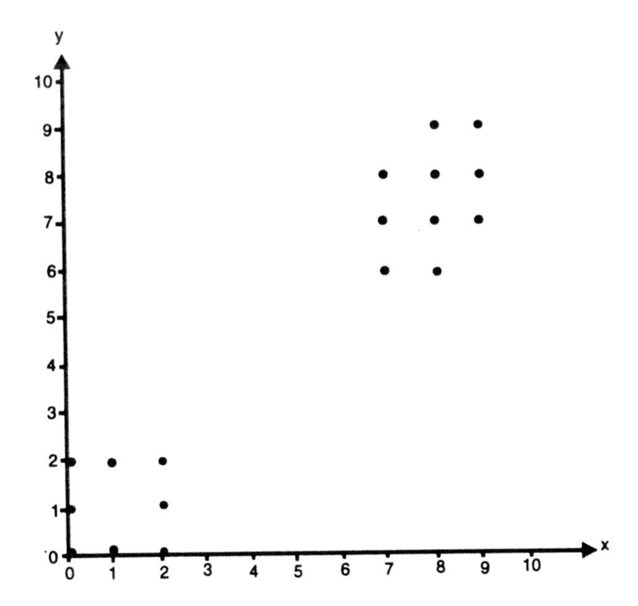

FIGURE 8.2 Example data to be clustered using the K-means algorithm.

The two-dimensional pattern vectors corresponding to the points in Figure 8.2 are as follows:

Pattern[0]=(0.000,0.000)
Pattern[1]=(1.000,0.000)
Pattern[2]=(0.000,1.000)
Pattern[3]=(2.000,1.000)
Pattern[4]=(1.000,2.000)
Pattern[5]=(2.000,2.000)
Pattern[6]=(2.000,0.000)
Pattern[7]=(0.000,2.000)
Pattern[8]=(7.000,6.000)
Pattern[9]=(7.000,7.000)
Pattern[10]=(7.000,8.000)
Pattern[11]=(8.000,6.000)
Pattern[12]=(8.000,7.000)
Pattern[13]=(8.000,8.000)
Pattern[14]=(8.000,9.000)
Pattern[15]=(9.000,7.000)
Pattern[16]=(9.000,8.000)
Pattern[17]=(9.000,9.000)

The initial cluster centers are taken from the first two pattern vectors. Thus initially the cluster centers are

ClusterCenter[0]=(0.000000,0.000000)
ClusterCenter[1]=(1.000000,0.000000)

We are now ready to apply the algorithm. The calculations for pass 1 begin with the association of each of the input vectors with the nearest cluster center as follows:

PASS=1

The distance from pattern 0 to cluster 0 is calculated as:
d=sqrt(.0000+ .0000) =0.000000

The distance from pattern 0 to cluster 1 is calculated as:
d=sqrt(1.0000+ .0000)=1.000000

} pattern 0 assigned to cluster 0

The distance from pattern 1 to cluster 0 is calculated as:
d=sqrt(1.0000+ .0000)=1.000000

The distance from pattern 1 to cluster 1 is calculated as:
d=sqrt(.0000+ .0000)=0.000000

} pattern 1 assigned to cluster 1

The distance from pattern 2 to cluster 0 is calculated as:
d=sqrt(.0000+ 1.0000)=1.000000

The distance from pattern 2 to cluster 1 is calculated as:
d=sqrt(1.0000+ 1.0000)=1.414214

pattern 2 assigned to cluster 0

The distance from pattern 3 to cluster 0 is calculated as:
d=sqrt(4.0000+ 1.0000)=2.236068

The distance from pattern 3 to cluster 1 is calculated as:
d=sqrt(1.0000+ 1.0000)=1.414214

pattern 3 assigned to cluster 1

The distance from pattern 4 to cluster 0 is calculated as:
d=sqrt(1.0000+ 4.0000)=2.236068

The distance from pattern 4 to cluster 1 is calculated as:
d=sqrt(.0000+ 4.0000)=2.000000

pattern 4 assigned to cluster 1

The distance from pattern 5 to cluster 0 is calculated as:
d=sqrt(4.0000+ 4.0000)=2.828427

The distance from pattern 5 to cluster 1 is calculated as:
d=sqrt(1.0000+ 4.0000)=2.236068

pattern 5 assigned to cluster 1

The distance from pattern 6 to cluster 0 is calculated as:
d=sqrt(4.0000+ .0000)=2.000000

The distance from pattern 6 to cluster 1 is calculated as:
d=sqrt(1.0000+ .0000)=1.000000

pattern 6 assigned to cluster 1

The distance from pattern 7 to cluster 0 is calculated as:
d=sqrt(.0000+ 4.0000)=2.000000

The distance from pattern 7 to cluster 1 is calculated as:
d=sqrt(1.0000+ 4.0000)=2.236068

pattern 7 assigned to cluster 0

The distance from pattern 8 to cluster 0 is calculated as:
d=sqrt(49.0000+ 36.0000)=9.219544

The distance from pattern 8 to cluster 1 is calculated as:
d=sqrt(36.0000+ 36.0000)=8.485281

pattern 8 assigned to cluster 1

The distance from pattern 9 to cluster 0 is calculated as:
d=sqrt(49.0000+ 49.0000)=9.899495

The distance from pattern 9 to cluster 1 is calculated as:
d=sqrt(36.0000+ 49.0000)=9.219544

pattern 9 assigned to cluster 1

The distance from pattern 10 to cluster 0 is calculated as:
d=sqrt(49.0000+ 64.0000)=10.630146

The distance from pattern 10 to cluster 1 is calculated as:
d=sqrt(36.0000+ 64.0000)=10.000000

pattern 10 assigned to cluster 1

The distance from pattern 11 to cluster 0 is calculated as:
d=sqrt(64.0000+ 36.0000)=10.000000

The distance from pattern 11 to cluster 1 is calculated as:
d=sqrt(49.0000+ 36.0000)=9.219544

pattern 11 assigned to cluster 1

The distance from pattern 12 to cluster 0 is calculated as:
d=sqrt(64.0000+ 49.0000)=10.630146

The distance from pattern 12 to cluster 1 is calculated as:
d=sqrt(49.0000+ 49.0000)=9.899495

pattern 12 assigned to cluster 1

The distance from pattern 13 to cluster 0 is calculated as:
d=sqrt(64.0000+ 64.0000)=11.313708

The distance from pattern 13 to cluster 1 is calculated as:
d=sqrt(49.0000+ 64.0000)=10.630146

pattern 13 assigned to cluster 1

The distance from pattern 14 to cluster 0 is calculated as:
d=sqrt(64.0000+ 81.0000)=12.041595

The distance from pattern 14 to cluster 1 is calculated as:
d=sqrt(49.0000+ 81.0000)=11.401754

pattern 14 assigned to cluster 1

The distance from pattern 15 to cluster 0 is calculated as:
d=sqrt(81.0000+ 49.0000)=11.401754

The distance from pattern 15 to cluster 1 is calculated as:
d=sqrt(64.0000+ 49.0000)=10.630146

pattern 15 assigned to cluster 1

The distance from pattern 16 to cluster 0 is calculated as:
d=sqrt(81.0000+ 64.0000)=12.041595

The distance from pattern 16 to cluster 1 is calculated as:
d=sqrt(64.0000+ 64.0000)=11.313708

pattern 16 assigned to cluster 1

The distance from pattern 17 to cluster 0 is calculated as:
d=sqrt(81.0000+ 81.0000)=12.727922

The distance from pattern 17 to cluster 1 is calculated as:
d=sqrt(64.0000+ 81.0000)=12.041595

pattern 17 assigned to cluster 1

All pattern vectors have now been associated with a cluster center. The next task is to recalculate the cluster centers consistant with the new membership list. The new cluster centers are now calculated as follows:

Cluster Center0 =(1/3)(0.000+0.000+0.000),
(1/3)(0.000+1.000+2.000))
Cluster Center1 = (1/15)(1.000+2.000+1.000+2.000+2.000+7.000+7.000+7.000+
8.000+8.000+8.000+8.000+9.000+9.000+9.000),
(1/15)(0.000+1.000+2.000+2.000+0.000+6.000+7.000+8.000+
6.000+7.000+8.000+9.000+7.000+8.000+9.000))

So, the new cluster centers become:

ClusterCenter[0]=(0.000000,1.000000)

ClusterCenter[1]=(5.866667,5.333333)

Pass 1 is now complete. Because the cluster centers have changed, we must proceed to pass 2. The calculations for pass 2 are as follows:

PASS=2

Distance calculations	Assignment
The distance from pattern 5 to cluster 0 is calculated as: d=sqrt(4.0000+ 1.0000)=2.236068 The distance from pattern 5 to cluster 1 is calculated as: d=sqrt(14.9511+ 11.1111)=5.10511	pattern 5 assigned to cluster 0
The distance from pattern 6 to cluster 0 is calculated as: d=sqrt(4.0000+ 1.0000)=2.236068 The distance from pattern 6 to cluster 1 is calculated as: d=sqrt(14.9511+ 28.4444)=6.587530	pattern 6 assigned to cluster 0
The distance from pattern 7 to cluster 0 is calculated as: d=sqrt(.0000+ 1.0000)=1.000000 The distance from pattern 7 to cluster 1 is calculated as: d=sqrt(34.4178+ 11.1111)=6.747510	pattern 7 assigned to cluster 0
The distance from pattern 8 to cluster 0 is calculated as: d=sqrt(49.0000+ 25.0000)=8.602325 The distance from pattern 8 to cluster 1 is calculated as: d=sqrt(1.2844+ .4444)=1.314872	pattern 8 assigned to cluster 1
The distance from pattern 9 to cluster 0 is calculated as: d=sqrt(49.0000+ 36.0000)=9.219544 The distance from pattern 9 to cluster 1 is calculated as: d=sqrt(1.2844+ 2.7778)=2.015496	pattern 9 assigned to cluster 1
The distance from pattern 10 to cluster 0 is calculated as: d=sqrt(49.0000+ 49.0000)=9.899495 The distance from pattern 10 to cluster 1 is calculated as: d=sqrt(1.2844+ 7.1111)=2.8975	pattern 10 assigned to cluster 1
The distance from pattern 11 to cluster 0 is calculated as: d=sqrt(64.0000+ 25.0000)=9.433981 The distance from pattern 11 to cluster 1 is calculated as: d=sqrt(4.5511+ .4444)=2.235074	pattern 11 assigned to cluster 1
The distance from pattern 12 to cluster 0 is calculated as: d=sqrt(64.0000+ 36.0000)=10.000000 The distance from pattern 12 to cluster 1 is calculated as: d=sqrt(4.5511+ 2.7778)=2.707192	pattern 12 assigned to cluster 1
The distance from pattern 13 to cluster 0 is calculated as: d=sqrt(64.0000+ 49.0000)=10.630146 The distance from pattern 13 to cluster 1 is calculated as: d=sqrt(4.5511+ 7.1111)=3.415000	pattern 13 assigned to cluster 1

The distance from pattern 14 to cluster 0 is calculated as:
d=sqrt(64.0000+ 64.0000)=11.313708

The distance from pattern 14 to cluster 1 is calculated as:
d=sqrt(4.5511+ 13.4444)=4.242117

} pattern 14 assigned to cluster 1

The distance from pattern 15 to cluster 0 is calculated as:
d=sqrt(81.0000+ 36.0000)=10.816654

The distance from pattern 15 to cluster 1 is calculated as:
d=sqrt(9.8178+ 2.7778)=3.549022

} pattern 15 assigned to cluster 1

The distance from pattern 16 to cluster 0 is calculated as:
d=sqrt(81.0000+ 49.0000)=11.401754

The distance from pattern 16 to cluster 1 is calculated as:
d=sqrt(9.8178+ 7.1111)=4.114473

} pattern 16 assigned to cluster 1

The distance from pattern 17 to cluster 0 is calculated as:
d=sqrt(81.0000+ 64.0000)=12.041595

The distance from pattern 17 to cluster 1 is calculated as:
d=sqrt(9.8178+ 13.4444)=4.823093

} pattern 17 assigned to cluster 1

The new cluster centers are now calculated as:

Cluster Center0 = (1/8)(.000+ 1.000+ .000+ 2.000+ 1.000+ 2.000+ 2.000+ .000),
(1/8)(.000+ .000+ 1.000+ 1.000+ 2.000+ 2.000+ .000+ 2.000))
Cluster Center1 = (1/10)(7.000+ 7.000+ 7.000+ 8.000+ 8.000+ 8.000+ 8.000+
9.000+ 9.000+ 9.000),
(1/10)(6.000+ 7.000+ 8.000+ 6.000+ 7.000+ 8.000+ 9.000+
7.000+ 8.000+ 9.000))

Thus the new cluster centers become:

ClusterCenter[0]=(1.000000,1.000000)
ClusterCenter[1]=(8.000000,7.500000)

Pass 2 is now complete. The cluster centers have changed so that we must proceed to pass 3.

PASS=3

The distance from pattern 0 to cluster 0 is calculated as:
d=sqrt(1.0000+ 1.0000)=1.414214

The distance from pattern 0 to cluster 1 is calculated as:
d=sqrt(64.0000+ 56.2500)=10.965856

} pattern 0 assigned to cluster 0

The distance from pattern 1 to cluster 0 is calculated as:
d=sqrt(.0000+ 1.0000)=1.000000

The distance from pattern 1 to cluster 1 is calculated as:
d=sqrt(49.0000+ 56.2500)=10.259142

} pattern 1 assigned to cluster 0

The distance from pattern 2 to cluster 0 is calculated as:
d=sqrt(1.0000+ .0000)=1.000000

The distance from pattern 2 to cluster 1 is calculated as:
d=sqrt(64.0000+ 42.2500)=10.307764

} pattern 2 assigned to cluster 0

The distance from pattern 3 to cluster 0 is calculated as:
d=sqrt(1.0000+ .0000)=1.000000

The distance from pattern 3 to cluster 1 is calculated as:
d=sqrt(36.0000+ 42.2500)=8.845903

} pattern 3 assigned to cluster 0

The distance from pattern 4 to cluster 0 is calculated as:
d=sqrt(.0000+ 1.0000)=1.000000

The distance from pattern 4 to cluster 1 is calculated as:
d=sqrt(49.0000+ 30.2500)8.902247

} pattern 4 assigned to cluster 0

The distance from pattern 5 to cluster 0 is calculated as:
d=sqrt(1.0000+ 1.0000)=1.414214

The distance from pattern 5 to cluster 1 is calculated as:
d=sqrt(36.0000+ 30.2500)=8.139410

} pattern 5 assigned to cluster 0

The distance from pattern 6 to cluster 0 is calculated as:
d=sqrt(1.0000+ 1.0000)=1.414214

The distance from pattern 6 to cluster 1 is calculated as:
d=sqrt(36.0000+ 56.2500)=9.604686

} pattern 6 assigned to cluster 0

The distance from pattern 7 to cluster 0 is calculated as:
d=sqrt(1.0000+ 1.0000)=1.414214

The distance from pattern 7 to cluster 1 is calculated as:
d=sqrt(64.0000+ 30.2500)=9.708244

} pattern 7 assigned to cluster 0

The distance from pattern 8 to cluster 0 is calculated as:
d=sqrt(36.0000+ 25.0000)=7.810250

The distance from pattern 8 to cluster 1 is calculated as:
d=sqrt(1.0000+ 2.2500)=1.802776

} pattern 8 assigned to cluster 1

The distance from pattern 9 to cluster 0 is calculated as:
d=sqrt(36.0000+ 36.0000)=8.485281

The distance from pattern 9 to cluster 1 is calculated as:
d=sqrt(1.0000+ .2500)=1.118034

} pattern 9 assigned to cluster 1

The distance from pattern 10 to cluster 0 is calculated as:
d=sqrt(36.0000+ 49.0000)=9.219544

The distance from pattern 10 to cluster 1 is calculated as:
d=sqrt(1.0000+ .2500)=1.118034

} pattern 10 assigned to cluster 1

The distance from pattern 11 to cluster 0 is calculated as:
d=sqrt(49.0000+ 25.0000)=8.602325

The distance from pattern 11 to cluster 1 is calculated as:
d=sqrt(.0000+ 2.2500)=1.500000

} pattern 11 assigned to cluster 1

The distance from pattern 12 to cluster 0 is calculated as:
d=sqrt(49.0000+ 36.0000)=9.219544

The distance from pattern 12 to cluster 1 is calculated as:
d=sqrt(.0000+ .2500)=0.500000

} pattern 12 assigned to cluster 1

The distance from pattern 13 to cluster 0 is calculated as:
d=sqrt(49.0000+ 49.0000)=9.899495

The distance from pattern 13 to cluster 1 is calculated as:
d=sqrt(.0000+ .2500)=0.500000

} pattern 13 assigned to cluster 1

The distance from pattern 14 to cluster 0 is calculated as:
d=sqrt(49.0000+ 64.0000)=10.630146

The distance from pattern 14 to cluster 1 is calculated as:
d=sqrt(.0000+ 2.2500)=1.500000

} pattern 14 assigned to cluster 1

The distance from pattern 15 to cluster 0 is calculated as:
d=sqrt(64.0000+ 36.0000)=10.000000

The distance from pattern 15 to cluster 1 is calculated as:
d=sqrt(1.0000+ .2500)=1.118034

} pattern 15 assigned to cluster 1

The distance from pattern 16 to cluster 0 is calculated as:
d=sqrt(64.0000+ 49.0000)=10.630146

The distance from pattern 16 to cluster 1 is calculated as:
d=sqrt(1.0000+ .2500)=1.118034

} pattern 16 assigned to cluster 1

The distance from pattern 17 to cluster 0 is calculated as:
d=sqrt(64.0000+ 64.0000)=11.313708

The distance from pattern 17 to cluster 1 is calculated as:
d=sqrt(1.0000+ 2.2500)=1.802776

} pattern 17 assigned to cluster 1

The new cluster centers are now calculated as:

Cluster Center0 = (1/8)(0.000+1.000+0.000+2.000+1.000+2.000+2.000+0.000),
(1/8)(0.000+0.000+1.000+1.000+2.000+2.000+0.000+2.000))
Cluster Center1 = (1/10)(7.000+7.000+7.000+8.000+8.000+8.000+8.000+9.000+
9.000+9.000),
(1/10)(6.000+7.000+8.000+6.000+7.000+8.000+9.000+7.000+
8.000+9.000))

Thus, the new cluster centers become:

ClusterCenter[0]=(1.000000,1.000000)
ClusterCenter[1]=(8.000000,7.500000)

Since the cluster centers did not change, the algorithm has converged. Thus the final cluster centers are as follows:

CLUSTER 0 = (1.000000,1.000000)
CLUSTER 1 = (8.000000,7.500000)

Figure 8.3 illustrates the cluster membership results along with the corresponding cluster centers detected in this example. Clearly the clusters and cluster centers are intuitively reasonable given the data presented.

8.3 AN INTRODUCTION TO THE KOHONEN MODEL

In section 8.2 we studied the clustering (i.e., self-organizing capabilities inherent in the K-means algorithm. Now we begin our investigation of self-organization in neural networks. We will begin timidly by illustrating the ability of the Kohonen model architecture to identify cluster centers just as the K-means algorithm did. This simplified introduction omits significant characteristics of the Kohonen model (in particular, localized lateral feedback) which contribute to the power of the SOFM. The choice of winning neuron in this simplified presentation is accomplished by taking a MAXNET approach. That is, the neuron with the biggest activation, net_j, becomes the winner.

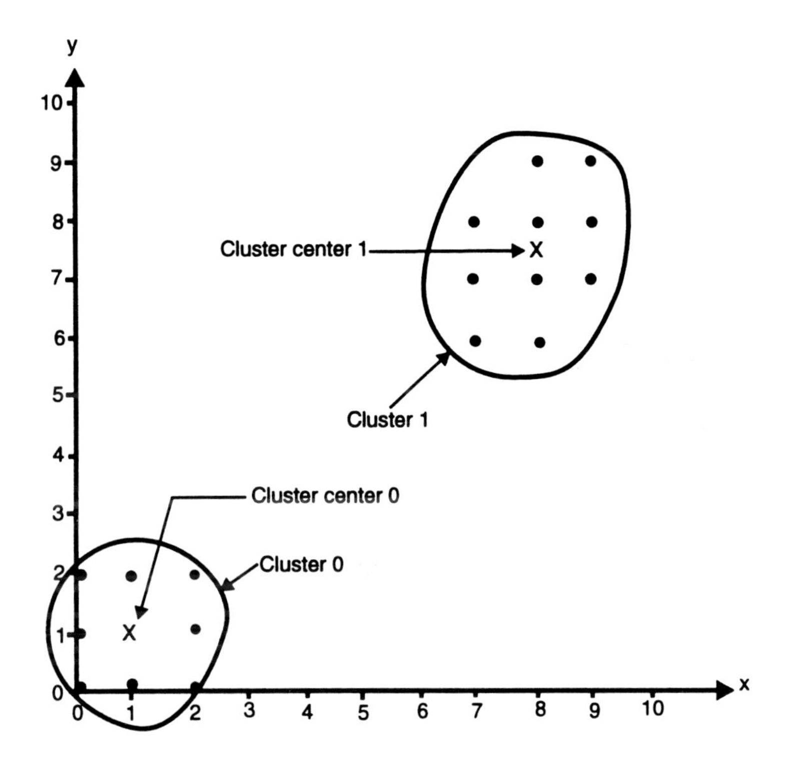

FIGURE 8.3 Clusters and cluster centers resulting from the application of the K-means algorithm.

Once we have observed clustering in context of this simplified model we shall proceed to discuss lateral feedback and ultimately the full SOFM.

The Kohonen net architecture consists of two layers, an input layer and a Kohonen layer. These two layers are fully connected. Each input layer neuron has a feed-forward connection to each output layer neuron. Figure 8.4 illustrates the Kohonen network architecture for the one-dimensional case (Figure 8.4[a]) and the two-dimensional case (Figure 8.4[b]). Higher dimensional cases are possible, albeit rare, and very difficult to draw. Let us assume that the inputs are normalized (i.e., $\|\mathbf{X}\| = 1$). Inputs to the Kohonen layer (i.e., the output layer) can be calculated conventionally using equation 7.

$$I_j = \sum_{i=1}^{n} \left(\mathbf{w}_{ij} \mathbf{x}_i \right) \tag{7}$$

Applying a winner-take-all paradigm, the winning output layer neuron will simply be the neuron with the biggest I_j. The output of the wining neuron will be a + 1. All other neurons in the kohonen layer will output nothing. In effect equation 7 is

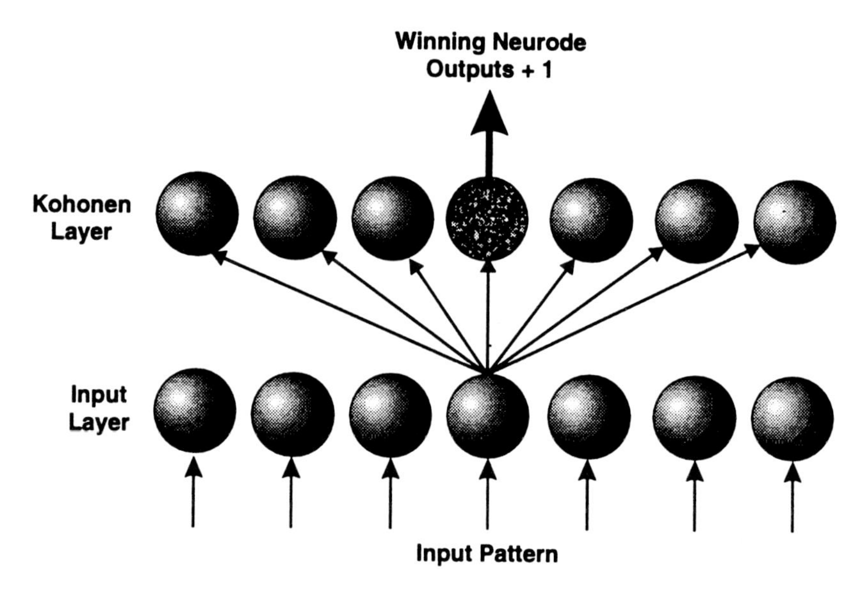

FIGURE 8.4(a) Architecture of the Kohonen network, one-dimensional case.

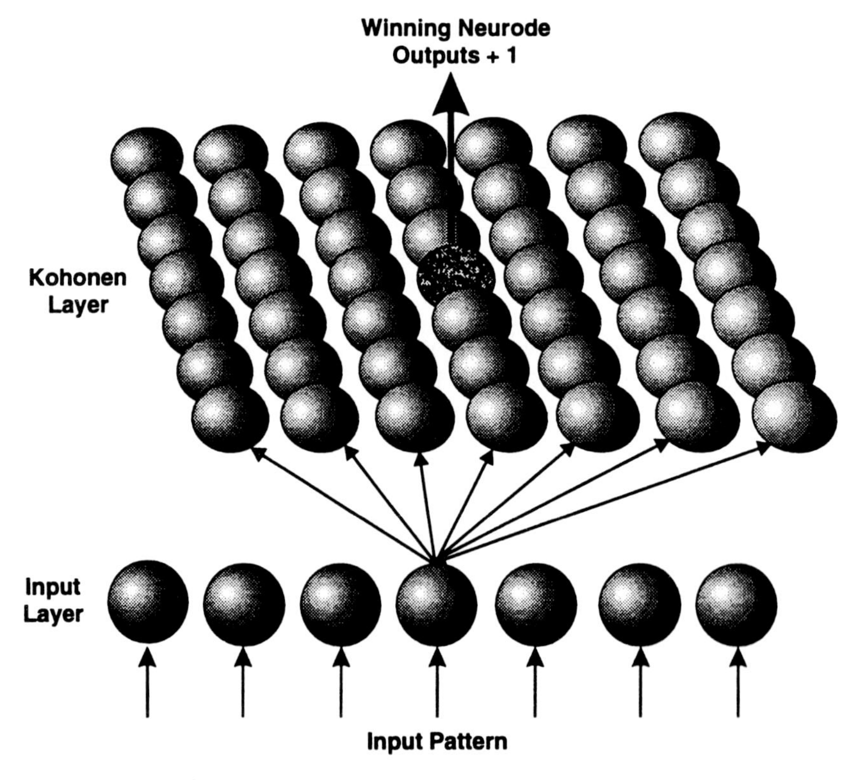

FIGURE 8.4(b) Architecture of the Kohonen network, two-dimensional case.

the dot product between a neuron weight vector and the input vector. Thus this method chooses a winning neuron such that the angle between the winning neuron weight vector and the input vector will be smaller than the corresponding dot product for all other neurons (see Figure 8.5).

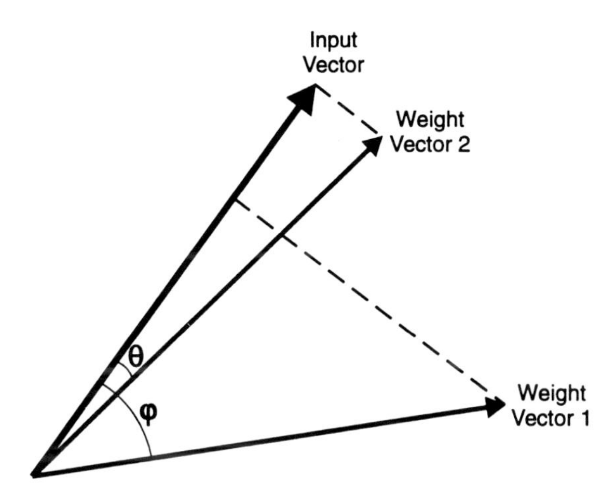

FIGURE 8.5 Dot products between pattern vector and weight vector. Notice that the smallest angle occurs between the pattern vector and the most similar weight vector.

An alternative method of choosing the winning neuron simply selects the neuron whose weight vector has a minimum of the Euclidean norm distance from the input vector (i.e., $d_j = \|\mathbf{w}_j - \mathbf{x}\|$). For vectors of unit length this method is equivalent to the method just described in that the same neuron will be chosen as the winner. The use of Euclidean distance to select a winner, however, may be advantageous in that it does not require weights or input vectors to be normalized.

The Kohonen nets train by competitive (unsupervised) learning. Neurons within the Kohonen layer compete when an input vector is presented to the network. A winner is chosen by one of the above methods. (Later we will see that the competition can be beneficially mediated by interneuron lateral feedback connections.) The winning neuron is trained according to the following equation:

$$\mathbf{W}_{ij}^{\text{new}} = \mathbf{W}_{ij}^{\text{old}} + \eta\left(\mathbf{x}_i - \mathbf{W}_{ij}^{\text{old}}\right) \tag{8}$$

where η is the learning parameter or gain and typical initial values are less than 0.25. The learning described by equation 8 has been referred to as Kohonen's learning.

Let us choose to normalize the inputs to network (each vector component is divided by the vector length). In this case the weight vectors are initialized randomly but they are also normalized. Training of the winning neuron can then be viewed as illustrated in Figure 8.6. As can be seen from the figure, the result of training is to nudge the weight vector closer to the input vector. Notice that the weight which results from the training is not necessarily of unit length. Renormalization of the weight vector is considered to be optional in that the effect can be expected to cancel over the ensemble of input vectors.

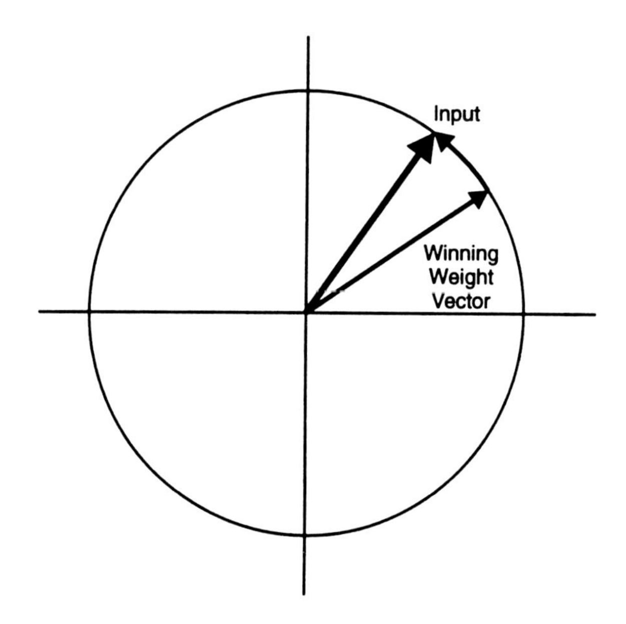

FIGURE 8.6 Training of the winning neuron moves its weight vector toward the pattern vector.

Listing 8.2

```
******************************
*KOHONEN NET                 *
******************************
#include <stdlib.h>
#include <stdio.h>
#include <math.h>

#define  MAXPATS      100
#define  MAXNEURONSIN 10
#define  MAXNEURONS   15

#define  MAXEPOCHS   1000
#define  ETAMIN      .001

unsigned int Random(int N) {
 unsigned int j;
j= (N*rand())/RAND_MAX;
if (j>=N) j=N;
return j;
}
class PATTERN {
 friend class KNET;
private:
 double      P[MAXPATS][MAXNEURONSIN];
```

```
 int        NumPatterns;
 int        Shuffle[MAXPATS];
 int        SizeVector;
public:
 PATTERN();
 int GetPatterns(char *);           //load pattern from file
 int GetRandPats(int,int);          //random patterns arg1=# of patterns, arg2=dimension
 double Query(int,int);             //returns P[arg1][arg2]
 double QueryR(int,int);            //returns P[Shuffle[arg1]][arg2]
 void ReShuffle(int N);
};

PATTERN::PATTERN(){
int i;
for (i=0; i<MAXPATS; i++)
  Shuffle[i]=i;
}

int PATTERN::GetPatterns(char *fname) {
 FILE *fp;
 int i,j;
 double x;
fp=fopen(fname,"r");
if (fp==NULL) return 0;                     // Test for failure.
fscanf(fp,"%d",&NumPatterns);
fscanf(fp,"%d",&SizeVector);
for (i=0; i<NumPatterns; i++) {             // For each vector
  for (j=0; j<SizeVector; j++) {            // create a pattern
    fscanf(fp,"%lg",&x);                    // consisting of all elements
    P[i][j]=x;
    } /* endfor */
  } /* endfor */
fclose(fp);
return 1;
}

int PATTERN::GetRandPats(int n1,int n2) {
  int i,j;
  double x;
NumPatterns=n1;
SizeVector=n2;
for (i=0; i<NumPatterns; i++) {             // For each vector
  for (j=0; j<SizeVector; j++) {            // create a pattern
    x=(double)rand()/RAND_MAX;              // consisting of random elements
     P[i][j]=x;                             // between 0 and 1
     } /* endfor */
  } /* endfor */
return 1;
}
void PATTERN::ReShuffle(int N) {
int i,a1,a2,tmp;
for (i=0; i<N ;i++) {
  a1=Random(NumPatterns);
  a2=Random(NumPatterns);
  tmp=Shuffle[a1];
```

```
  Shuffle[a1]=Shuffle[a2];
  Shuffle[a2]=tmp;
  }
}

double PATTERN::Query(int pat,int j) {
return P[pat][j];
}

double PATTERN::QueryR(int pat,int j) {
return P[Shuffle[pat]][j];
}

class KNET {
private:
 double W[MAXNEURONSIN][MAXNEURONS];  // The weight matrix
 double Yout[MAXNEURONS];             // The output layer neurons
 double Yin[MAXNEURONSIN];            //The input layer neurons
 int    YinSize;                      //input layer dimensions
 int    YoutSize;                     //outlayer dimensions

 int    epoch;
 double eta;                          //The learning rate
 double delta_eta;                    //Amount to change l.r. each epoch
 int    StochFlg;                     //Present vectors in rand order if 1
 PATTERN *Pattern;

 int    LoadInLayer(int);             //pattern->input layer
 double EucNorm(int);                 //Calc Euclidean distance
 int    FindWinner();                 //get coords of winning neuron
 void   Train(int);
 void   AdaptParms();
public:
 KNET();
 void SetPattern(PATTERN *);
 void SetParms(int, double);
 void PrintWeights();
 void PrintWinner();
 void RunTrn();
 void Run();
};

KNET::KNET(){
StochFlg=0;
}

void KNET::SetPattern(PATTERN *p) {
  Pattern=p;
  YinSize=p->SizeVector;
}
void KNET::SetParms(int X, double LR){
 int i,k;
YoutSize=X;
eta=LR;
```

```
delta_eta=0.005;
for (i=0; i<YoutSize; i++) {
  for (k=0; k<YinSize; k++) {
    W[k][i]= (double)rand()/(10.0 * (double)RAND_MAX);
     } /* endfor */
  } /* endfor */
}

int KNET::LoadInLayer(int P){
  int i;
for (i=0; i<YinSize; i++){
  if (StochFlg){
    Yin[i]=Pattern->QueryR(P,i);
     }
    else {
    Yin[i]=Pattern->Query(P,i);
     }
  }
return 1;
}


void KNET::AdaptParms(){
eta=eta-delta_eta;
if (eta<ETAMIN)
  eta=ETAMIN;
printf(" New eta=%f\n",eta);
}

void KNET::PrintWeights() {
  int i,k;
for (i=0; i<YoutSize; i++) {
    for (k=0; k<YinSize; k++) {
       printf("W[%d][%d]=%f ",k,i,W[k][i]);
       } /* endfor */
     printf("\n");
  } /* endfor */
}

void KNET::RunTrn(){
int i,np;
int Winner;
epoch=0;
np=Pattern->NumPatterns;
while (epoch<=MAXEPOCHS){
  for (i=0; i<np; i++){
   LoadInLayer(i);
   Winner=FindWinner();
   Train(Winner);
   }
  if(5*(epoch/5)==epoch) {
   printf("Epoch=%d\n",epoch);
   PrintWeights();
   }
  epoch++;
```

```
  if (StochFlg)
   Pattern->ReShuffle(np);
  AdaptParms();
   }
}

void KNET::Train(int Winner){
 int k;
for (k=0; k<YinSize; k++){
  W[k][Winner]=W[k][Winner]+eta*(Yin[k]-W[k][Winner]);
  } /*endfor*/
}

int KNET::FindWinner(){
 int i;
 double d,best;
 int Winner;
best=1.0e99;
Winner=-1;
for (i=0; i<YoutSize; i++){
 d=EucNorm(i);
 if (d<best) {
    best=d;
    Winner=i;
    } // endif
 } // endfor
return Winner;
}

double KNET::EucNorm(int x){                     // Calc Euclidean norm of vector dif
int i;
double dist;
dist=0;
for (i=0; i< YinSize;i++){
  dist += (W[i][x]-Yin[i]) * (W[i][x]-Yin[i]);
  } /* endfor */
dist=sqrt(dist);
return dist;
}

//==================================================================
// GLOBAL OBJECTS
//==================================================================

PATTERN InPat;
KNET    net;

//==================================================================
// Main()
//==================================================================

main(int argc, char *argv[]) {
//srand(17);
if (argc>1) {
  InPat.GetPatterns(argv[1]);                  //Establish pattern
  net.SetPattern(&InPat);                      //Inform the feature map about the pattern
```

```
    net.SetParms(3, 0.500);                    //Init fm parms
    net.RunTrn();                              //Run the FM w/ training enabled
    }
  else {
    printf("USAGE: KNET PATTERN_FILE");
    }
}
```

8.3.1 KOHONEN EXAMPLE

The following example illustrates the operation of the basic Kohonen net as described above. Figure 8.7 illustrates the pattern which will be applied to the network. We have chosen a one-dimensional network with three output layer neurons. After the first full epoch with $\eta = 0.5$, the network has already achieved a fairly good approximation of the cluster centers. The state of the network after the first epoch is as follows:

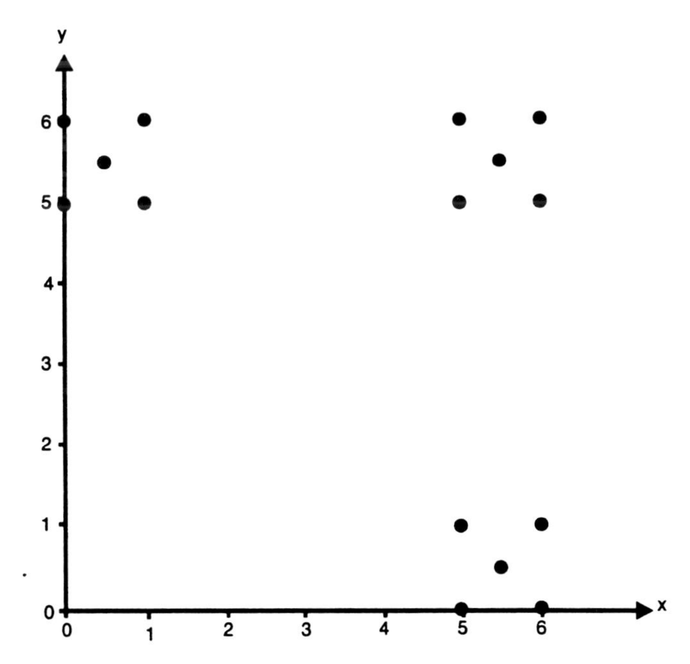

FIGURE 8.7 Example two-dimensional input pattern to be clustered by the Kohonen network.

W[0][0]=5.470356 W[1][0]=0.563049

W[0][1]=0.563464 W[1][1]=5.470420

W[0][2]=5.409211 W[1][2]=5.281787

After 250 epochs η has been reduced to 0.001. The state of the network is as follows:

W[0][0]=5.507603 W[1][0]=0.503805
W[0][1]=0.503805 W[1][1]=5.507603
W[0][2]=5.503805 W[1][2]=5.499810

After 500 epochs η remains at its minimum value of 0.001. The network weights (representing the cluster centers discovered by the network) are as follows:

W[0][0]=5.502463 W[1][0]=0.501232
W[0][1]=0.501232 W[1][1]=5.502463
W[0][2]=5.501232 W[1][2]=5.499945

The network can be seen to have accomplished the task of properly clustering the input patterns as shown below in Figure 8.8.

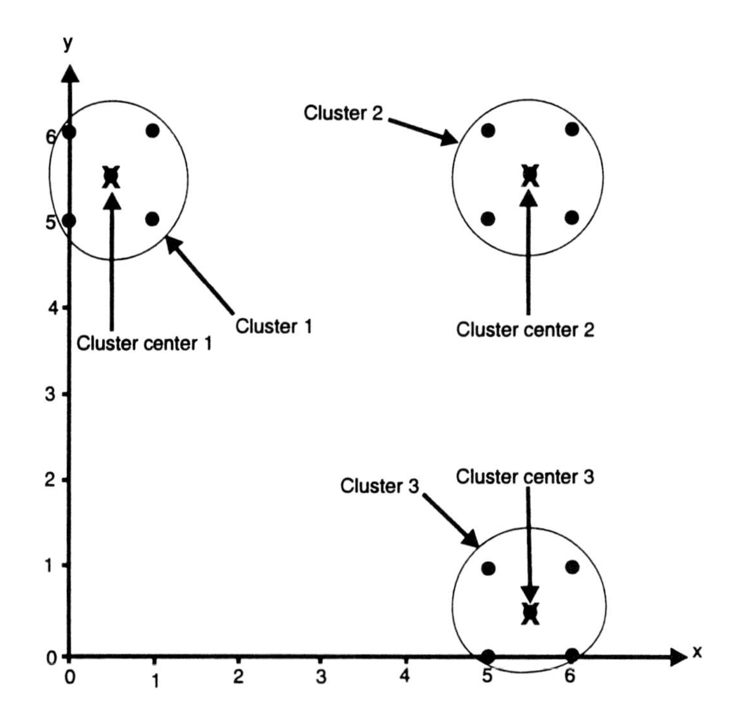

FIGURE 8.8 Clusters and cluster centers detected by the Kohonen network.

8.4 THE ROLE OF LATERAL FEEDBACK

The operation of the Kohonen SOFM requires a somewhat more sophisticated approach to lateral feedback than was taken in section 8.3. In particular, there was nothing in the implementation found in section 8.3 to motivate the cluster centers to order themselves in any particular way. As a result, although the network was self-organizing, the features did not organize in any geometrically relevant way within the output layer. If we desire features to congregate within the output layer such that relatively like features may be found nearby one another in the lattice structure, we must change our approach to lateral feedback in such a manner as will induce this behavior.

To accomplish this task, feedback connections are established within the output layer as illustrated for a one-dimensional lattice in Figure 8.9 and a two-dimensional lattice in Figure 8.10. The magnitude and type (excitatory or inhibitory) of feedback, expressed in the lateral weights, will be a function of the geometric distance between the neurons within the lattice. It remains now to determine what sort of lateral connections will produce the desired result. In this case we are fortunate in that we may draw inspiration from biological systems.

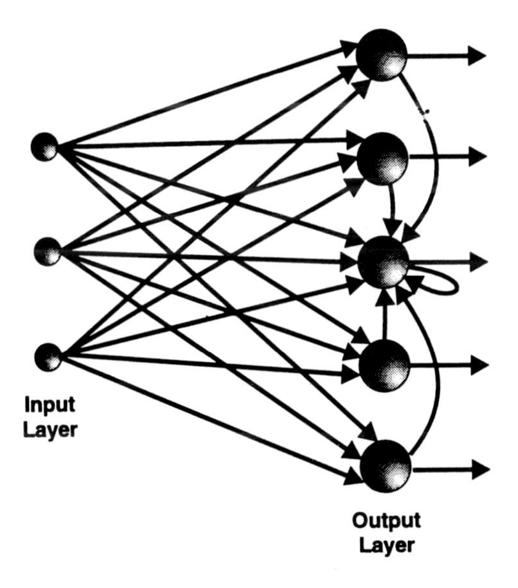

FIGURE 8.9 Lateral feedback in a one-dimensional lattice Kohonen layer.

A geometric mapping of features like the one we are seeking can be found in the visual cortex. Thus the short-range lateral feedback found within the visual cortex provides the desired model which we may incorporate into our network. Equation 9, frequently referred to as the Mexican hat function, thus becomes our model for lateral feedback.

$$\nabla^2 h = \left(\frac{r^2 - \sigma^2}{\sigma^4}\right) e^{\frac{r^2}{2\sigma^2}} \tag{9}$$

The graph of equation 9 may be found in Figure 8.11.

Referring to Figure 8.11 above it can be seen that the Mexican hat function contains three distinct regions of lateral interaction as a function of the distance between

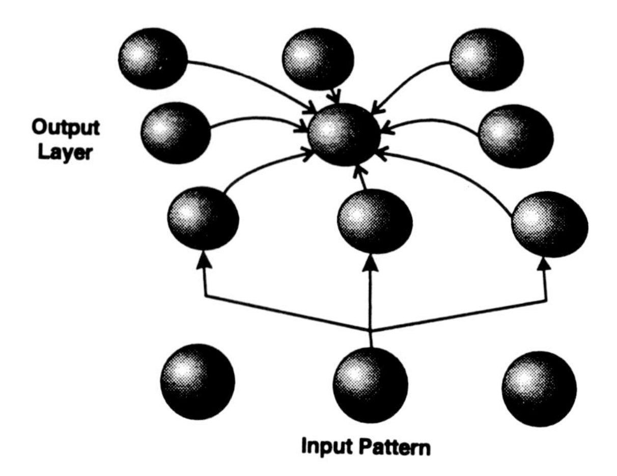

FIGURE 8.10 Lateral feedback in a two-dimensional lattice Kohonen layer.

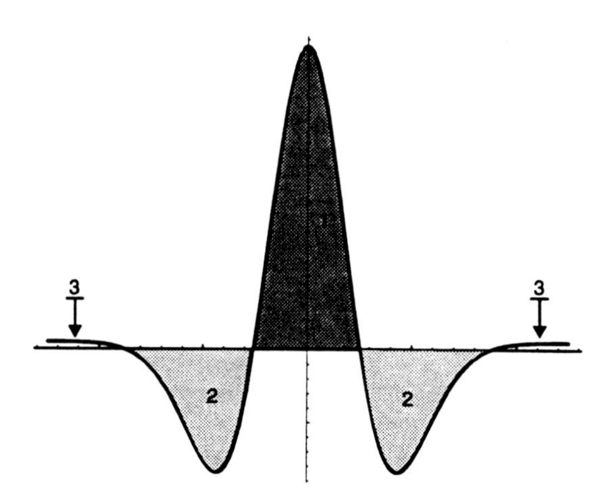

FIGURE 8.11 Mexican hat function. Note the regions of positive and negative reinforcement.

neurons. At distances less than R_0, the lateral feedback is excitatory. In the region between R_0 and R_1 there is a penumbra of inhibitory feedback. Beyond R_1 there is weak excitation. For additional detail regarding correspondence of this model with the visual cortex see Kohonen (1982).

We now examine the operation of the Kohonen layer in the presence of lateral feedback described above. The net input to the jth neuron of the output layer may now be expressed as:

$$\text{net}_j = \left(I_j + \sum_{k=-K}^{K} c_{j,j+k} y_{j+k} \right), \quad j = 1, 2, \ldots, n \tag{10}$$

where K defines a maximum region over which lateral feedback is active and I_j is given by:

$$I_j = \sum_{i=1}^{p} w_{ji} x_i \tag{11}$$

The output of the jth neuron is obtained by applying a nonlinear function, $\phi()$, to net_j as follows: $y_j = \phi(net_j)$. ϕ is chosen so that it constrains $a > y_j \geq 0$, where a is an arbitrary constant. The lateral feedback in the equation is carried in the weights c_{jk}. These interlayer weights are fixed, meaning that they are not trained but rather are derived from equation 9 or at least a suitable approximation thereof. Such lateral connections serve to assist or excite neurons in close physical proximity and to inhibit those farther away. Most frequently an approximation to the Mexican hat function is adequate for establishing these weights. Figure 8.12 shows such an approximation.

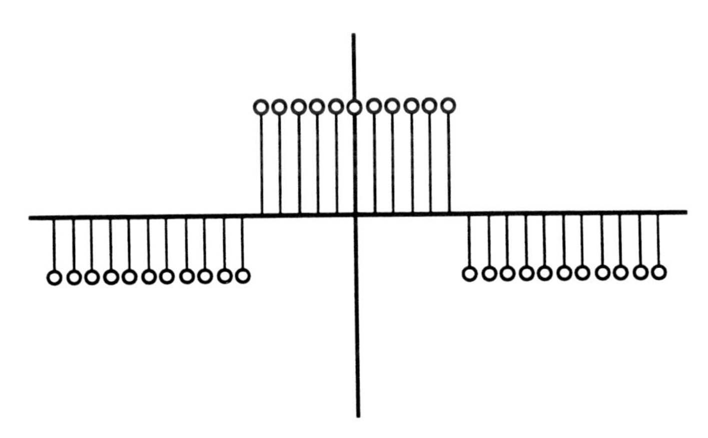

FIGURE 8.12 An approximation of the Mexican hat function.

The solution to the equations 9 and 10 above requires an iterative approach in which the neurons of the output layer in effect settle down into an equilibrium state over time. The iterative equation is as follows:

$$y_j(n+1) = \varphi\left(I_j + \beta \sum_{k=-K}^{K} c_{j,j+k} y_{j+k}(n)\right), \quad j = 1,2, \ldots, N \tag{12}$$

where n represents a discrete time step and β is a constant governing the rate of convergence. Where $\phi()$ and β have been appropriately chosen, equation 11 will converge to produce a bounded spatial region or activity bubble. Within the bubble the response is maximum, $y_j \to a$. Outside the bubble, response falls off to zero. The width of the bubble is governed by the feedback connections. Positive feedback causes the bubble to be wider. Negative feedback causes the bubble to be sharper.

The bubble is formed centered around the neuron for which $y_j(0)$ (due to the stimulus I_j alone) is maximum. These results suggest that the effect of lateral feedback due to the Mexican hat function can thus be simulated in a computationally efficient way. Lateral weights are eliminated and replaced by a region in which the response y_j is maximized. The region corresponds to the activity bubble. The adjustment of the lateral can then be emulated by simply adjusting the size of this neighborhood: larger for more positive feedback, smaller for more negative feedback. This is exactly the approach that we will encounter when we study the SOFM in section 8.5.

8.5 KOHONEN SELF-ORGANIZING FEATURE MAP

The Kohonen SOFM utilizes the Kohonen architecture and Kohonen learning as described in section 8.3. By augmenting this implementation with lateral feedback, as described in section 8.4, the goals of the self-organizing feature mapping algorithm are achieved. That goal is the mapping of an input space of n-dimensions into a one- or two-dimensional lattice (of output layer neurons) which comprises the output space such that a meaningful topological ordering exists within the output space.

As before the weight vector associated with each output layer neuron is regarded as an exemplar of the kind of input vector the neuron will respond to. Let us denote the input vector, **x**, as follow:

$$\mathbf{x} = \left[x_i, x_2, \ldots, x_p\right]^T \tag{13}$$

The weight vector, $\mathbf{w}_j$ corresponding to output layer neuron j can be written:

$$\mathbf{w}_j = \left[w_{ji}, w_{j2}, \ldots, w_{jp}\right]^T \quad j = 1, 2, \ldots, N \tag{14}$$

Determination of the winning output layer neuron amounts to selecting the output layer neuron whose weight vector $\mathbf{w}_j$ best matches the input vector **x**. Recall from our earlier discussion that this may be done in two ways. We may select the output layer neuron whose stimulus $I_j = \mathbf{W}_j^T\mathbf{x}$ is maximum. Alternatively, we may select the output layer neuron whose weight vector lies at a minimum Euclidean norm from the input vector. If $i(\mathbf{x})$ is used to designate the index of the winning neuron, this latter approach can be expressed as:

$$i(\mathbf{x}) = k \text{ where } \left\|\mathbf{W}_k - \mathbf{x}\right\| < \left\|\mathbf{W}_j - \mathbf{x}\right\| \quad j = 1, 2, \ldots, n \tag{15}$$

The manner in which lateral feedback is moderated remains to be defined. To do this a function is introduced which defines the size of the neighborhood (corresponding to the activity bubble) surrounding the winning neuron. This function

$\Lambda_{i(x)}(n)$ is a function of discrete time (i.e., iteration). By this it is meant that the amount of lateral feedback (i.e., the width of the activity bubble) can be varied over the course of training the network. Larger neighborhoods mean more positive feedback, a larger activity bubble, and training that takes place at a more global level. It is through a large value of the neighborhood function during the early training of the network that the topological ordering of the network is achieved. Subsequent reduction of the neighborhood then makes the clusters sharper so that the cluster response may be refined. Typically it is convenient to express the neighborhood function in terms of a rectangular lattice as shown in Figure 8.13. Notice that a radius of zero includes only the winning neuron. At a radius of one the eight nearest-neighbor neurons are also included. Other arrangements such as the hexagonal lattice illustrated in Figure 8.14 are possible as well.

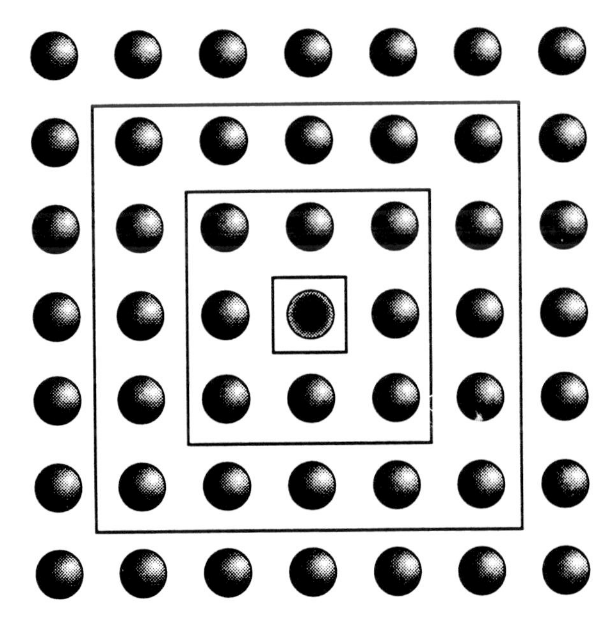

FIGURE 8.13 Neighborhoods on a rectangular lattice.

The neighborhood function may now be utilized to modify the learning process as indicated in equation 15:

$$w_j(n+1) = \begin{cases} w_j(n) + \eta(n)\left[x - w_j(n)\right] & j \in \Lambda_{i(x)}(n) \\ w_j(n) & \text{otherwise} \end{cases} \tag{16}$$

Note that the modified training method of equation 15 leads or motivates neurons within a neighborhood to contain similar synaptic weight vectors.

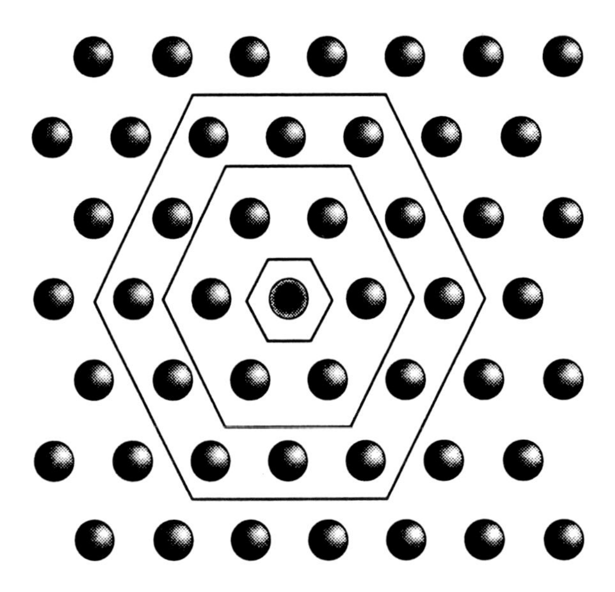

FIGURE 8.14 Neighborhoods on a hexagonal lattice.

The Kohonen self-organizing feature map algorithm can be expressed in the following steps.

Step 1. Initialization:
Initialize the weight vectors, $\mathbf{w}_j(0)$ to random values. It may be beneficial to choose these values such that the magnitude is small. Initialize the learning rate η (0) and the neighborhood function $\Lambda_i(\mathbf{x})(0)$. Both learning rate and neighborhood function should be large initially. (We will have more to say about choosing these initial values later on.)

Step 2. For each vector, $\mathbf{x}$ in the of samples perform steps 2a, 2b and 2c.

Step 2a. Place the sensory stimulus vector, $\mathbf{x}$ onto the input layer of the network.

Step 2b. Similarity matching:
Select the neuron whose weight vector best matches $\mathbf{x}$ as the winning neuron. Using the Euclidean criteria, the index of the winning neuron will be

$$i(x) = k \text{ where } \left\|\mathbf{W}_k - \mathbf{x}\right\| < \left\|\mathbf{W}_j - \mathbf{x}\right\| \quad j = 1, 2, \ldots, n \tag{17}$$

Step 2c. Training:
Train the weight vectors such that neurons within the activity bubble are moved toward the input vector as follows:

$$\mathbf{w}_j(n+1) = \begin{cases} \mathbf{w}_j(n) + \eta(n)\left[x - w_j(n)\right] & j \in \Lambda_{i(x)}(n) \\ \mathbf{w}_j(n) & \text{otherwise} \end{cases} \tag{18}$$

Step 3. Update the learning rate, $\eta(n)$:
A linear decrease of the learning rate should produce satisfactory results.

Step 4. Reduce the neighborhood function, $\Lambda_{i(x)}(n)$.

Step 5. Check stopping condition:
Exit when no noticeable change to the feature map has occurred. Otherwise go to step 2.

Learning in the SOFM can be seen to take place in two principal phases: the ordering phase and the convergence phase. The ordering phase takes place first. During this phase the (global) topological ordering of the weight vectors $\mathbf{w}_j$ takes place. During this phase η is maintained relatively large. Initially, η may be chosen close to 1.0. During the ordering phase it will typically not decrease to below 0.1. Additionally during the ordering phase, the neighborhood function begins relatively large, frequently encompassing all neurons in the network. Over the course of the algorithm it will shrink to produce neighborhoods limited to a few neurons (or perhaps even only the winning neuron). The ordering phase may last about a thousand or so iterations and thus is typically much shorter than the convergence phase. During the convergence phase, output neurons home in on precise exemplar weight vector values. To this end, during the convergence phase of the algorithm a much smaller value is desired for both η and neighborhood function. The value of η at this point would typically be held to 0.01 or less.

Listing 8.3 below provides complete source code for the self-organizing feature map.

Listing 8.3

```
/*************************************************
* Kohonen's self organizing feature map          *
*************************************************/

#include <stdlib.h>
#include <stdio.h>
#include <math.h>

#define MAXPATS       100
#define MAXNEURONSIN 10
#define MAXNEURONSX  15
#define MAXNEURONSY  15

#define MAXEPOCHS  2000
#define ETAMIN     .005

unsigned int Random(int N) {
 unsigned int j;
j= (N*rand())/RAND_MAX;
if (j>=N) j=N;
return j;
}
/*******************************************************
```

```
* Pattern class definition                              *
********************************************************/
class PATTERN {
 friend class SOFM;
private:
 double        P[MAXPATS][MAXNEURONSIN];
 int           NumPatterns;
 int           Shuffle[MAXPATS];
 int           SizeVector;
public:
 PATTERN();
 int GetPatterns(char *);               //load pattern form file
 int GetRandPats(int,int);              //random patterns arg1=# of patterns, arg2=dimension
 double Query(int,int);                 //returns P[arg1][arg2]
 double QueryR(int,int);                //returns P[Shuffle[arg1]][arg2]
 void ReShuffle(int N);
};

/*******************************************************
* Pattern class method implementation                  *
********************************************************/

//*******************************************************
// Pattern class constructor
//*******************************************************

PATTERN::PATTERN(){
int i;
for (i=0; i<MAXPATS; i++)
  Shuffle[i]=i;
}

//*******************************************************
// GetPatterns
//    Read pattern data from file
//*******************************************************

int PATTERN::GetPatterns(char *fname) {
 FILE *fp;
 int i,j;
 double x;
fp=fopen(fname,"r");
if (fp==NULL) return 0;                          // Test for failure.
fscanf(fp,"%d",&NumPatterns);
fscanf(fp,"%d",&SizeVector);
for (i=0; i<NumPatterns; i++) {                  // For each vector
  for (j=0; j<SizeVector; j++) {                 // create a pattern
    fscanf(fp,"%lg",&x);                         // consisting of all elements
    P[i][j]=x;
    } /* endfor */
  } /* endfor */
fclose(fp);
return 1;
}
//*******************************************************
```

```
// GetRandPats
//    Creates n1 random pattern vectors
//    with n2 elements
//*******************************************************

int PATTERN::GetRandPats(int n1,int n2) {
  int i,j;
  double x;
NumPatterns=n1;
SizeVector=n2;
for (i=0; i<NumPatterns; i++) {             // For each vector
  for (j=0; j<SizeVector; j++) {            // create a pattern
    x=(double)rand()/RAND_MAX;              // consisting of random elements
     P[i][j]=x;                             // between 0 and 1
     } /* endfor */
  } /* endfor */
return 1;
}


//*******************************************************
// Reshuffle
//    Randomize an array consisting of all patterns
//*******************************************************

void PATTERN::ReShuffle(int N) {
int i,a1,a2,tmp;
for (i=0; i<N ;i++) {
  a1=Random(NumPatterns);
  a2=Random(NumPatterns);
  tmp=Shuffle[a1];
  Shuffle[a1]=Shuffle[a2];
  Shuffle[a2]=tmp;
   }
}

//*******************************************************
// Query
//    Returns the jth element of the
//    specified pattern vector
//*******************************************************

double PATTERN::Query(int pat,int j) {
return P[pat][j];
}


//*******************************************************
// QueryR
//    Returns the jth element of the pattern whose
//    index is obtained from the shuffle array.
//    (Used in conjunction with reshuffle to train
//      patterns in random order
//*******************************************************
double PATTERN::QueryR(int pat,int j) {
```

```
return P[Shuffle[pat]][j];
}

struct iPair {
 int x,y;
};

/*****************************************************
* SOFM class definition                          *
*****************************************************/

class SOFM {
private:
 double W[MAXNEURONSIN][MAXNEURONSX][MAXNEURONSY]; // The weight
matrix
 double Yout[MAXNEURONSX][MAXNEURONSY];        // The output layer neurons
 double Yin[MAXNEURONSIN];                     //The input layer neurons
 int    Lattice;                               //Square Vs triangular lattice
 int    YinSize;                               //input layer dimensions
 iPair  YoutSize;                              //outlayer dimensions

 int    R;                                     //update neighborhood radius
 int    MaxEpoch;
 int    epoch;
 double eta;                                   //The learning rate
 double delta_eta;                   //Amount to change l.r. each epoch
 double Erosion;                     //Urban decay metric..Neighborhoods shrink
 int    StochFlg;                    //Present vectors in rand order if 1
 PATTERN *Pattern;

 int    LoadInLayer(int);            //pattern->input layer
 double EucNorm(int, int);           //Calc Euclidean distance
 iPair  FindWinner();                //get coords of winning neuron
 void   Train(iPair);
 void   AdaptParms();
public:
 SOFM();
 void SetPattern(PATTERN *);
 void SetParms(int, int, double);
 void PrintWeights();
 void PrintWinner();
 void RunTrn();
 void Run();
};

/*****************************************************
* Pattern method implementations                 *
*****************************************************/

//*****************************************************
// SOFM class constructor
//*****************************************************

SOFM::SOFM(){
StochFlg=1;
```

```
Erosion=0;
}

//*********************************************************
// SetPattern
//    Establish linkage w/ the pattern
//*********************************************************

void SOFM::SetPattern(PATTERN *p) {
  Pattern=p;
  YinSize=p->SizeVector;
}

//*********************************************************
// SetParms
//    Establish needed parameters
//*********************************************************

void SOFM::SetParms(int X, int Y, double LR){
  int ix,iy,k;
YoutSize.x=X;
YoutSize.y=Y;
R=(X+Y)/4;
eta=LR;
delta_eta=0.005;
for (ix=0; ix<X; ix++) {
   for (iy=0; iy<Y; iy++) {
     for (k=0; k<YinSize; k++) {
       W[k][ix][iy]= (double)rand()/(10.0 * (double)RAND_MAX);
         } /* endfor */
       } /* endfor */
   } /* endfor */
}

//*********************************************************
// LoadInLayer
//    Initialize input layer w/ the designated pattern
//*********************************************************

int SOFM::LoadInLayer(int P){
  int i;
for (i=0; i<YinSize; i++){
   if (StochFlg){
     Yin[i]=Pattern->QueryR(P,i);
      }
    else {
     Yin[i]=Pattern->Query(P,i);
      }
   }
 return 1;
 }

//*********************************************************
// AdaptParms
//    Adjust learning rate and neighborhood size //
```

```
*****************************************************

void SOFM::AdaptParms(){
Erosion += .01;
if (Erosion>=1.0) {
  Erosion=0.0;
  if (R>0)
    R—;
  printf("New neighborhood.  Radius=%d", R);
  } /* endif */
if (epoch<500) {                    //Reduce learning rate more slowly over 1st 1k epochs
  eta=eta-delta_eta/10.0;
  }
else {
  eta=eta-delta_eta;
  } /* endif */
if (eta<ETAMIN)
  eta=ETAMIN;
printf(" New eta=%f\n",eta);
}

//*****************************************************
// PrintWeights
//   Display the weight matrix
//*****************************************************

void  SOFM::PrintWeights() {
  int ix,iy,k;
for (ix=0; ix<YoutSize.x; ix++) {
  for (iy=0; iy<YoutSize.y; iy++) {
    for (k=0; k<YinSize; k++) {
      printf("W[%d][%d][%d]=%f  ",k,ix,iy,W[k][ix][iy]);
      } /* endfor */
      printf("\n");
    } /* endfor */
  } /* endfor */
}


//*****************************************************
// RunTrn
//   Run the network w/ training enabled.
//*****************************************************

void SOFM::RunTrn(){
int i,np;
iPair Winner;
epoch=0;
np=Pattern->NumPatterns;
while (epoch<=MAXEPOCHS){
  for (i=0; i<np; i++){
    LoadInLayer(i);
    Winner=FindWinner();
    Train(Winner);
    }
```

```
 if(20*(epoch/20)==epoch) {
  printf("Epoch=%d\n",epoch);
   PrintWeights();
   }
 epoch++;
 if (StochFlg)
  Pattern->ReShuffle(np);
 AdaptParms();
  }
}


//*******************************************************
// Train
//  Update the weights of the winning neuron and
//  update the weights of neurons within the winning
//  neurons neighborhood
//*******************************************************


void SOFM::Train(struct iPair Winner){
 int ix,iy,k;
for (ix=Winner.x-R; ix<=Winner.x+R; ix++){
 if ((ix>=0) && (ix<YoutSize.x)){
   for (iy=Winner.y-R; iy<=Winner.y+R; iy++){
    if ((iy>=0) && iy<YoutSize.y) {
      for (k=0; k<YinSize; k++){
       W[k][ix][iy]=W[k][ix][iy]+eta*(Yin[k]-W[k][ix][iy]);
        } /*endfor*/
      } /*endif*/
    } /*endfor*/
   } /*endif*/
 } /*endfor*/
}


//*******************************************************
// FindWinner
//  Discover winning neuron based on min Euclidean
//  distance
//*******************************************************


iPair SOFM::FindWinner(){
 int ix,iy;
 double d,best;
 iPair Winner;
best=1.0e99;
Winner.x=-1;
Winner.y=-1;
for (ix=0; ix<YoutSize.x; ix++){
 for (iy=0; iy<YoutSize.y; iy++){
   d=EucNorm(ix,iy);
   if (d<best) {
     best=d;
```

```
      Winner.x=ix;
      Winner.y=iy;
      } // endif
   } // endfor
 } // endfor
return Winner;
}


//*******************************************************
// EucNorm
//   Calculate Euclidean distance between a pattern and
//   the neuron at x,y in the output layer lattice
//*******************************************************

double SOFM::EucNorm(int x, int y){             // Calc Euclidean norm of vector dif
int i;
double dist;
dist=0;
for (i=0; i< YinSize;i++){
  dist += (W[i][x][y]-Yin[i]) * (W[i][x][y]-Yin[i]);
  } /* endfor */
dist=sqrt(dist);
return dist;
}


//==================================================================
// GLOBAL OBJECTS
//==================================================================

PATTERN InPat;
SOFM    FMap;


//*******************************************************
// main
//*******************************************************

main(int argc, char *argv[]) {
//srand(17);
if (argc>1) {
  InPat.GetPatterns(argv[1]);                 //Establish pattern
  FMap.SetPattern(&InPat);                    //Inform the feature map about the pattern
  FMap.SetParms(5,5,0.900);                   //Init fm parms
  FMap.RunTrn();                              //Run the FM w/ training enabled
  }
 else {
  printf("USAGE: SOFM PATTERN_FILE");
  }
}
```

8.5.1 SOFM EXAMPLE

We now illustrate the operation of the self-organizing feature map. Figure 8.15 illustrates the input pattern to which the SOFM will be applied. The input vectors corresponding to Figure 8.15 are as follows:

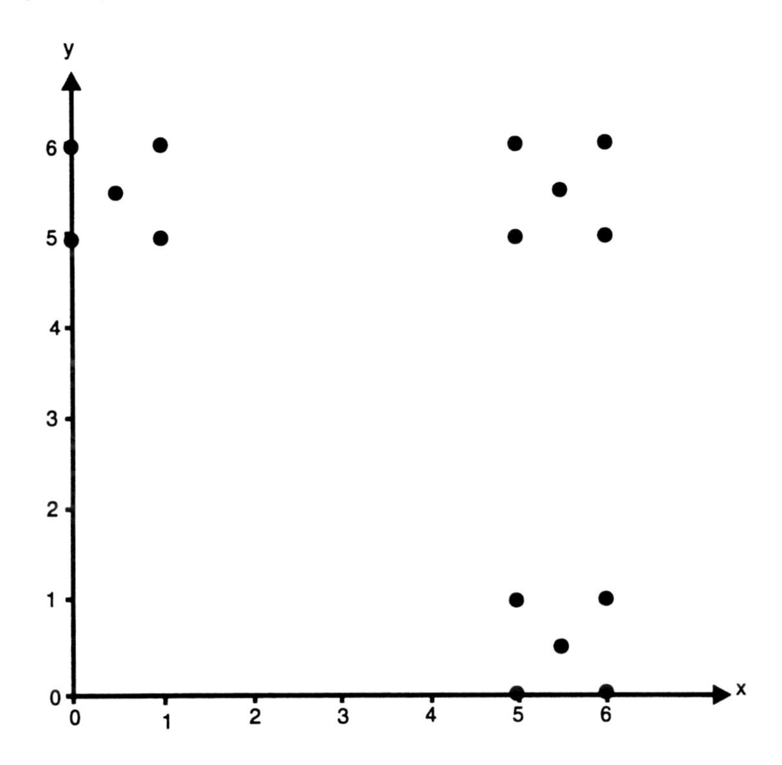

FIGURE 8.15 Example input pattern for the SOFM.

{(5.0, 5.0), (6.0, 6.0), (5.0, 6.0), (6.0, 5.0), (5.5, 5.5)
(5.0, 0.0), (5.0, 1.0), (6.0, 0.0), (6.0, 1.0), (5.5, 0.5)
(0.0, 5.0), (1.0, 5.0), (0.0, 6.0), (1.0, 6.0), (0.5, 5.5)}

The network topology for the SOFM chosen for this example is a two-dimensional 5 × 5 neuron rectangular array. Initially the neighborhood consists of all neurons within a radius of three of the winning neuron. The radius is decreased as training progresses. The initial learning rate is $\eta = 0.9$. The learning rate also decreases as learning progresses.

After one epoch, the weight matrix is as follows:

W[0][0][0]=0.995085 W[1][0][0]=5.009985
W[0][0][1]=1.004001 W[1][0][1]=4.995960
W[0][0][2]=1.004001 W[1][0][2]=4.995960
W[0][0][3]=1.004001 W[1][0][3]=4.995960
W[0][0][4]=1.003950 W[1][0][4]=4.995900

```
W[0][1][0]=0.540955  W[1][1][0]=5.548956
W[0][1][1]=0.540955  W[1][1][1]=5.548955
W[0][1][2]=0.540955  W[1][1][2]=5.548955
W[0][1][3]=0.540955  W[1][1][3]=5.548955
W[0][1][4]=1.004900  W[1][1][4]=5.004095
W[0][2][0]=0.540955  W[1][2][0]=5.548956
W[0][2][1]=0.540955  W[1][2][1]=5.548955
W[0][2][2]=0.540955  W[1][2][2]=5.548955
W[0][2][3]=0.540955  W[1][2][3]=5.548955
W[0][2][4]=1.004900  W[1][2][4]=5.004095
W[0][3][0]=0.540955  W[1][3][0]=5.548956
W[0][3][1]=0.540955  W[1][3][1]=5.548955
W[0][3][2]=0.540955  W[1][3][2]=5.548955
W[0][3][3]=0.540955  W[1][3][3]=5.548955
W[0][3][4]=1.004900  W[1][3][4]=5.004095
W[0][4][0]=0.540960  W[1][4][0]=5.548964
W[0][4][1]=0.540960  W[1][4][1]=5.548960
W[0][4][2]=0.540960  W[1][4][2]=5.548960
W[0][4][3]=0.540960  W[1][4][3]=5.548960
W[0][4][4]=1.049491  W[1][4][4]=5.045401
```

where $W[x][y][i]$ designates the ith element of the weights incident on the output layer neuron at position (x,y) in the output layer lattice.

After 100 epochs the learning rate has decreased to 0.85 and the radius of the neighborhood is decreased to 2. The weight matrix is as follows:

```
W[0][0][0]=5.083817  W[1][0][0]=5.066834
W[0][0][1]=5.862818  W[1][0][1]=1.609674
W[0][0][2]=5.552790  W[1][0][2]=0.650543
W[0][0][3]=5.555808  W[1][0][3]=0.554776
W[0][0][4]=5.555326  W[1][0][4]=0.552985
W[0][1][0]=5.083817  W[1][1][0]=5.066836
W[0][1][1]=5.775423  W[1][1][1]=1.601451
W[0][1][2]=5.541011  W[1][1][2]=0.662873
W[0][1][3]=5.473923  W[1][1][3]=0.650328
W[0][1][4]=5.473912  W[1][1][4]=0.650287
W[0][2][0]=5.083004  W[1][2][0]=5.067118
W[0][2][1]=5.775368  W[1][2][1]=1.601509
W[0][2][2]=5.537867  W[1][2][2]=0.666115
W[0][2][3]=5.454642  W[1][2][3]=0.672364
W[0][2][4]=5.454642  W[1][2][4]=0.672363
```

W[0][3][0]=5.064609 W[1][3][0]=5.086207
W[0][3][1]=4.499728 W[1][3][1]=5.012878
W[0][3][2]=4.348711 W[1][3][2]=5.065260
W[0][3][3]=0.555089 W[1][3][3]=5.428316
W[0][3][4]=0.554743 W[1][3][4]=5.425703
W[0][4][0]=5.064617 W[1][4][0]=5.086217
W[0][4][1]=4.499690 W[1][4][1]=5.012925
W[0][4][2]=4.348711 W[1][4][2]=5.065260
W[0][4][3]=0.555051 W[1][4][3]=5.428363
W[0][4][4]=0.553039 W[1][4][4]=5.427828

After 200 epochs the learning rate has decreased to 0.80 and the radius of the neighborhood is decreased to 1. The weight matrix is as follows:

W[0][0][0]=0.176284 W[1][0][0]=5.022428
W[0][0][1]=0.209436 W[1][0][1]=4.984109
W[0][0][2]=4.957143 W[1][0][2]=0.241577
W[0][0][3]=5.118544 W[1][0][3]=0.080189
W[0][0][4]=5.118544 W[1][0][4]=0.080189
W[0][1][0]=0.323467 W[1][1][0]=4.876737
W[0][1][1]=0.329895 W[1][1][1]=4.868832
W[0][1][2]=4.953037 W[1][1][2]=0.246759
W[0][1][3]=5.823735 W[1][1][3]=0.816265
W[0][1][4]=5.823737 W[1][1][4]=0.816262
W[0][2][0]=4.033593 W[1][2][0]=1.992169
W[0][2][1]=5.767417 W[1][2][1]=4.200818
W[0][2][2]=5.887439 W[1][2][2]=4.150783
W[0][2][3]=5.983738 W[1][2][3]=1.654697
W[0][2][4]=5.918693 W[1][2][4]=1.073483
W[0][3][0]=0.321344 W[1][3][0]=5.839687
W[0][3][1]=1.132862 W[1][3][1]=5.801589
W[0][3][2]=5.560012 W[1][3][2]=5.427188
W[0][3][3]=5.592318 W[1][3][3]=4.766708
W[0][3][4]=5.567962 W[1][3][4]=4.799698
W[0][4][0]=0.204840 W[1][4][0]=5.992078
W[0][4][1]=1.129061 W[1][4][1]=5.807937
W[0][4][2]=5.561553 W[1][4][2]=5.433309
W[0][4][3]=5.561591 W[1][4][3]=5.433542
W[0][4][4]=5.439809 W[1][4][4]=5.598542

After 300 epochs the learning rate has decreased to 0.75 and the radius of the neighborhood is decreased to 0. Thus from this point forward, only the winning neuron will be trained. The weight matrix at epoch 300 is as follows:

W[0][0][0]=0.152532 W[1][0][0]=5.141537
W[0][0][1]=0.919860 W[1][0][1]=5.140855
W[0][0][2]=5.210362 W[1][0][2]=5.038421
W[0][0][3]=5.097558 W[1][0][3]=5.856369
W[0][0][4]=5.847307 W[1][0][4]=5.858987
W[0][1][0]=0.131883 W[1][1][0]=5.879134
W[0][1][1]=0.919861 W[1][1][1]=5.890854

```
W[0][1][2]=5.052640  W[1][1][2]=5.008963
W[0][1][3]=5.847556  W[1][1][3]=5.106356
W[0][1][4]=5.472323  W[1][1][4]=5.483821
W[0][2][0]=1.713121  W[1][2][0]=4.239573
W[0][2][1]=2.114362  W[1][2][1]=4.810238
W[0][2][2]=5.193171  W[1][2][2]=4.807382
W[0][2][3]=5.243454  W[1][2][3]=4.807361
W[0][2][4]=5.225576  W[1][2][4]=4.228176
W[0][3][0]=4.745697  W[1][3][0]=1.301179
W[0][3][1]=5.002715  W[1][3][1]=0.953236
W[0][3][2]=5.387721  W[1][3][2]=0.565731
W[0][3][3]=5.577927  W[1][3][3]=0.563596
W[0][3][4]=5.905813  W[1][3][4]=0.894870
W[0][4][0]=4.751555  W[1][4][0]=1.295121
W[0][4][1]=5.008562  W[1][4][1]=0.802011
W[0][4][2]=5.096930  W[1][4][2]=0.141378
W[0][4][3]=5.519473  W[1][4][3]=0.515830
W[0][4][4]=5.905812  W[1][4][4]=0.143951
```

Since the neighborhood radius is now zero, we have entered the convergence phase. After 2000 iterations the learning rate has reached its minimum value of 0.005 and the state of the network is as follows:

```
W[0][0][0]=0.000000  W[1][0][0]=5.000000
W[0][0][1]=0.750000  W[1][0][1]=5.250000
W[0][0][2]=5.210362  W[1][0][2]=5.038421
W[0][0][3]=5.000000  W[1][0][3]=6.000000
W[0][0][4]=6.000000  W[1][0][4]=6.000000
W[0][1][0]=0.000000  W[1][1][0]=6.000000
W[0][1][1]=1.000000  W[1][1][1]=6.000000
W[0][1][2]=5.000000  W[1][1][2]=5.000000
W[0][1][3]=6.000000  W[1][1][3]=5.000000
W[0][1][4]=5.500000  W[1][1][4]=5.500000
W[0][2][0]=1.320044  W[1][2][0]=3.266798
W[0][2][1]=2.114362  W[1][2][1]=4.810238
W[0][2][2]=5.193171  W[1][2][2]=4.807382
W[0][2][3]=5.243454  W[1][2][3]=4.807361
W[0][2][4]=5.225576  W[1][2][4]=4.228176
W[0][3][0]=4.745697  W[1][3][0]=1.301179
W[0][3][1]=5.000000  W[1][3][1]=1.000000
W[0][3][2]=5.387721  W[1][3][2]=0.565731
W[0][3][3]=5.577927  W[1][3][3]=0.563596
W[0][3][4]=6.000000  W[1][3][4]=1.000000
W[0][4][0]=4.751555  W[1][4][0]=1.295121
W[0][4][1]=5.008562  W[1][4][1]=0.802011
W[0][4][2]=5.000000  W[1][4][2]=0.000000
W[0][4][3]=5.500000  W[1][4][3]=0.500000
W[0][4][4]=6.000000  W[1][4][4]=0.000000
```

Figure 8.16 shows the distribution of weight vectors corresponding to the clusters formed by the feature maps. To emphasize that the weight array has arranged itself into cluster centers that are geometrically meaningful in terms of the coordinates of the weight array itself, Figure 8.17 shows the weight array partitioned according to the cluster centers represented by the corresponding weight vector.

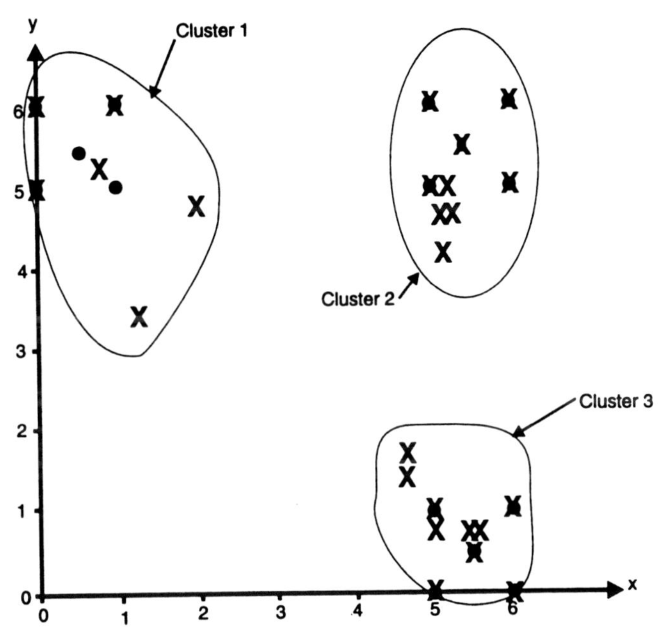

FIGURE 8.16 Clusters and cluster centers detected by the SOFM.

8.6 LEARNING VECTOR QUANTIZATION

Learning vector quantization (LVQ) [Kohonen, 1989,1990a] is a supervised learning extension of the Kohonen network methods we have studied thus far. It allows specification of the categories into which inputs will be classified. The designated categories for the training set are known in advance and are part of the training set. LVQ network architecture is exactly the same one we have become familiar with for the SOFM with the single exception that each neuron in the output layer is designated as belonging to one of the several classification categories. This is illustrated in Figure 8.18. In general, several output neurons will be assigned to each class. As before the weight vector to a given output unit represents an exemplar of the input vectors to which it will most strongly respond. In the context of LVQ such weight vectors are sometimes referred to as a reference or codebook vector. When an input pattern, **x**, is input to the network, the neuron with the closest (Euclidean

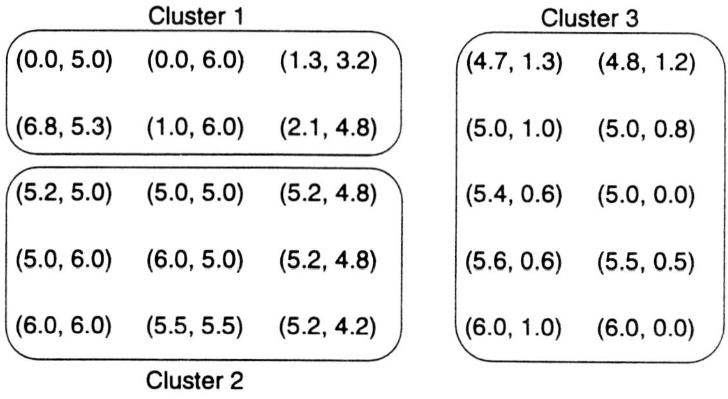

Cluster 1

(0.0, 5.0) (0.0, 6.0) (1.3, 3.2)

(6.8, 5.3) (1.0, 6.0) (2.1, 4.8)

(5.2, 5.0) (5.0, 5.0) (5.2, 4.8)

(5.0, 6.0) (6.0, 5.0) (5.2, 4.8)

(6.0, 6.0) (5.5, 5.5) (5.2, 4.2)

Cluster 2

Cluster 3

(4.7, 1.3) (4.8, 1.2)

(5.0, 1.0) (5.0, 0.8)

(5.4, 0.6) (5.0, 0.0)

(5.6, 0.6) (5.5, 0.5)

(6.0, 1.0) (6.0, 0.0)

FIGURE 8.17 Arrangement of clusters on the SOFM two-dimensional lattice as indicated by the weight matrix.

norm) weight vector is declared to be the winner. The training procedure utilizes a rule that is again similar to the earlier Kohonen learning. Only the winning neuron is modified. Equation 19 indicates the training adjustment to be

$$\mathbf{W}_{ij}^{\text{new}} = \mathbf{W}_{ij}^{\text{old}} + \eta\left(x_i - \mathbf{W}_{ij}^{\text{old}}\right) \quad \text{Category is correct}$$

$$\mathbf{W}_{ij}^{\text{new}} = \mathbf{W}_{ij}^{\text{old}} - \eta\left(x_i - \mathbf{W}_{ij}^{\text{old}}\right) \quad \text{Category is incorrect} \tag{19}$$

It is evident that this training method rewards a winning neuron if it belongs to the correct category by moving it toward the input vector. Conversely, if the winning neuron does not belong to the correct category, it is punished in that it is forced to move away from the input. This behavior is illustrated in Figure 8.19 below. Before we present the LVQ algorithm let us introduce some needed nomenclature. Let $\mathbf{x}^{(p)}$ represent the pth training pattern vector. Let $T^{(p)}$ represent the desired class or category into which $\mathbf{x}^{(p)}$ falls (according to the training set). Finally let C_j be the class or category represented by the jth output neuron. The LVQ algorithm may now be expressed in the following steps.

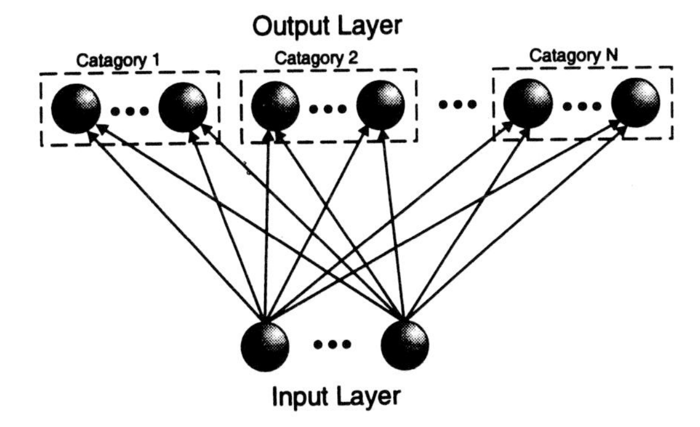

FIGURE 8.18 Architecture of the learning vector quantization network.

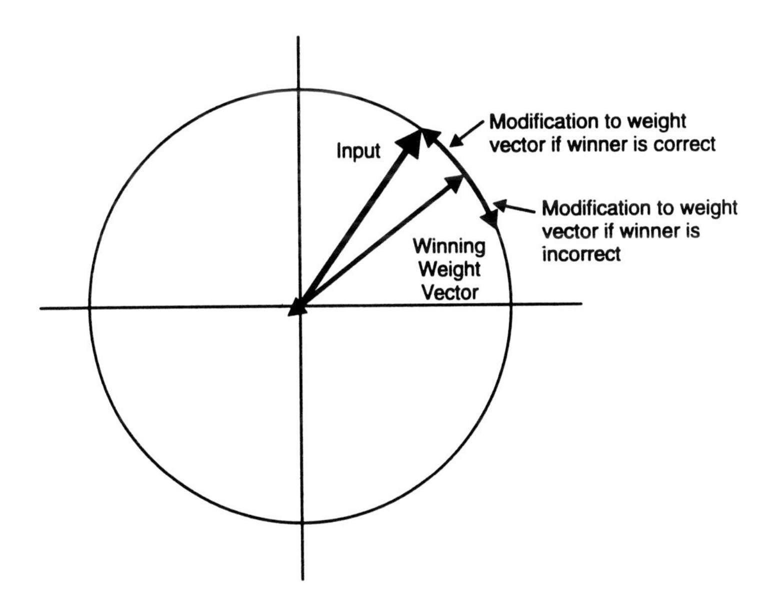

FIGURE 8.19 Training of an LVQ network.

Step 1. Initialization:
Initialize the weight vectors. The weight vectors may be initialized randomly. However, there are some other choices which we shall discuss shortly. Also, initialize the learning rate.

Step 2. For each vector $\mathbf{x}^{(p)}$ in the training set follow steps 2a and 2b.

Step 2a. Find the winning neuron k such that:

$$i\left(\mathbf{x}^{(p)}\right) = k \text{ where } \left\|\mathbf{W}_k - \mathbf{x}^{(p)}\right\| < \left\|\mathbf{W}_j - \mathbf{x}^{(p)}\right\| \quad j = 1, 2, \ldots, n \tag{20}$$

Step 2b. Update the weights $\mathbf{w}_k$ as follows:

$$\mathbf{W}_k^{\text{new}} = \begin{cases} \mathbf{W}_k^{\text{old}} + \eta\left(\mathbf{x}^{(p)} - \mathbf{W}_k^{\text{old}}\right) & \text{if } T = C_j \\ \mathbf{W}_k^{\text{old}} - \eta\left(\mathbf{x}^{(p)} - \mathbf{W}_k^{\text{old}}\right) & \text{if } T \neq C_j \end{cases} \tag{21}$$

Step 3. Adjust the learning rate:
The learning rate is reduced as a function of iteration.

Step 4. Check for termination:
Exit if termination conditions are met. Otherwise go to step 2.

A complete C++ implementation of the learning vector quantization algorithm may be found in listing 8.4

LVQ networks have been applied to the problem of character recognition. Baykal and Yalabki [1992] have utilized a Kohonen LVQ net in conjunction with a feed-forward net in the recognition of multifont characters. They report a recognition rate was 87% even with distorted, shifted, and rotated characters.

Listing 8.4

```
//*********************************************************
// LEARNING VECTOR QUANTIZATION
//*********************************************************

#include <stdlib.h>
#include <stdio.h>
#include <math.h>

#define  MAXPATS      100
#define  MAXNEURONSIN 50
#define  MAXNEURONS   15

#define  MAXEPOCHS    1500
#define  ETAMIN       .001

unsigned int Random(int N) {
  unsigned int j;
j= (N*rand())/RAND_MAX;
if (j>=N) j=N;
return j;
}

class TPATTERN {
  friend class LVQ;
```

```
private:
 double       P[MAXPATS][MAXNEURONSIN];
 int          PClass[MAXPATS];
 int          NumPatterns;
 int          NumClasses;
 int          Shuffle[MAXPATS];
 int          SizeVector;
public:
 TPATTERN();
 int GetPatterns(char *);             //load pattern form file
 int GetRandPats(int,int);            //random patterns arg1=# of patterns, arg2=dimension
 double Query(int,int);               //returns P[arg1][arg2]
 double QueryR(int,int);              //returns P[Shuffle[arg1]][arg2]
 int  QueryClass(int);
 void ReShuffle(int N);
};

TPATTERN::TPATTERN(){
int i;
for (i=0; i<MAXPATS; i++)
  Shuffle[i]=i;
}

int TPATTERN::GetPatterns(char *fname) {
 FILE *fp;
 int i,j,k;
 double x;
fp=fopen(fname,"r");
if (fp==NULL) return 0;  // Test for failure.
fscanf(fp,"%d",&NumPatterns);
fscanf(fp,"%d",&SizeVector);
fscanf(fp,"%d",&NumClasses);
for (i=0; i<NumPatterns; i++) {              // For each vector
  for (j=0; j<SizeVector; j++) {             // create a pattern
    fscanf(fp,"%lg",&x);                     // consisting of all elements
    P[i][j]=x;
    } /* endfor */
  fscanf(fp,"%d",&k);
  PClass[i]=k;
  } /* endfor */
fclose(fp);
return 1;
}

int TPATTERN::GetRandPats(int n1,int n2) {
  int i,j;
  double x;
NumPatterns=n1;
SizeVector=n2;
for (i=0; i<NumPatterns; i++) {              // For each vector
  for (j=0; j<SizeVector; j++) {             // create a pattern
   x=(double)rand()/RAND_MAX;                // consisting of random elements
    P[i][j]=x;                               // between 0 and 1
    } /* endfor */
  } /* endfor */
```

```
return 1;
}

void TPATTERN::ReShuffle(int N) {
int i,a1,a2,tmp;
for (i=0; i<N ;i++) {
  a1=Random(NumPatterns);
  a2=Random(NumPatterns);
  tmp=Shuffle[a1];
  Shuffle[a1]=Shuffle[a2];
  Shuffle[a2]=tmp;
  }
}

double TPATTERN::Query(int pat,int j) {
return P[pat][j];
}

double TPATTERN::QueryR(int pat,int j) {
return P[Shuffle[pat]][j];
}

int TPATTERN::QueryClass(int pat) {
return PClass[pat];
}

class LVQ {
private:
 double W[MAXNEURONSIN][MAXNEURONS];      //The weight matrix
 int    zClass[MAXNEURONS];               //Neuron Class assignment
 double Yout[MAXNEURONS];                 //The output layer neurons
 double Yin[MAXNEURONSIN];                //The input layer neurons
 int    YinSize;                          //input layer dimensions
 int    YoutSize;                         //outlayer dimensions

 int    epoch;
 double eta;                              //The learning rate
 double delta_eta;                        //Amount to change l.r. each epoch
 int    StochFlg;                         //Present vectors in rand order if 1
 TPATTERN *Pattern;

 int    LoadInLayer(int);                 //pattern->input layer
 double EucNorm(int);                     //Calc Euclidean distance
 int    FindWinner();                     //get coords of winning neuron
 void   Train(int,int);
 void   AdaptParms();
public:
 LVQ();
 void SetPattern(TPATTERN *);
 void SetParms(int, double);
 void PrintWeights();
 void PrintWinner();
 void RunTrn();
 void Run();
};
```

```
LVQ::LVQ(){
StochFlg=0;
}

void LVQ::SetPattern(TPATTERN *p) {
  Pattern=p;
  YinSize=p->SizeVector;
}

void LVQ::SetParms(int X, double LR){
 int i,k,m;
YoutSize=X;
eta=LR;
delta_eta=0.002;
for (i=0; i<YoutSize; i++) {
  for (k=0; k<YinSize; k++) {
    W[k][i]= (double)rand()/(10.0 * (double)RAND_MAX);
    //W[k][i]= (double)rand()/((double)RAND_MAX);
     } /* endfor */
  m=YoutSize/Pattern->NumClasses;
  zClass[i]= i/m;
  } /* endfor */
}

int LVQ::LoadInLayer(int P){
 int i;
for (i=0; i<YinSize; i++){
  if (StochFlg){
    Yin[i]=Pattern->QueryR(P,i);
     }
    else {
     Yin[i]=Pattern->Query(P,i);
     }
  }
return 1;
}

void LVQ::AdaptParms(){
eta=eta-delta_eta;
if (eta<ETAMIN)
  eta=ETAMIN;
}

void  LVQ::PrintWeights() {
  int i,k;
for (i=0; i<YoutSize; i++) {
   printf("W[%d]=",i);
   for (k=0; k<YinSize; k++) {
      printf("%f ",W[k][i]);
      } /* endfor */
   printf("\n");
   } /* endfor */
}

void LVQ::RunTrn(){
```

```
  int pat,np;
  int k,z;
  int Winner;
  epoch=0;
  np=Pattern->NumPatterns;
  while (epoch<=MAXEPOCHS){
    if( (epoch<=50) || (25*(epoch/25)==epoch) ) {          //output control
      printf("EPOCH=%d\n",epoch);
      printf("eta=%f\n",eta);
      }
    for (pat=0; pat<np; pat++){                          //Traverse all patterns
     LoadInLayer(pat);
      Winner=FindWinner();
     if( (epoch<=50) || (25*(epoch/25)==epoch) ) {         //output control
       printf("winner=%d/pat=%d\n",Winner,pat);
       printf("winner class=%d/pat class=%d\n",zClass[Winner],Pattern->QueryClass(pat));
       }
     Train(Winner,pat);
     if( (epoch<=50) || (25*(epoch/25)==epoch) ) {         //output control
        printf("W[%d]=",Winner);
        z=1;
       for (k=0; k<YinSize; k++) {
          printf("%f ",W[k][Winner]);
          if (z>4){
            printf("\n    ");
            z=1;
            }
          else
            z++;
         } /* endfor */
       printf("\n\n");
       }
     //printf("\n");
     }
// if(1*(epoch/1)==epoch) {                               //output control
//   printf("Epoch=%d\n",epoch);
//   PrintWeights();
//   }
  epoch++;                                                //keep track of epochs
  if (StochFlg)                                           // if desired
   Pattern->ReShuffle(np);                                //  reorder training patterns
  AdaptParms();                                           //Adjust the learning rate
  }
}
void LVQ::Run(){
 int pat,np,i;
 int Winner;
printf("\n");
np=Pattern->NumPatterns;
for (pat=0; pat<np; pat++){                               //Traverse all patterns
  LoadInLayer(pat);
  Winner=FindWinner();
  printf("Responding neuron %d is of class %d \n",Winner,zClass[Winner]);
  printf("The desired class for pattern %d is: %d\n",pat,Pattern->QueryClass(pat));
  printf("The distances to each of the output layer neurons are:\n");
```

```
for (i=0;i<YoutSize ; i++) {
  printf("distance from pattern %d to neuron %d is: %f\n",pat,i,EucNorm(i));
   } /* endfor */
 printf("\n");
 }
}

void LVQ::Train(int Winner, int pat){
 int c,k;
c=Pattern->QueryClass(pat);
if (c==zClass[Winner]) {
  for (k=0; k<YinSize; k++){
    W[k][Winner]=W[k][Winner]+eta*(Yin[k]-W[k][Winner]);
    } /*endfor*/
  }
 else {
  for (k=0; k<YinSize; k++){
    W[k][Winner]=W[k][Winner]-eta*(Yin[k]-W[k][Winner]);
    } /*endfor*/
  } /* endif */
 }

int LVQ::FindWinner(){
  int i;
  double d,best;
  int Winner;
 best=1.0e99;
 Winner=-1;
 for (i=0; i<YoutSize; i++){
  d=EucNorm(i);
  if (d<best) {
     best=d;
     Winner=i;
     } // endif
   } // endfor
  return Winner;
  }

 double LVQ::EucNorm(int x){                    // Calc Euclidean norm of vector dif
 int i;
 double dist;
 dist=0;
 for (i=0; i< YinSize;i++){
   dist += (W[i][x]-Yin[i]) * (W[i][x]-Yin[i]);
    } /* endfor */
  dist=sqrt(dist);
  return dist;
  }

  //=================================================================
  // GLOBAL OBJECTS
  //=================================================================
  TPATTERN InPat;
  TPATTERN InPat2;
  LVQ    net;
```

```
//===================================================================
// Main()
//===================================================================

main(int argc, char *argv[]) {
//srand(17);
if (argc>1) {
  InPat.GetPatterns(argv[1]);           //Establish training pattern
  net.SetPattern(&InPat);               //Inform the net about the pattern
// net.SetParms(8, 0.2500);             //Init fm parms
  net.SetParms(4, 0.2500);              //Init fm parms
// net.SetParms(4, 0.3000);             //Init fm parms
  net.RunTrn();                         //Run the FM w/ training enabled
  InPat2.GetPatterns(argv[2]);          //Establish test pattern
  net.SetPattern(&InPat2);              //Inform the net about the new pattern
  net.Run();
  }
else {
  printf("USAGE: LVQ PATTERN_FILE");
  }
}
```

8.6.1 LVQ Example

To illustrate learning vector quantization we have chosen a character recognition example. Eight training patterns corresponding to four distinct classes will be applied to the LVQ network. The training patterns and their corresponding classes are illustrated in Figure 8.20. After training is complete the network will be exposed to the two test vectors shown in Figure 8.21. The network topology for this example utilizes an input layer of x neurons as required by the dimensionality of the input. The output layer consists of a one-dimensional array of four neurons, one neuron per class. LVQ networks would typically contain more than one neuron per class. In our case one is sufficient due to the limited scope of the example. The input training vectors corresponding to Figure 8.20 follow.

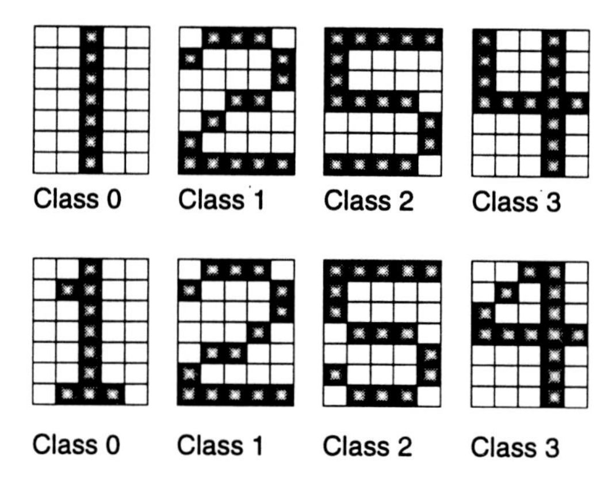

FIGURE 8.20 Example training patterns for an LVQ network with corresponding class designations.

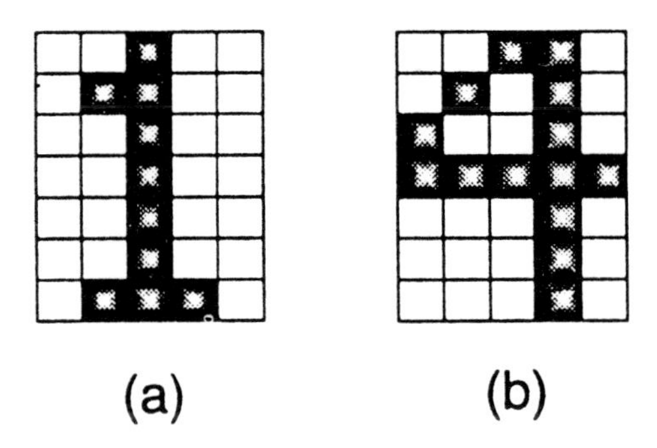

FIGURE 8.21 Example test vectors. The corresponding class designations are the anticipated test result.

We are now ready to train the network. For each pattern, the winning output layer neuron is shown. Next, the winning neuron class is shown along side of the pattern class. Finally the weight or codebook vector is shown for the winning neuron after training. The calculations for the first epoch go as follows:

```
EPOCH=0 η=0.250000

winner=1/pat=0
winner class=1/pat class=0
W[1]=0.072737 0.048654 -0.205546 0.025029 0.103366
     0.051988 0.057939 -0.127602 0.015805 0.026578
     0.119808 0.092185 -0.198870 0.097514 0.094737
     0.119606 0.003510 -0.210158 0.094619 0.030374
     0.073695 0.005425 -0.130493 0.039892 0.007420
     0.055235 0.114380 -0.178468 0.014855 0.071223
     0.031507 0.061983 -0.220408 0.059622 0.050760


winner=3/pat=1
winner class=3/pat class=1
W[3]=  0.025998 -0.232505 -0.213176 -0.149648  0.027367
      -0.179613  0.089450  0.024693  0.123730 -0.218745
       0.053827  0.094409  0.107616  0.111850 -0.127739
       0.049425  0.054025 -0.234107 -0.192789  0.029729
       0.123257 -0.168397  0.075529  0.030236  0.056860
      -0.151254  0.009850  0.059549  0.019074  0.030717
      -0.131874 -0.173246 -0.126476 -0.190340 -0.150037
```

```
winner=0/pat=2
winner class=0/pat class=2
W[0]=- 0.185766 -0.228034 -0.211421 -0.183184 -0.131546
       -0.228534  0.087779  0.028302  0.061846  0.015587
       -0.239513  0.048704  0.034654  0.046007  0.122932
       -0.183077 -0.154290 -0.169191 -0.154107  0.097530
        0.102870  0.018990  0.078185  0.039335 -0.206637
        0.114650  0.064970  0.050146  0.075846 -0.151822
       -0.133557 -0.141259 -0.141682 -0.165685  0.094802


winner=2/pat=3
winner class=2/pat class=3
W[2]=-0.140873  0.053369  0.044778 -0.202254  0.005394
     -0.229926  0.065294  0.087073 -0.237865  0.050104
     -0.153321  0.030602  0.042852 -0.221252  0.037233
     -0.211932 -0.139100 -0.245418 -0.168607 -0.200175
      0.084536  0.091575  0.117225 -0.220840  0.104812
      0.120903  0.097331  0.053938 -0.165738  0.101173
      0.019845  0.034986  0.016915 -0.141976  0.093776


winner=2/pat=4
winner class=2/pat class=0
W[2]=-0.176092  0.066712 -0.194027 -0.252817  0.006743
     -0.287408 -0.168382 -0.141159 -0.297331  0.062630
     -0.191651  0.038253 -0.196435 -0.276564  0.046541
     -0.264915 -0.173874 -0.556773 -0.210759 -0.250218
      0.105670  0.114468 -0.103468 -0.276049  0.131015
      0.151129  0.121664 -0.182578 -0.207173  0.126466
      0.024806 -0.206268 -0.228856 -0.427470  0.117220


winner=1/pat=5
winner class=1/pat class=1
W[1]= 0.054553  0.286491  0.095841  0.268772  0.077525
      0.288991  0.043455 -0.095701  0.011854  0.269933
      0.089856  0.069139 -0.149153  0.073136  0.321053
      0.089704  0.002632 -0.157619  0.320964  0.022780
      0.055271  0.254068  0.152130  0.029919  0.005565
      0.291426  0.085785 -0.133851  0.011141  0.053417
      0.273630  0.296487  0.084694  0.294716  0.288070


winner=1/pat=6
winner class=1/pat class=2
W[1]=-0.181809  0.108113 -0.130199  0.085965 -0.153094
      0.111239  0.054318 -0.119627  0.014817  0.337417
     -0.137680  0.086423 -0.186441  0.091420  0.401316
      0.112130 -0.246710 -0.447023  0.151205  0.028475
      0.069089  0.317586  0.190163  0.037398 -0243044
      0.114283  0.107231 -0.167314  0.013926 -0.183229
      0.342037  0.120609 -0.144133  0.118395  0.360087
```

```
winner=3/pat=7
winner class=3/pat class=3
W[3]= 0.019498 -0.174379  0.090118  0.137764  0.020526
     -0.134710  0.317087  0.018520  0.342797 -0.164059
      0.290370  0.070807  0.080712  0.333888 -0.095804
      0.287069  0.290519  0.074419  0.105408  0.272297
      0.092442 -0.126298  0.056647  0.272677  0.042645
     -0.113440  0.007387  0.044662  0.264306  0.023038
     -0.098906 -0.129934 -0.094857  0.107245 -0.112527
```

Notice that during this first epoch the class of the winning neuron frequently does not match the pattern training class. In such cases the weights are trained such that the winning neuron weight vector is made more dissimilar to the pattern vector as specified in equation 21.

By epoch 10 the training has progressed to the point where the winning neuron class most frequently agrees with the training vector class. The calculations for epoch 10 go as follows:

```
EPOCH=10 η=0.230000

winner=0/pat=0
winner class=0/pat class=0
W[0]=-0.182456 -0.183187  0.979025 -0.182411 -0.181517
     -0.183196  0.532081  1.162414  0.001071  0.000270
     -0.183386  0.000843  1.162524  0.000797  0.002129
     -0.003170 -0.181911  0.979756 -0.181907  0.001689
      0.001781  0.000329  1.163278  0.000681 -0.182817
     -0.177254  0.001125  1.162793  0.001313 -0.181868
     -0.002312  0.348876  0.980232  0.348453  0.001641


winner=1/pat=1
winner class=1/pat class=1
W[1]=-0.326132  0.977788  0.971853  0.977236 -0.325417
      0.977866  0.001353 -0.002979  0.000369  1.305103
     -0.325033  0.002152 -0.004643  0.002277  1.306695
     -0.224235 -0.327749  0.389607  0.978861  0.000709
      0.001721  1.304609  0.579091  0.000931 -0.327657
      1.204969  0.002671 -0.004167  0.000347 -0.326168
      1.078190  0.978099  0.971506  0.978044  1.305668


winner=1/pat=2
winner class=1/pat class=2
W[1]=-0.631143  0.972679  0.965379  0.972001 -0.630263
      0.972775  0.001664 -0.003664  0.000454  1.605277
     -0.629791  0.002647 -0.005711  0.002800  1.607234
     -0.505809 -0.633131  0.249216  0.973999  0.000872
      0.002116  1.604669  0.712283  0.001146 -0.633019
      1.482112  0.003285 -0.005125  0.000427 -0.631186
      1.096174  0.973062  0.964952  0.972994  1.605971
```

```
  winner=3/pat=3
winner class=3/pat class=3
W[3]=  0.385936 -0.341704  0.250249  0.979510 -0.335685
       0.381174  0.599714 -0.005565  1.322161 -0.005066
       0.984223  0.002187 -0.003644  1.321885 -0.002959
       1.158810  0.984227  0.971417  0.978511  1.319983
       0.002855 -0.003900 -0.004387  1.319995 -0.335002
      -0.178193  0.000228 -0.004757  1.319737 -0.335608
      -0.164684 -0.340332 -0.345385  0.978567 -0.003475


winner=0/pat=4
winner class=0/pat class=0
W[0]=-0.140491 -0.141054  0.983849 -0.140456 -0.139768
     -0.141061  0.639702  1.125059  0.000825  0.000208
     -0.141207  0.000649  1.125144  0.000613  0.001639
     -0.002441 -0.140071  0.984412 -0.140069  0.001300
      0.001371  0.000253  1.125724  0.000524 -0.140769
     -0.136486  0.000866  1.125350  0.001011 -0.140038
     -0.001781  0.498635  0.984779  0.498309  0.001264

  winner=1/pat=5
winner class=1/pat class=1
W[1]=-0.485980  0.978963  0.973342  0.978441 -0.485303
      0.979037  0.001281 -0.002822  0.000349  1.466063
     -0.484939  0.002038 -0.004398  0.002156  1.467570
     -0.389473 -0.487511  0.191897  0.979979  0.000672
      0.001630  1.465595  0.778458  0.000882 -0.487424
      1.371227  0.002529 -0.003946  0.000328 -0.486014
      1.074054  0.979258  0.973013  0.979205  1.466598

  winner=0/pat=6
winner class=0/pat class=2
W[0]=-0.402804 -0.403497 0.980134 -0.402761 -0.401914
     -0.403505  0.786834  1.383823  0.001014  0.000256
     -0.403685  0.000799  1.383927  0.000754  0.002016
     -0.003002 -0.402287  0.980827 -0.402284  0.001599
      0.001687  0.000311  1.384641  0.000645 -0.403146
     -0.397877  0.001065  1.384181  0.001244 -0.402247
     -0.002190  0.383321  0.981278  0.382920  0.001555

  winner=3/pat=7
winner class=3/pat class=3
W[3]=0.297171 -0.263112  0.422692  0.984223 -0.258478
     0.293504  0.691780 -0.004285  1.248064 -0.003901
     0.987851  0.001684 -0.002806  1.247852 -0.002278
     1.122284  0.987855  0.977991  0.983453  1.246387
     0.002198 -0.003003 -0.003378  1.246396 -0.257952
    -0.137209  0.000176 -0.003663  1.246197 -0.258418
    -0.126806 -0.262055 -0.265946  0.983497 -0.002676
```

By epoch 20 the winning neuron and pattern vector classes agree throughout the epoch. Training during the 20th epoch is as follows:

```
EPOCH=20 η=0.210000

winner=0/pat=0
winner class=0/pat class=0
W[0]=-0.065621 -0.065633  0.999673 -0.065621 -0.065607
     -0.065633  0.475598  1.065309  0.000017  0.000004
     -0.065636  0.000013  1.065311  0.000012  0.000033
     -0.031805 -0.065613  0.999684 -0.065613  0.000026
      0.000028  0.000005  1.065322  0.000011 -0.065627
     -0.033785  0.000018  1.065315  0.000020 -0.065612
     -0.031792  0.409964  0.999692  0.409958  0.000026

winner=1/pat=1
winner class=1/pat class=1
W[1]=-0.121791  0.999482  0.999344  0.999469 -0.121774
      0.999484  0.000032 -0.000069  0.000009  1.121301
     -0.121765  0.000050 -0.000108  0.000053  1.121338
     -0.040615 -0.121829  0.493460  0.999507  0.000017
      0.000040  1.121289  0.505810  0.000022 -0.121827
      1.040166  0.000062 -0.000097  0.000008 -0.121792
      1.080624  0.999489  0.999336  0.999488  1.121314

winner=2/pat=2
winner class=2/pat class=2
W[2]= 0.720564  0.778254  0.716303  0.702335  0.764005
      0.694116 -0.040007 -0.033539 -0.070645  0.014881
      0.716867  0.009089 -0.046672 -0.065711  0.011058
      0.361543  0.721091  0.630116  0.712327 -0.059451
      0.025107  0.027197 -0.024584 -0.065588  0.793532
      0.373825  0.028907 -0.043380 -0.049224  0.792451
      0.430380  0.713395  0.708028  0.660838  0.027851

winner=3/pat=3
winner class=3/pat class=3
W[3]=0.510088 -0.086869  0.402623  0.999611 -0.086754
     0.509998  0.489639 -0.000106  1.086498 -0.000096
     0.999700  0.000042 -0.000069  1.086492 -0.000056
     1.013832  0.999701  0.999457  0.999592  1.086456
     0.000054 -0.000074 -0.000083  1.086456 -0.086741
    -0.014200  0.000004 -0.000090  1.086452 -0.086753
    -0.072690 -0.086843 -0.086938  0.999593 -0.000066

winner=0/pat=4
winner class=0/pat class=0
W[0]=-0.051841 -0.051850  0.999742 -0.051840 -0.051829
     -0.051850  0.585722  1.051594  0.000013  0.000003
     -0.051852  0.000010  1.051595  0.000010  0.000026
     -0.025126 -0.051834  0.999751 -0.051834  0.000021
      0.000022  0.000004  1.051605  0.000008 -0.051845
     -0.026690  0.000014  1.051599  0.000016 -0.051834
     -0.025115  0.533872  0.999757  0.533867  0.000020
```

```
winner=1/pat=5
winner class=1/pat class=1
W[1]=-0.096215  0.999591  0.999482  0.999581 -0.096202
      0.999592  0.000025 -0.000055  0.000007  1.095828
     -0.096195  0.000040 -0.000086  0.000042  1.095857
     -0.032086 -0.096245  0.389834  0.999611  0.000013
      0.000032  1.095818  0.609590  0.000017 -0.096243
      1.031731  0.000049 -0.000077  0.000006 -0.096216
      1.063693  0.999597  0.999475  0.999596  1.095838


winner=2/pat=6
winner class=2/pat class=2
W[2]= 0.779246  0.824820  0.775879  0.764844  0.813564
      0.758352 -0.031606 -0.026496 -0.055810  0.011756
      0.776325  0.007180 -0.036871 -0.051912  0.008736
      0.285619  0.779662  0.707791  0.772739 -0.046966
      0.019834  0.021486 -0.019421 -0.051815  0.836890
      0.505322  0.022837 -0.034270 -0.038887  0.836036
      0.340000  0.773582  0.769342  0.732062  0.022002


winner=3/pat=7
winner class=3/pat class=3
W[3]= 0.402970 -0.068626  0.528072  0.999693 -0.068536
      0.402898  0.596815 -0.000083  1.068333 -0.000076
      0.999763  0.000033 -0.000055  1.068329 -0.000044
      1.010928  0.999763  0.999571  0.999678  1.068300
      0.000043 -0.000058 -0.000066  1.068301 -0.068526
     -0.011218  0.000003 -0.000071  1.068297 -0.068535
     -0.057425 -0.068606 -0.068681  0.999679 -0.000052
```

At this point training must proceed until it has advanced sufficiently that the cluster centers for each class have been formed. Notice that as training progresses η is reduced. The calculations for the final epoch are:

```
EPOCH=1500  eta=0.001000


winner=0/pat=0
winner class=0/pat class=0
W[0] =-0.000000 -0.000000  1.000000 -0.000000 -0.000000
      -0.000000  0.500377  1.000000  0.000000  0.000000
      -0.000000  0.000000  1.000000  0.000000  0.000000
      -0.000000 -0.000000  1.000000 -0.000000  0.000000
       0.000000  0.000000  1.000000  0.000000 -0.000000
      -0.000000  0.000000  1.000000  0.000000 -0.000000
      -0.000000  0.500377  1.000000  0.500377  0.000000
winner=1/pat=1
winner class=1/pat class=1
W[1]=-0.000000  1.000000  1.000000  1.000000 -0.000000
      1.000000  0.000000 -0.000000  0.000000  1.000000
     -0.000000  0.000000 -0.000000  0.000000  1.000000
     -0.000000 -0.000000  0.499623  1.000000  0.000000
      0.000000  1.000000  0.500377  0.000000 -0.000000
      1.000000  0.000000 -0.000000  0.000000 -0.000000
      1.000000  1.000000  1.000000  1.000000  1.000000
```

```
winner=2/pat=2
winner class=2/pat class=2
W[2]= 1.000000 1.000000 1.000000 1.000000 1.000000
      1.000000 -0.000000 -0.000000 -0.000000 0.000000
      1.000000 0.000000 -0.000000 -0.000000 0.000000
      0.499623 1.000000 1.000000 1.000000 -0.000000
      0.000000 0.000000 -0.000000 -0.000000 1.000000
      0.500377 0.000000 -0.000000 -0.000000 1.000000
      0.499623 1.000000 1.000000 1.000000 0.000000

winner=3/pat=3
winner class=3/pat class=3
W[3]= 0.499623 -0.000000 0.500377 1.000000 -0.000000
      0.499623 0.500377 -0.000000 1.000000 -0.000000
      1.000000 0.000000 -0.000000 1.000000 -0.000000
      1.000000 1.000000 1.000000 1.000000 1.000000
      0.000000 -0.000000 -0.000000 1.000000 -0.000000
      -0.000000 0.000000 -0.000000 1.000000 -0.000000
      -0.000000 -0.000000 -0.000000 1.000000 -0.000000

winner=0/pat=4
winner class=0/pat class=0
W[0]=-0.000000 -0.000000 1.000000 -0.000000 -0.000000
      -0.000000 0.500877 1.000000 0.000000 0.000000
      -0.000000 0.000000 1.000000 0.000000 0.000000
      -0.000000 -0.000000 1.000000 -0.000000 0.000000
      0.000000 0.000000 1.000000 0.000000 -0.000000
      -0.000000 0.000000 1.000000 0.000000 -0.000000
      -0.000000 0.500877 1.000000 0.500877 0.000000

winner=1/pat=5
winner class=1/pat class=1
W[1]=-0.000000 1.000000 1.000000 1.000000 -0.000000
      1.000000 0.000000 -0.000000 0.000000 1.000000
      -0.000000 0.000000 -0.000000 0.000000 1.000000
      -0.000000 -0.000000 0.499123 1.000000 0.000000
      0.000000 1.000000 0.500877 0.000000 -0.000000
      1.000000 0.000000 -0.000000 0.000000 -0.000000
      1.000000 1.000000 1.000000 1.000000 1.000000
winner=2/pat=6
winner class=2/pat class=2
W[2]= 1.000000 1.000000 1.000000 1.000000 1.000000
      1.000000 -0.000000 -0.000000 -0.000000 0.000000
      1.000000 0.000000 -0.000000 -0.000000 0.000000
      0.499123 1.000000 1.000000 1.000000 -0.000000
      0.000000 0.000000 -0.000000 -0.000000 1.000000
      0.500877 0.000000 -0.000000 -0.000000 1.000000
      0.499123 1.000000 1.000000 1.000000 0.000000
```

```
winner=3/pat=7
winner class=3/pat class=3
W[3]= 0.499123 -0.000000  0.500877  1.000000 -0.000000
      0.499123  0.500877 -0.000000  1.000000 -0.000000
      1.000000  0.000000 -0.000000  1.000000 -0.000000
      1.000000  1.000000  1.000000  1.000000  1.000000
      0.000000 -0.000000 -0.000000  1.000000 -0.000000
     -0.000000  0.000000 -0.000000  1.000000 -0.000000
     -0.000000 -0.000000 -0.000000  1.000000 -0.000000
```

Having trained the network we may now apply test vectors. We begin by applying the following pattern corresponding to Figure 8.21 (a):

```
0.0 0.0 1.0 0.0 0.0
0.0 1.0 1.0 0.0 0.0
0.0 0.0 1.0 0.0 0.0
0.0 0.0 1.0 0.0 0.0
0.0 0.0 1.0 0.0 0.0
0.0 0.0 1.0 0.0 0.0
0.0 1.0 1.0 1.0 0.0
```

The network calculates its response to this test vector below. The distances to each of the output layer neurons are

```
The distance from pattern 0 to neuron 0 is: 0.864507
The distance from pattern 0 to neuron 1 is: 3.807887
The distance from pattern 0 to neuron 2 is: 3.968517
The distance from pattern 0 to neuron 3 is: 4.122681
```

Neuron 0 (representing class 0) responds to the pattern since it has the shortest Euclidean distance. The desired class for pattern 0 is also 0. The network has thus correctly classified the pattern as belonging to class 0.

We now apply the second test vector corresponding to Figure 8.21 (b):

```
0.0 0.0 1.0 1.0 0.0
0.0 1.0 0.0 1.0 0.0
1.0 0.0 0.0 1.0 0.0
1.0 1.0 1.0 1.0 1.0
0.0 0.0 0.0 1.0 0.0
0.0 0.0 0.0 1.0 0.0
0.0 0.0 0.0 1.0 0.0
```

The network calculates its response to this test vector as follows:

```
Responding neuron 3 is of class 3
The desired class for pattern 1 is 3
The distances to each of the output layer neurons are
```

distance from pattern 1 to neuron 0 is: 3.968517
distance from pattern 1 to neuron 1 is: 4.416079
distance from pattern 1 to neuron 2 is: 3.840687
distance from pattern 1 to neuron 3 is: 0.998247

Neuron 3 (representing class 3) responds to the pattern since it has the shortest Euclidean distance. The desired class for pattern 2 is class 3. The network has thus correctly classified the pattern as belonging to class 3. Again, the test vector is correctly evaluated.

8.7 VARIATIONS ON LVQ

We will now look at some variations on the basic LVQ method we have just studied. Specifically, we will cover several improvements to the algorithm proposed by Kohonen [1990a] termed LVQ2, LVQ2.1, and LVQ3. All of these devolve around the manner in which the network is trained. Otherwise the network structure remains in tact. In particular, training will no longer be exclusively restricted to the winning neuron.

8.7.1 LVQ2

Learning vector quantization can be extended by training both the winning vector and the first runner-up under appropriate conditions. The principal idea is that when the winning neuron does not represent the correct category and the first runner-up does represent the correct category, we may wish to train both. The winner is moved away from the exemplar pattern vector and the runner-up is trained to be closer. The winner is punished and the closest vector in the correct category is rewarded. LVQ2 imposes the further condition that the distance, d_c, from the input vector to the winner not be too different from the distance, d_r, between the input vector and the runner-up. (When the above conditions are not met, training proceeds exactly as before for LVQ.)

The condition that d_r and d_c not be too different can be expressed in terms of a window within which the condition is considered to be met. The window is defined as follows:

$$\frac{d_c}{d_r} = 1 - \varepsilon$$

$$\frac{d_r}{d_c} = 1 + \varepsilon \tag{22}$$

When all of the above conditions are met, training under LVQ2 proceeds as follows:

$$\mathbf{W}_R^{\text{new}} = \mathbf{W}_R^{\text{old}} + \eta\left(\mathbf{x}^{(p)} - \mathbf{W}_R^{\text{old}}\right)$$

$$\mathbf{W}_C^{\text{new}} = \mathbf{W}_C^{\text{old}} - \eta\left(\mathbf{x}^{(p)} - \mathbf{W}_C^{\text{old}}\right) \tag{23}$$

where $\mathbf{W}_R$ is the reference vector for the runner-up belonging to the same class as the training vector, $\mathbf{x}^{(p)}$, and $\mathbf{W}_C$ is the reference vector for the winning neuron (which does not belong to the correct class for $\mathbf{x}^{(p)}$).

8.7.2 LVQ2.1

In a further modification of learning vector quantization [Kohonen, 1990a] the requirement (in LVQ2) that the winning neuron not be of the correct class for the training vector is dropped. Here the best two reference vectors are trained, provided that one of them belongs to the correct class and the other does not, without regard for which, best or runner-up, represents the correct class. LCQ2.1 does keep a windowing requirement in a somewhat modified form. The windowing requirement is as follows:

$$\min\left[\frac{d_{C_1}}{d_{C_2}}, \frac{d_{C_2}}{d_{C_1}}\right] > 1 - \varepsilon \tag{24}$$

and

$$\max\left[\frac{d_{C_1}}{d_{C_2}}, \frac{d_{C_2}}{d_{C_1}}\right] < 1 + \varepsilon \tag{25}$$

where d_{c1} is the distance from $\mathbf{x}^{(p)}$ to $\mathbf{W}_{C1}$ and d_{C2} is the distance from $\mathbf{x}^{(p)}$ to $\mathbf{W}_{C2}$. $\mathbf{W}_{C1}$ and $\mathbf{W}_{C2}$ are the two best reference vectors. With the above conditions met, training proceeds as follows:

$$\mathbf{W}_{C_1}^{\text{new}} = \mathbf{W}_{C_1}^{\text{old}} + \eta\left(\mathbf{x}^{(p)} - \mathbf{W}_{C_1}^{\text{old}}\right)$$

$$\mathbf{W}_{C_2}^{\text{new}} = \mathbf{W}_{C_2}^{\text{old}} - \eta\left(\mathbf{x}^{(p)} - \mathbf{W}_{C_2}^{\text{old}}\right) \tag{26}$$

where we have taken $\mathbf{W}_{C1}$ as the neuron representing the correct category.

8.7.3 LVQ3

Another variation on LVQ due to Kohonen [1990a] is termed LVQ3. The LVQ3 algorithm again varies the windowing condition and extends the circumstances under which both the winner and the runner-up neurons are trained. Where the winner and runner-up neurons belong to different classes, training proceeds as in LVQ2.1. When both winner and runner-up neurons belong to the same class and the windowing criteria are met, both winner and runner-up are trained. The windowing criteria in LVQ3 is as follows:

$$\min\left[\frac{d_{C_1}}{d_{C_2}}, \frac{d_{C_2}}{d_{C_1}}\right] > \frac{1-\varepsilon}{1+\varepsilon} \tag{27}$$

When the above criteria are met, training is accomplished as follows:

$$\mathbf{W}_C^{\text{new}} = \mathbf{W}_C^{\text{old}} + \beta\left(\mathbf{x}^{(p)} - \mathbf{W}_C^{\text{old}}\right) \tag{28}$$

where

$$\beta = m\eta \tag{29}$$

The multiplier, m, is in the range $0.1 < m < 0.5$. A typical value of ε is about 0.2. The multiplier, m, may also be varied such that a narrower window leads to a smaller multiplier.

8.7.4 A Final Variation of LVQ

Yizhak and Chevalier [1991] have proposed a supervised learning variation of the Kohonen network (that is different from the LVQ learning) for handwritten digit recognition. In this variation two sets of neurons are connected to a Kohonen-type layer. The first is the conventional input layer, I. The second, T, serves to distribute the desired output/target pattern to the Kohonen layer during training and then later will serve as the output layer of the network. There are two distinct groups of weights, one for the I layer and one for the T layer.

The I layer always receives the input pattern vector, and its associated weights are designated $\mathbf{W}_{ij}$. Weights associated with the I layer, $\mathbf{W}_{ij}$, are trained relative to the input pattern vector. During training the T layer behaves as an input and receives the desired output pattern. Thus, the T layer weights, designated $\mathbf{W}_{tk}$, are trained relative to the desired output.

After training, during the recognition phase, the sense of the T layer is reversed and the T layer neurons become outputs with weights given by $\mathbf{W}_{kt}$. Training proceeds with neurons in the neighborhood of the winning Kohonen layer neuron being updated along with the winner. A radius defining this neighborhood is established by the strength of the winning neuron response. The more strongly the winner has responded to the input vector, the smaller the neighborhood (because there already exists a neuron which is classifying the vector well).

A further enhancement presented in this implementation is a criterion for uncertainty in the network results during operational mode. Three parameters are established, these being σ_1, σ_2, and σ_3. The network is only considered to have recognized a character if criteria involving all of these parameters or thresholds are met. (Otherwise the network is said to be in the confused state.) σ_1 is a threshold such that the strength of the winning output neuron must be greater than σ_1. The strength of the second best neuron must be less than σ_2. Finally, the difference in strength of the best and runner-up neurons must be greater than σ_3.

The sample size used in this study is small, 735 training patterns and 265 test patterns. In the best reported configuration recognition for the training set and test set was 90.7 and 75.5%, respectively. For the same configuration the network was confused by 4.9 and 14.3% of the input patterns for the training set and the test set, respectively.

REFERENCES AND BIBLIOGRAPHY

Caudill, M., "A little knowledge is a dangerous thing", *AI Expert,* pp. 16–22, June 1993.

Baykal, N. and Yalabik, N., "Object orientation detection and character recognition using optimal feed-forward network and Kohonen's feature map," *SPIE,* vol. 1709, pp. 292–303, 1992.

Kohonen, T., "Self-organized formation of topographically correct feature maps." *Biol. Cybern.*, Vol. 43, pp. 59–69, 1982.

Kohonen, T., *Self-organization and Associative Memory,* 3rd ed., Springer-Verlag, Berlin, 1989.

Kohonen, T., "Improved versions of learning vector quantizations," *Int. Joint Conf. Neural Networks,* I, pp. 545–550, 1990(a).

Kohonen, T., "The self-organizing map," *Proc. IEEE,* vol. 78 no. 9, 1464–1480, 1990(b).

Yizhak, I. and Chevalier, R. C., "Handwritten digit recognition by a supervised Kohonen-like learning algorithm," *1991 IEEE Joint Conf. Neural Networks IJCNN,* pp. 1576–1581, 1991.

9 Neural Associative Memories and Hopfield Networks

9.1 GENERAL

In this chapter we address the subject of neural associative memories. As the name implies neural associative memories operate by learning an association between pattern pairs. That is, when an input pattern is presented to these networks, the network provides the associated response. As such we can see that such memories will be content addressable. Moreover, there is a sense in which we may think of these associative neural networks as learning (in some sense) as humans do. When the network sees a pattern that is different from all learned exemplar patterns, the network will yield a response associated with the exemplar pattern to which the input pattern was (by some criteria) most similar. It is in this property that we find an analogy between the function of such networks and biological learning. When a human sees a pattern representing, for example, the letter "A", the pattern does not need to be identical to some specimen pattern of the letter "A" that he or she has previously learned. To be recognized, it is enough that the pattern is sufficiently similar to our experience to remind us of other A's we have encountered (i.e., it appeals to our concept of A-ness). Beyond this rough and tenuous analogy we do not address here the extent to which such network associative memories may be plausible as simplified models of either human or biological memory. Readers interested in more discussion on biological plausibility of such models may refer to Hopfield [1982, 1984], Kohonen [1977], or Levine [1990].

Pioneering work on associative memory networks was done by several researchers including Nakano [1972], Kohonen [1977, 1980], and Anderson and Bower [1973]. Hopfield's memory models [Hopfield, 1982, 1984] played an important role in the current resurgence of interest in artificial neural networks. His discrete and continuous models along with applications to optimization problems of neural computation by connecting like the traveling salesman problem (TSP) have had a profound, stimulating effect on the scientific community in the field of neural network models [Hopfield and Tank, 1985; Tank and Hopfield, 1987]. Several textbooks on neural networks (for example, see Zurada [1992], Freeman and Skapura [1991], Fausett [1994], etc.) deal with all these topics in great detail. In this text we restrict the discussion to a detailed view of the discrete Hopfield memory model and its performance as an associative memory.

In typical digital computer memories, data are accessed by providing the relevant addresses in the memory. In contrast, usable addressing schemes do not exist for associate memories because the stored information is spatially distributed and superimposed throughout the network. Thus the digital computer memories are called address addressable memories, while associative memories are referred to as content addressable memories (CAMs).

By way of taxonomy, neural associative memories may be either autoassociative or heteroassociative. For an autoassociative network memory, the training input and the target output are identical. The network may thus be thought of as memorizing a pattern by associating that pattern with itself. We can obtain a matrix autoassociative memory by forming an outerproduct of the pattern vector with itself. If we impose many such arrays on top of each other, we can design a memory that will store several patterns, and provide an autoassociative recall. By contrast, a heteroassociative network memory associates an input pattern with a different and distinct training pattern provided by a teacher.

Alternatively, networks can be classified as feed-forward or recurrent. In feed-forward networks information flows only from input to output. Recurrent networks contain connectors among the neurons which form loops. Therefore they are iterative and converge to a minimum energy state corresponding to the desired association.

In this chapter several of the best known associative network memories will be considered. We begin the discussion with the linear associative memory (LAM) which is a single-layer feed-forward network. We also provide an example of an autoassociative LAM.

Hopfield networks, perhaps the best known of the associative neural network memories, are discussed in section 9.3. Hopfield networks are recurrent and autoassociative. In such autoassociative memories a number of different patterns can be stored, such that if any one of them is presented (i.e., memory is set in one of the stored states), it will remain stable in that state. When a distorted version of a stored pattern is presented, it will evolve from that state to the stable stored state. The convergence properties and storage capacity of the Hopfield networks are examined. Finally, we present the bidirectional associative memory (BAM) in sections 9.7 and 9.8 to provide an example of recurrent, heteroassociative type of memory.

The LAM model is also called feed-forward type, while Hopfield networks and BAM are classified as feedback type, since they have feedback connections that facilitate recurrent operations. The recurrent models, unlike the feed-forward models, require many iterations before retrieving a final pattern.

The presentation of associative memories in this chapter is by no means exhaustive. The intention is to present the basic concepts along with the potential benefits and limitations of such memories.

9.2 LINEAR ASSOCIATIVE MEMORY (LAM)

The linear associative memory is a single-layer feed-forward network providing a mapping from an input space of dimension K to an output space of dimension N. The objective of the mapping is to recover the output pattern from full or partial information in the input pattern. The LAM may be autoassociative or heteroassociative.

(By definition, in the autoassociative case the input and output training set patterns will be identical; thus the dimensionality of the input and output space must therefore be identical, $K = N$.) The linear associative memory was invented by several researchers working independently during the period from 1968 through 1972. The articles by Amari [1972], Anderson [1972], and Kohonen [1972] are considered classical references on the subject.

A linear associative memory is depicted in Figure 9.1.

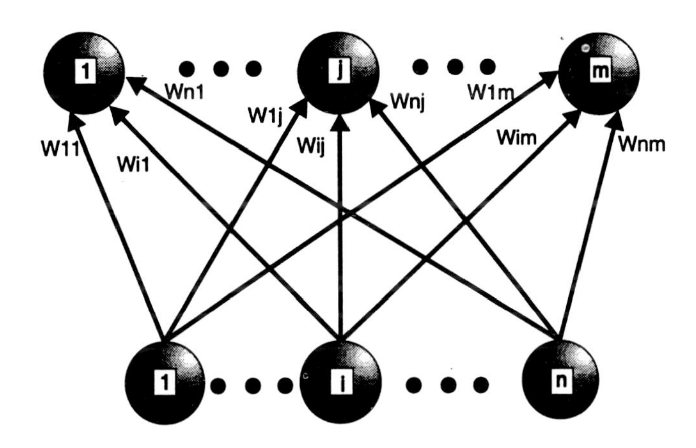

FIGURE 9.1 Architecture of a linear associative memory.

As usual the mapping is carried in the weights, **W**, and thresholds, θ, Equations 1 through 4 specify how the LAM network output is determined.

$$\hat{\mathbf{a}}^{(m)} = \mathbf{W}\mathbf{b}^{(m)} \tag{1}$$

or

$$\hat{a}_i = \sum_j W_{ij} b_j \tag{2}$$

Here **a** represents a vector, a_i represents the ith component of **a**. $\hat{\mathbf{a}}$ represents a real value. a_i is a binary value.

$$a_i^{(m)} = \begin{cases} 1 & \hat{a}_i^{(m)} > \theta_i \\ 0 & \hat{a}_i^{(m)} \geq \theta_i \end{cases} \tag{3}$$

where the input vector **b** and the output vector **a** for the mth pattern are as follows:

$$a^{(m)} = \begin{pmatrix} a_1^{(m)} \\ a_2^{(m)} \\ \vdots \\ a_N^{(m)} \end{pmatrix} \qquad b^{(m)} = \begin{pmatrix} b_1^{(m)} \\ b_2^{(m)} \\ \vdots \\ b_K^{(m)} \end{pmatrix} \tag{4}$$

The LAM network is easily trained in a single pass. Training the LAM network simply involves taking the correlation between the input and output vectors. As shown by equation 5 for bipolar inputs (i.e., inputs of [1, –1].

$$W_{ij} = \sum_m a_i^{(m)} b_j^{(m)} \tag{5}$$

For binary valued inputs [0, 1], the network must be trained by equation 6 which properly takes into account the bias of 0.5 introduced by this representation.

$$W_{ij} = \sum_m \left(2a_i^{(m)} - 1\right)\left(2b_j^{(m)} - 1\right) \tag{6}$$

To produce the binary-valued output, the threshold values, θ, (see equation 3), are calculated as shown in equation 7.

$$\theta_i = \sum_{j=1}^{K} W_{ij} \tag{7}$$

where K is the dimension of the input vector, $\mathbf{b}^{(m)}$.

9.2.1 An Autoassociative LAM Example

To illustrate the operation of the linear associative memory consider the patterns for the digits 1 through 4 shown in Figure 9.2.

These 3 × 5 images in Figure 9.2 can be mapped to 15 dimensional vector representations. Thus, we can design an autoassociative network with 15 input neurons and 15 output neurons. There will be 225 (15 × 15) weights connecting them. Fifteen threshold values will be associated with the output neurons.

In viewing this as a matrix memory, the cross products obtained in formation of outer products become the elements of the matrix. From the neural networks perspective, the cross products can be viewed as the values of the synaptic strengths (weights) linking two respective neurons. Thus, two different but related perspectives exist in two different communities. Kohonen [1977] describes their relationship very well in his book on associative memory.

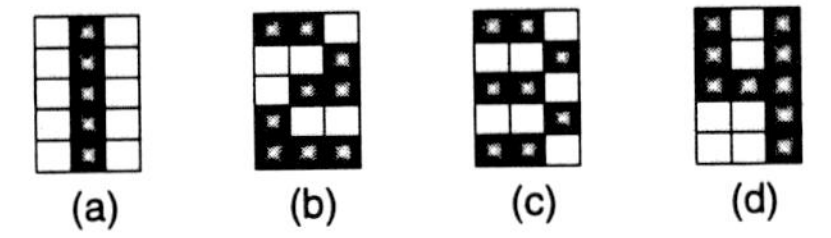

FIGURE 9.2 LAM input patterns.

For the digits shown in Figure 9.2, the corresponding binary-valued vectors are as follows:

$$\mathbf{b}^{(0)} = \begin{pmatrix} 0\\1\\0\\0\\1\\0\\0\\1\\0\\0\\1\\0\\0\\1\\0 \end{pmatrix} \quad \mathbf{b}^{(1)} = \begin{pmatrix} 1\\1\\0\\0\\0\\1\\0\\1\\1\\1\\0\\0\\1\\1\\1 \end{pmatrix} \quad \mathbf{b}^{(2)} = \begin{pmatrix} 1\\1\\0\\0\\0\\1\\1\\1\\0\\0\\0\\1\\1\\1\\0 \end{pmatrix} \quad \mathbf{b}^{(3)} = \begin{pmatrix} 1\\0\\1\\1\\0\\1\\1\\1\\1\\0\\0\\1\\0\\0\\1 \end{pmatrix} \tag{8}$$

The weight matrix corresponding to these training patterns can be calculated using equation 6 as shown below. (Note that the training set for the autoassociative case constrains $a_i = b_i$.)

```
W[0,0]=-1*-1+1*1+1*1+1*1=4          W[0,1]=-1*1+1*1+1*1+1*-1=0
W[0,2]=-1*-1+1*-1+1*-1+1*1=0        W[0,3]=-1*-1+1*-1+1*-1+1*1=0
W[0,4]=-1*1+1*-1+1*-1+1*-1=-4       W[0,5]=-1*-1+1*1+1*1+1*1=4
W[0,6]=-1*-1+1*-1+1*1+1*1=2         W[0,7]=-1*1+1*1+1*1+1*1=2
W[0,8]=-1*-1+1*1+1*-1+1*1=2         W[0,9]=-1*-1+1*1+1*-1+1*-1=0
W[0,10]=-1*1+1*-1+1*-1+1*-1=-4      W[0,11]=-1*-1+1*-1+1*1+1*1=2
W[0,12]=-1*-1+1*1+1*1+1*-1=2        W[0,13]=-1*1+1*1+1*1+1*-1=0
W[0,14]=-1*-1+1*1+1*-1+1*1=2        W[1,0]=1*-1+1*1+1*1+-1*1=0
W[1,1]=1*1+1*1+1*1+-1*-1=4          W[1,2]=1*-1+1*-1+1*-1+-1*1=-4
W[1,3]=1*-1+1*-1+1*-1+-1*1=-4       W[1,4]=1*1+1*-1+1*-1+-1*-1=0
W[1,5]=1*-1+1*1+1*1+-1*1=0          W[1,6]=1*-1+1*-1+1*1+-1*1=-2
W[1,7]=1*1+1*1+1*1+-1*1=2           W[1,8]=1*-1+1*1+1*-1+-1*1=-2
W[1,9]=1*-1+1*1+1*-1+-1*-1=0        W[1,10]=1*1+1*-1+1*-1+-1*-1=0
W[1,11]=1*-1+1*-1+1*1+-1*1=-2       W[1,12]=1*-1+1*1+1*1+-1*-1=2
W[1,13]=1*1+1*1+1*1+-1*-1=4         W[1,14]=1*-1+1*1+1*-1+-1*1=-2
```

```
W[2,0]=-1*-1+-1*1+-1*1+1*1=0
W[2,1]=-1*1+-1*1+-1*1+1*-1=-4
W[2,2]=-1*-1+-1*-1+-1*-1+1*1=4
W[2,3]=-1*-1+-1*-1+-1*-1+1*1=4
W[2,4]=-1*1+-1*-1+-1*-1+1*-1=0
W[2,5]=-1*-1+-1*1+-1*1+1*1=0
W[2,6]=-1*-1+-1*-1+-1*1+1*1=2
W[2,7]=-1*1+-1*1+-1*1+1*1=-2
W[2,8]=-1*-1+-1*1+-1*-1+1*1=2
W[2,9]=-1*-1+-1*1+-1*-1+1*-1=0
W[2,10]=-1*1+-1*-1+-1*-1+1*-1=0
W[2,11]=-1*-1+-1*-1+-1*1+1*1=2
W[2,12]=-1*-1+-1*1+-1*1+1*-1=-2
W[2,13]=-1*1+-1*1+-1*1+1*-1=-4
W[2,14]=-1*-1+-1*1+-1*-1+1*1=2
W[3,0]=-1*-1+-1*1+-1*1+1*1=0
W[3,1]=-1*1+-1*1+-1*1+1*-1=-4
W[3,2]=-1*-1+-1*-1+-1*-1+1*1=4
W[3,3]=-1*-1+-1*-1+-1*-1+1*1=4
W[3,4]=-1*1+-1*-1+-1*-1+1*-1=0
W[3,5]=-1*-1+-1*1+-1*1+1*1=0
W[3,6]=-1*-1+-1*-1+-1*1+1*1=2
W[3,7]=-1*1+-1*1+-1*1+1*1=-2
W[3,8]=-1*-1+-1*1+-1*-1+1*1=2
W[3,9]=-1*-1+-1*1+-1*-1+1*-1=0
W[3,10]=-1*1+-1*-1+-1*-1+1*-1=0
W[3,11]=-1*-1+-1*-1+-1*1+1*1=2
W[3,12]=-1*-1+-1*1+-1*1+1*-1=-2
W[3,13]=-1*1+-1*1+-1*1+1*-1=-4
W[3,14]=-1*-1+-1*1+-1*-1+1*1=2
W[4,0]=1*-1+-1*1+-1*1+-1*1=-4
W[4,1]=1*1+-1*1+-1*1+-1*-1=0
W[4,2]=1*-1+-1*-1+-1*-1+-1*1=0
W[4,3]=1*-1+-1*-1+-1*-1+-1*1=0
W[4,4]=1*1+-1*-1+-1*-1+-1*-1=4
W[4,5]=1*-1+-1*1+-1*1+-1*1=-4
W[4,6]=1*-1+-1*-1+-1*1+-1*1=-2
W[4,7]=1*1+-1*1+-1*1+-1*1=-2
W[4,8]=1*-1+-1*1+-1*-1+-1*1=-2
W[4,9]=1*-1+-1*1+-1*-1+-1*-1=0
W[4,10]=1*1+-1*-1+-1*-1+-1*-1=4
W[4,11]=1*-1+-1*-1+-1*1+-1*1=-2
W[4,12]=1*-1+-1*1+-1*1+-1*-1=-2
W[4,13]=1*1+-1*1+-1*1+-1*-1=0
W[4,14]=1*-1+-1*1+-1*-1+-1*1=-2`
W[5,0]=-1*-1+1*1+1*1+1*1=4
W[5,1]=-1*1+1*1+1*1+1*-1=0
W[5,2]=-1*-1+1*-1+1*-1+1*1=0
W[5,3]=-1*-1+1*-1+1*-1+1*1=0
W[5,4]=-1*1+1*-1+1*-1+1*-1=-4
W[5,5]=-1*-1+1*1+1*1+1*1=4
W[5,6]=-1*-1+1*-1+1*1+1*1=2
W[5,7]=-1*1+1*1+1*1+1*1=2
W[5,8]=-1*-1+1*1+1*-1+1*1=2
W[5,9]=-1*-1+1*1+1*-1+1*-1=0
W[5,10]=-1*1+1*-1+1*-1+1*-1=-4
W[5,11]=-1*-1+1*-1+1*1+1*1=2
W[5,12]=-1*-1+1*1+1*1+1*-1=2
W[5,13]=-1*1+1*1+1*1+1*-1=0
W[5,14]=-1*-1+1*1+1*-1+1*1=2
W[6,0]=-1*-1+-1*1+1*1+1*1=2
W[6,1]=-1*1+-1*1+1*1+1*-1=-2
W[6,2]=-1*-1+-1*-1+1*-1+1*1=2
W[6,3]=-1*-1+-1*-1+1*-1+1*1=2
W[6,4]=-1*1+-1*-1+1*-1+1*-1=-2
W[6,5]=-1*-1+-1*1+1*1+1*1=2
W[6,6]=-1*-1+-1*-1+1*1+1*1=4
W[6,7]=-1*1+-1*1+1*1+1*1=0
W[6,8]=-1*-1+-1*1+1*-1+1*1=0
W[6,9]=-1*-1+-1*1+1*-1+1*-1=-2
W[6,10]=-1*1+-1*-1+1*-1+1*-1=-2
W[6,11]=-1*-1+-1*-1+1*1+1*1=4
W[6,12]=-1*-1+-1*1+1*1+1*-1=0
W[6,13]=-1*1+-1*1+1*1+1*-1=-2
W[6,14]=-1*-1+-1*1+1*-1+1*1=0
W[7,0]=1*-1+1*1+1*1+1*1=2
W[7,1]=1*1+1*1+1*1+1*-1=2
W[7,2]=1*-1+1*-1+1*-1+1*1=-2
W[7,3]=1*-1+1*-1+1*-1+1*1=-2
W[7,4]=1*1+1*-1+1*-1+1*-1=-2
W[7,5]=1*-1+1*1+1*1+1*1=2
W[7,6]=1*-1+1*-1+1*1+1*1=0
W[7,7]=1*1+1*1+1*1+1*1=4
W[7,8]=1*-1+1*1+1*-1+1*1=0
W[7,9]=1*-1+1*1+1*-1+1*-1=-2
W[7,10]=1*1+1*-1+1*-1+1*-1=-2
W[7,11]=1*-1+1*-1+1*1+1*1=0
W[7,12]=1*-1+1*1+1*1+1*-1=0
W[7,13]=1*1+1*1+1*1+1*-1=2
W[7,14]=1*-1+1*1+1*-1+1*1=0
W[8,0]=-1*-1+1*1+-1*1+1*1=2
W[8,1]=-1*1+1*1+-1*1+1*-1=-2
W[8,2]=-1*-1+1*-1+-1*-1+1*1=2
W[8,3]=-1*-1+1*-1+-1*-1+1*1=2
W[8,4]=-1*1+1*-1+-1*-1+1*-1=-2
W[8,5]=-1*-1+1*1+-1*1+1*1=2
W[8,6]=-1*-1+1*-1+-1*1+1*1=0
W[8,7]=-1*1+1*1+-1*1+1*1=0
W[8,8]=-1*-1+1*1+-1*-1+1*1=4
W[8,9]=-1*-1+1*1+-1*-1+1*-1=2
W[8,10]=-1*1+1*-1+-1*-1+1*-1=-2
W[8,11]=-1*-1+1*-1+-1*1+1*1=0
W[8,12]=-1*-1+1*1+-1*1+1*-1=0
W[8,13]=-1*1+1*1+-1*1+1*-1=-2
W[8,14]=-1*-1+1*1+-1*-1+1*1=4
W[9,0]=-1*-1+1*1+-1*1+-1*1=0
W[9,1]=-1*1+1*1+-1*1+-1*-1=0
W[9,2]=-1*-1+1*-1+-1*-1+-1*1=0
W[9,3]=-1*-1+1*-1+-1*-1+-1*1=0
W[9,4]=-1*1+1*-1+-1*-1+-1*-1=0
W[9,5]=-1*-1+1*1+-1*1+-1*1=0
W[9,6]=-1*-1+1*-1+-1*1+-1*1=-2
W[9,7]=-1*1+1*1+-1*1+-1*1=-2
W[9,8]=-1*-1+1*1+-1*-1+-1*1=2
W[9,9]=-1*-1+1*1+-1*-1+-1*-1=4
W[9,10]=-1*1+1*-1+-1*-1+-1*-1=0
W[9,11]=-1*-1+1*-1+-1*1+-1*1=-2
W[9,12]=-1*-1+1*1+-1*1+-1*-1=2
W[9,13]=-1*1+1*1+-1*1+-1*-1=0
W[9,14]=-1*-1+1*1+-1*-1+-1*1=2
```

```
W[10,0]=1*-1+-1*1+-1*1+-1*1=-4
W[10,1]=1*1+-1*1+-1*1+-1*-1=0
W[10,2]=1*-1+-1*-1+-1*-1+-1*1=0
W[10,3]=1*-1+-1*-1+-1*-1+-1*1=0
W[10,4]=1*1+-1*-1+-1*-1+-1*-1=4
W[10,5]=1*-1+-1*1+-1*1+-1*1=-4
W[10,6]=1*-1+-1*-1+-1*1+-1*1=-2
W[10,7]=1*1+-1*1+-1*1+-1*1=-2
W[10,8]=1*-1+-1*1+-1*-1+-1*1=-2
W[10,9]=1*-1+-1*1+-1*-1+-1*-1=0
W[10,10]=1*1+-1*-1+-1*-1+-1*-1=4
W[10,11]=1*-1+-1*-1+-1*1+-1*1=-2
W[10,12]=1*-1+-1*1+-1*1+-1*-1=-2
W[10,13]=1*1+-1*1+-1*1+-1*-1=0
W[10,14]=1*-1+-1*1+-1*-1+-1*1=-2
W[11,0]=-1*-1+-1*1+1*1+1*1=2
W[11,1]=-1*1+-1*1+1*1+1*-1=-2
W[11,2]=-1*-1+-1*-1+1*-1+1*1=2
W[11,3]=-1*-1+-1*-1+1*-1+1*1=2
W[11,4]=-1*1+-1*-1+1*-1+1*-1=-2
W[11,5]=-1*-1+-1*1+1*1+1*1=2
W[11,6]=-1*-1+-1*-1+1*1+1*1=4
W[11,7]=-1*1+-1*1+1*1+1*1=0
W[11,8]=-1*-1+-1*1+1*-1+1*1=0
W[11,9]=-1*-1+-1*1+1*-1+1*-1=-2
W[11,10]=-1*1+-1*-1+1*-1+1*-1=-2
W[11,11]=-1*-1+-1*-1+1*1+1*1=4
W[11,12]=-1*-1+-1*1+1*1+1*-1=0
W[11,13]=-1*1+-1*1+1*1+1*-1=-2
W[11,14]=-1*-1+-1*1+1*-1+1*1=0
W[12,0]=-1*-1+1*1+1*1+-1*1=2
W[12,1]=-1*1+1*1+1*1+-1*-1=2
W[12,2]=-1*-1+1*-1+1*-1+-1*1=-2
W[12,3]=-1*-1+1*-1+1*-1+-1*1=-2
W[12,4]=-1*1+1*-1+1*-1+-1*-1=-2
W[12,5]=-1*-1+1*1+1*1+-1*1=2
W[12,6]=-1*-1+1*-1+1*1+-1*1=0
W[12,7]=-1*1+1*1+1*1+-1*1=0
W[12,8]=-1*-1+1*1+1*-1+-1*1=0
W[12,9]=-1*-1+1*1+1*-1+-1*-1=2
W[12,10]=-1*1+1*-1+1*-1+-1*-1=-2
W[12,11]=-1*-1+1*-1+1*1+-1*1=0
W[12,12]=-1*-1+1*1+1*1+-1*-1=4
W[12,13]=-1*1+1*1+1*1+-1*-1=2
W[12,14]=-1*-1+1*1+1*-1+-1*1=0
W[13,0]=1*-1+1*1+1*1+-1*1=0
W[13,1]=1*1+1*1+1*1+-1*-1=4
W[13,2]=1*-1+1*-1+1*-1+-1*1=-4
W[13,3]=1*-1+1*-1+1*-1+-1*1=-4
W[13,4]=1*1+1*-1+1*-1+-1*-1=0
W[13,5]=1*-1+1*1+1*1+-1*1=0
W[13,6]=1*-1+1*-1+1*1+-1*1=-2
W[13,7]=1*1+1*1+1*1+-1*1=2
W[13,8]=1*-1+1*1+1*-1+-1*1=-2
W[13,9]=1*-1+1*1+1*-1+-1*-1=0
W[13,10]=1*1+1*-1+1*-1+-1*-1=0
W[13,11]=1*-1+1*-1+1*1+-1*1=-2
W[13,12]=1*-1+1*1+1*1+-1*-1=2
W[13,13]=1*1+1*1+1*1+-1*-1=4
W[13,14]=1*-1+1*1+1*-1+-1*1=-2
W[14,0]=-1*-1+1*1+-1*1+1*1=2
W[14,1]=-1*1+1*1+-1*1+1*-1=-2
W[14,2]=-1*-1+1*-1+-1*-1+1*1=2
W[14,3]=-1*-1+1*-1+-1*-1+1*1=2
W[14,4]=-1*1+1*-1+-1*-1+1*-1=-2
W[14,5]=-1*-1+1*1+-1*1+1*1=2
W[14,6]=-1*-1+1*-1+-1*1+1*1=0
W[14,7]=-1*1+1*1+-1*1+1*1=0
W[14,8]=-1*-1+1*1+-1*-1+1*1=4
W[14,9]=-1*-1+1*1+-1*-1+1*-1=2
W[14,10]=-1*1+1*-1+-1*-1+1*-1=-2
W[14,11]=-1*-1+1*-1+-1*1+1*1=0
W[14,12]=-1*-1+1*1+-1*1+1*-1=0
W[14,13]=-1*1+1*1+-1*1+1*-1=-2
W[14,14]=-1*-1+1*1+-1*-1+1*1=4
```

The weight matrix calculated above is as follows:

$$
\mathbf{W} = \begin{pmatrix}
4 & 0 & 0 & 0 & -4 & 4 & 2 & 2 & 2 & 0 & -4 & 2 & 2 & 0 & 2 \\
0 & 4 & -4 & -4 & 0 & 0 & -2 & 2 & -2 & 0 & 0 & -2 & 2 & 4 & -2 \\
0 & -4 & 4 & 4 & 0 & 0 & 2 & -2 & 2 & 0 & 0 & 2 & -2 & -4 & 2 \\
0 & -4 & 4 & 4 & 0 & 0 & 2 & -2 & 2 & 0 & 0 & 2 & -2 & -4 & 2 \\
-4 & 0 & 0 & 0 & 4 & -4 & -2 & -2 & -2 & 0 & 4 & -2 & -2 & 0 & -2 \\
4 & 0 & 0 & 0 & -4 & 4 & 2 & 2 & 2 & 0 & -4 & 2 & 2 & 0 & 2 \\
2 & -2 & 2 & 2 & -2 & 2 & 4 & 0 & 0 & -2 & -2 & 4 & 0 & -2 & 0 \\
2 & 2 & -2 & -2 & -2 & 2 & 0 & 4 & 0 & -2 & -2 & 0 & 0 & 2 & 0 \\
2 & -2 & 2 & 2 & -2 & 2 & 0 & 0 & 4 & 2 & -2 & 0 & 0 & -2 & 4 \\
0 & 0 & 0 & 0 & 0 & 0 & -2 & -2 & 2 & 4 & 0 & -2 & 2 & 0 & 2 \\
-4 & 0 & 0 & 0 & 4 & -4 & -2 & -2 & -2 & 0 & 4 & -2 & -2 & 0 & -2 \\
2 & -2 & 2 & 2 & -2 & 2 & 4 & 0 & 0 & -2 & -2 & 4 & 0 & -2 & 0 \\
2 & 2 & -2 & -2 & -2 & 2 & 0 & 0 & 0 & 2 & -2 & 0 & 4 & 2 & 0 \\
0 & 4 & -4 & -4 & 0 & 0 & -2 & 2 & -2 & 0 & 0 & -2 & 2 & 4 & -2 \\
2 & -2 & 2 & 2 & -2 & 2 & 0 & 0 & 4 & 2 & -2 & 0 & 0 & -2 & 4
\end{pmatrix} \tag{9}
$$

The corresponding thresholds, θ, as calculated by equation 10 are as follows:

$$\theta = \begin{pmatrix} 6 \\ -2 \\ 2 \\ 2 \\ -6 \\ 6 \\ 3 \\ 1 \\ 5 \\ 2 \\ -6 \\ 3 \\ 3 \\ -2 \\ 5 \end{pmatrix} \tag{10}$$

Notice that where the input pattern precisely matches the training pattern, the network output is precisely correct:

IN: 010010010010010
OUT: 010010010010010

IN: 110001011100111
OUT: 110001011100111

IN: 110001110001110
OUT: 110001110001110

IN: 101101111001001
OUT: 101101111001001

As seen below, a somewhat modified input vector is associated with a similar vector from the training set.

IN: 111101111011001
OUT: 101101111001001

The input and output patterns are depicted in Figure 9.3. Notice that while the input (Figure 9.3 [a]) is a modified version of an input, (Figure 9.2 [a]). The output fully matches this vector. See Figure 9.3 [b].

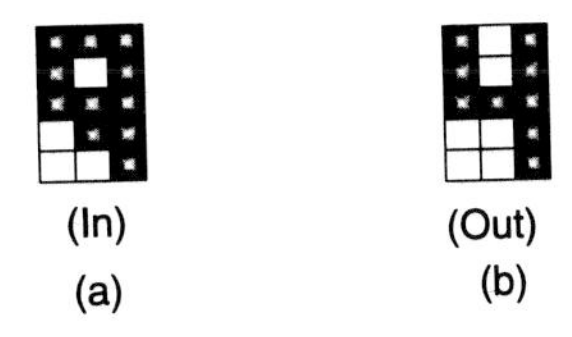

FIGURE 9.3 Example LAM input (a) and output (b) patterns for a successful retrieval.

However, an input vector that is more dissimilar from training set vectors may not map to any of the training vectors. This situation is illustrated below.

IN: 010101111101101
OUT: 101101111101101

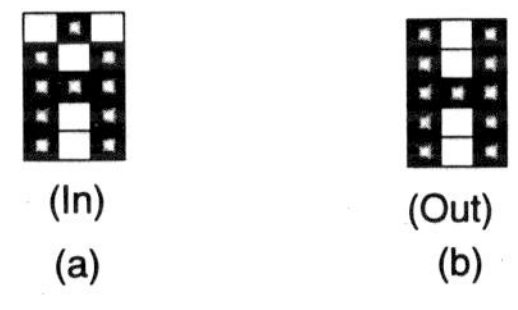

FIGURE 9.4 Example LAM input (a) and output (b) patterns illustrating an unsuccessful retrieval.

Notice that the output does not match any of the stored patterns. Therefore the retrieval has failed. The complete C++ implementation of a LAM network may be found in listing 9.1.

Listing 9.1

```
/********************************************************************************
*                                                                               *
* LAM                                                                           *
*                                                                               *
********************************************************************************/

#include <stdio.h>
#include <stdlib.h>
#include <string.h>
#include <conio.h>
#include <math.h>

// FUNCTION PROTOTYPES

//void  ShowResults(int, long int,double temp, double alpha, double eta);
```

```
// DEFINES
#define MAXNEURONS 100        // MAX NUMBER OF NEURONS PER LAYER
#define MAXPATTERNS 80        // MAX NUMBER OF PATTERNS IN A TRAINING SET

class LAME
{

private:
 int W[MAXNEURONS][MAXNEURONS];              // WEIGHTS MATRIX
 int Thresh[MAXNEURONS];
 int TrnSet[MAXPATTERNS][MAXNEURONS];        //TRAINING SET
 int TstSet[MAXPATTERNS][MAXNEURONS];        //TRAINING SET
 int inVect[MAXNEURONS];
 int outVect[MAXNEURONS];
 int  Neurons;
 int  TrnPatterns;                         // # of training set patterns
 int  TstPatterns;                         // # of test set patterns
public:
 LAME(void);
 void  GetTrnSet(char *Fname);
 int  GetTstSet(char *Fname);
 void  Train();
 void  Run(int i);
 void  ShowWeights();
 void  ShowThresholds();
 void  ShowInVect();
 void  ShowOutVect();
};

//------------------------------------------------------------------
// METHOD DEFINITIONS

LAME::LAME(){
Neurons=0;
TrnPatterns=0;
TstPatterns=0;
}

void LAME::GetTrnSet(char *Fname){
FILE *PFILE;
int  i,j,k;

PFILE = fopen(Fname,"r");                  // batch
if (PFILE==NULL){
  printf("\nUnable to open file %s\n",Fname);
  exit(0);
  }
fscanf(PFILE,"%d",&TrnPatterns);
fscanf(PFILE,"%d",&Neurons);
for (i=0; i<TrnPatterns; i++) {
  for (j=0; j<Neurons; j++) {
    fscanf(PFILE,"%d",&k);
    TrnSet[i][j]=k;
    } /* endfor */
  } /* endfor */
```

```
}

int LAME::GetTstSet(char *Fname){
  FILE *PFILE;
  int i,j,k;
PFILE = fopen(Fname,"r");                    // batch
if (PFILE==NULL){
  printf("\nUnable to open file %s\n",Fname);
  exit(0);
  }
fscanf(PFILE,"%d",&TstPatterns);
for (i=0; i<TstPatterns; i++) {
  for (j=0; j<Neurons; j++) {
    fscanf(PFILE,"%d",&k);
    TstSet[i][j]=k;
    } /* endfor */
  } /* endfor */
return(TstPatterns);
}

void LAME::Train(){
  int i,j,p;
//Calc weight matrix

for (i=0; i<Neurons; i++) {
  for (j=0; j<Neurons; j++) {
    W[i][j]=0;
    for (p=0; p<TrnPatterns; p++) {
      W[i][j] += (2*TrnSet[p][i]-1) *(2*TrnSet[p][j]-1);
      } /* endfor */
    } /* endfor */
  } /* endfor */

//Calc Thresholds
for (i=0; i<Neurons; i++) {
  for (j=0; j<Neurons; j++) {
    Thresh[i] += W[i][j];
    } /* endfor */
  Thresh[i]=Thresh[i]/2;
  } /* endfor */
}

void LAME::Run(int tp){
  int i,j;
  int RawOutVect[MAXNEURONS];
for (i=0; i<Neurons; i++) {
  inVect[i]=TstSet[tp][i];
  } /* endfor */
for (i=0; i<Neurons; i++) {
  RawOutVect[i] = 0;
  for (j=0; j<Neurons; j++) {
    RawOutVect[i]+=W[i][j] * inVect[j];        //Calc Raw output vect
    } /* endfor */
  } /* endfor */
//apply threshold
```

```
for (i=0; i<Neurons; i++) {
  if (RawOutVect[i]>Thresh[i]) {
    outVect[i]=1;
  } else {
    outVect[i]=0;
  } /* endif */
  } /* endfor */

}
void LAME::ShowWeights(){
  int i,j;
for (i=0; i<Neurons; i++) {
  for (j=0; j<Neurons; j++) {
    printf("%d ",W[i][j]);
    } /* endfor */
  printf("\n");
  } /* endfor */
}

void LAME::ShowThresholds(){
  int i;
for (i=0; i<Neurons; i++) {
  printf("%d ",Thresh[i]);
  } /* endfor */
printf("\n");
}

void LAME::ShowInVect(){
  int i;
printf("IN:  ");
for (i=0; i<Neurons; i++) {
  printf("%d ",inVect[i]);
  } /* endfor */
printf("\n");
}

void LAME::ShowOutVect(){
  int i;
printf("OUT: ");
for (i=0; i<Neurons; i++) {
  printf("%d ",outVect[i]);
  } /* endfor */
printf("\n");
}

//------------------------------------------------------------------------

LAME   LAM;

/*****************************************************************************
*
* MAIN                                                                      *
*****************************************************************************
/
```

```
int main(int argc, char *argv[])
{
  int TstSetSize;
  int i;
if (argc>2) {
  LAM.GetTrnSet(argv[1]);
  TstSetSize=LAM.GetTstSet(argv[2]);
  LAM.Train();
  LAM.ShowWeights();
  for (i=0; i<TstSetSize; i++) {
    LAM.Run(i);                          //Evaluate ith test pattern
    printf("\n");
    LAM.ShowInVect();
    LAM.ShowOutVect();
  } /* endfor */
  }
 else {
  printf("USAGE: LAM TRAINING_FILE TEST_FILE\n");
  exit(0);
  }


return 0;
}
```

9.3 HOPFIELD NETWORKS

Hopfield networks [Hopfield, 1982; 1984] are iterative autoassociative networks consisting of a single layer of fully connected processing elements which can function as an associative memory. Hopfield's contribution received considerable attention, since he presented the memory in terms of an energy (Liapunov) function and incorporated asynchronous processing at individual processing elements (neurons).

One of the remarkable features of neural networks, besides their adaptive nature, is that for many of the proposed architectures, both information processing and information storage are carried out in a parallel fashion. Thus, all the processing elements (neurons) perform computations synchronously (concurrently). Also their architecture is designed for parallel distributed memory, such that only all the storage elements (weights) together present meaningful information. These features provide a great deal of fault tolerance with regard to the storage and processing elements. With increasing availability of hardware suitable for connected, but independent, concurrent, elemental processors, these neural network models are receiving renewed attention.

Hopfield networks are quite interesting from the information processing system point of view. They do support the parallel distributed information storage aspect common to all neural network models, but information processing is not really performed in a parallel fashion. The processing elements (neurons) perform computations in an asynchronous fashion. This allows the network to avoid the difficulties encountered in propagating synchronization signals throughout the large network.

Another advantage is that this relieves the need of requiring that each processing element have global knowledge of the entire network at all times.

For the discrete Hopfield [1982] network processing, elements may take on either binary (1/0) or bipolar(1/–1) values. A pattern constituting N discrete binary u_i can be presented to the network, resulting in a state vector for the system. The network is fully connected in the sense that each processing element (or neuron) is connected to every other processing element. The weights, designated T_{ij}, are symmetric (see equation 11); and there is no connection between a neuron and itself (see equation 12):

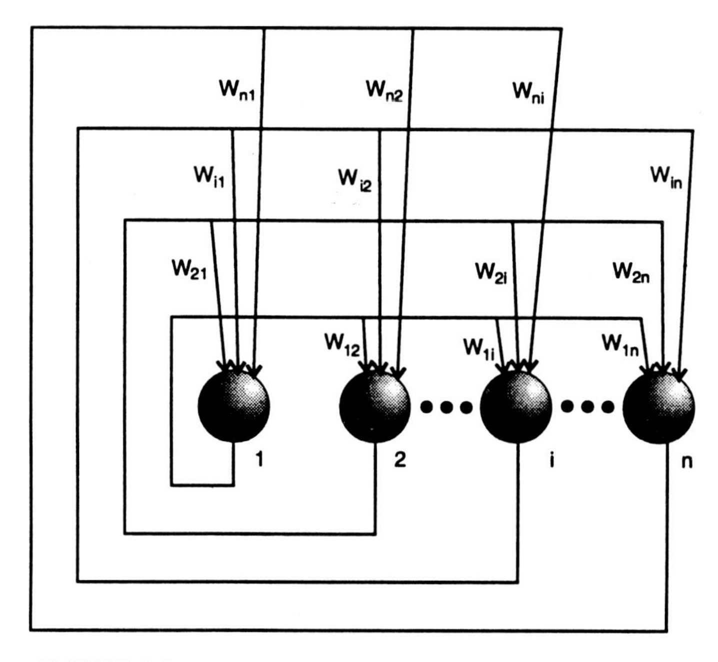

FIGURE 9.5 Architecture of a Hopfield network with η neurons.

$$T_{ij} = T_{ji} \tag{11}$$

$$T_{ii} = 0 \tag{12}$$

The structure of the network is shown in Figure 9.5 for the general case of N neurons. Figure 9.6 (a) shows a Hopfield network having 3 neurons and Figure 9.6 (b) shows a Hopfield network having 4 neurons. In Figure 9.7 (a) and (b) the energy flow cubes corresponding to each of these networks are shown to illustrate the possible state transitions within these networks.

The state of the network is the vector composed of the activity levels or states of the ordered processing elements. States have an associated energy (Liapunov) function given by:

$$E = -\frac{1}{2}\sum_{j}\sum_{\substack{i \\ i\neq j}} T_{ji}u_j u_i \tag{13}$$

where T_{ij} is the weight from unit i to unit j and u_i is the output of the ith unit in the network. As an iterative network, neurons are permitted to update, one at a time, until convergence occurs. The network is considered to have converged when a minimum of the energy function has been achieved and no individual neuron is motivated to change state when evaluated. It can be shown that the Hopfield network will always converge to a state of minimum energy when updated in the prescribed manner. Also, once such a state is achieved, none of the neurons would change their state.

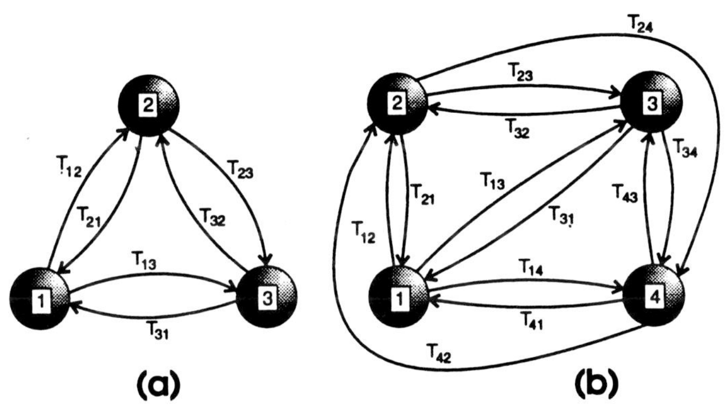

FIGURE 9.6 Architecture of a Hopfield network with (a) 3 neurons and (b) 4 neurons.

Figure 9.8 shows the energy landscape in two dimensions with several stored states represented by global minima. In general, the energy landscape will be a hyperspace where valid states will be represented by binary state vectors at the corners of a hypercube. The discrete Hopfield memory model computes by jumping from one such corner to another with lower energy, eventually ending up at a corner with the lowest energy. In contrast, the continuous Hopfield memory computes by traversing the hyperspace from a starting state to a global minimum (stored state) as shown in Figure 9.8.

Note that the neuronal states, u_i, may be considered to be short-term memory (STM). The memory can record and retain any pattern for a short while by modifying the u_i's. The synaptic strengths (weights), T_{ij}'s, can be thought of as being a long-term memory (LTM), since the network can learn by modifying these weight values.

We now describe the method by which a given neuron updates in the Hopfield model. The stimulus to the jth neuron will be a sum of products of the inputs u_i and weights T_{ij} as follows:

$$S_j = \sum_{\substack{i=1 \\ i\neq j}}^{n} u_i T_{ji} \tag{14}$$

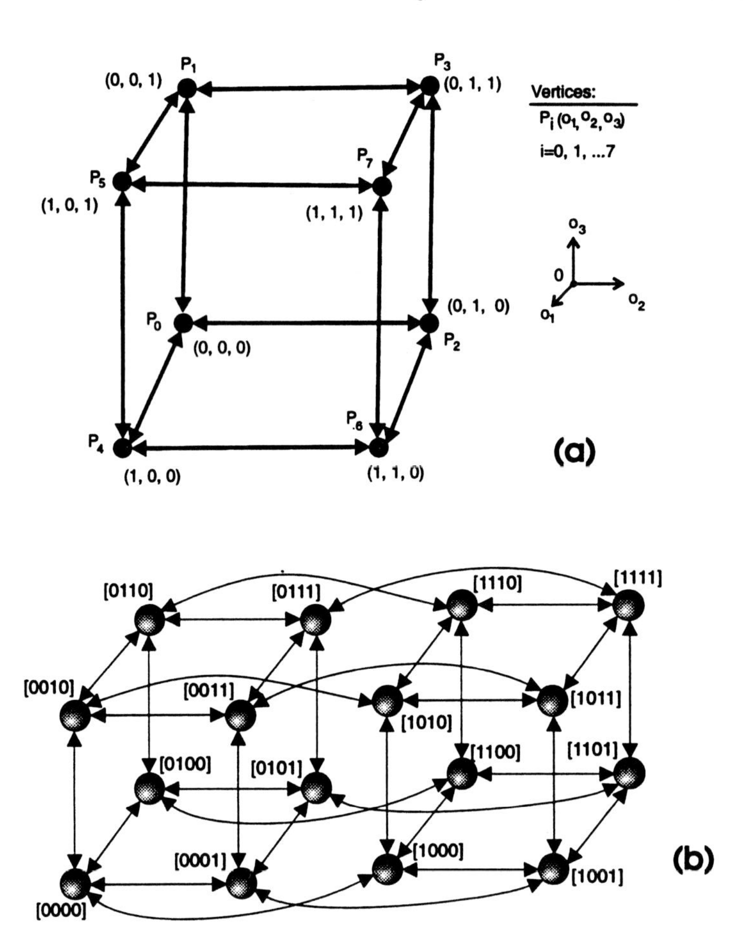

FIGURE 9.7 Energy flow cubes illustrating state transitions for (a) 3-neuron Hopfield networks and (b) 4-neuron Hopfield networks.

The resulting output of the jth neuron will be

$$u_j = \begin{cases} 1 & S_j \geq 0 \\ 0 & S_j < 0 \end{cases} \tag{15}$$

Let us now examine the implication of equations 14 and 15 in terms of the energy function. In particular, we demonstrate that when neurons are updated in this fashion, the network must converge to a state of minimum energy. Since the Hopfield network is an iterative network and only one neuron will be updated at a time, it is sufficient to consider the change in energy resulting from a single-neuron update. To begin, the energy due to a single neuron, j, is given by:

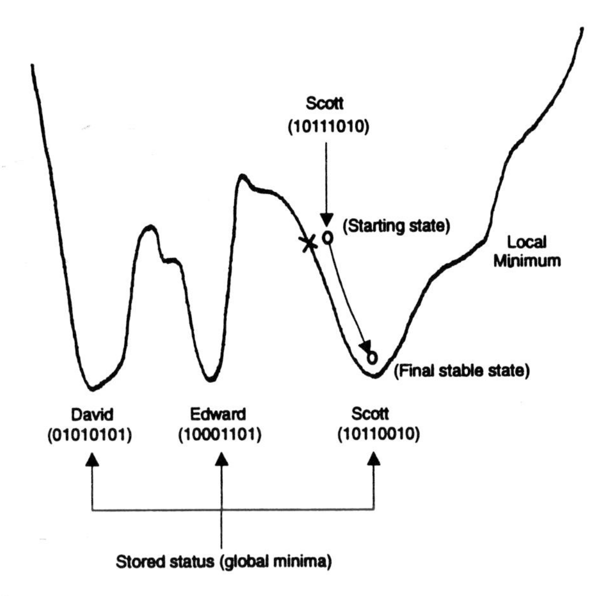

FIGURE 9.8 Two-dimensional energy landscape of a Hopfield network with multiple attractors.

$$E = -\frac{1}{2}\sum_{\substack{i \\ i \neq j}} T_{ij} u_i u_j \tag{16}$$

Since the sum is only over i, we may rewrite equation 16 as:

$$E_j = u_j \left(-\frac{1}{2}\sum_{\substack{i \\ i \neq j}} T_{ij} u_i \right) \tag{17}$$

Therefore when neuron j updates from a previous value, u_j^{old}, to a new value, u_j^{new}, the change in energy will be ΔE_j, where:

$$\Delta E_j = E_j^{new} - E_j^{old} \tag{18}$$

Thus by equation 17 we obtain:

$$\Delta E_j = u_j^{new} \left(-\frac{1}{2}\sum_{\substack{i \\ i \neq j}} T_{ij} u_i \right) - u_j^{old} \left(-\frac{1}{2}\sum_{\substack{i \\ i \neq j}} T_{ij} u_i \right) \tag{19}$$

$$\Delta E_j = \left(u_j^{\text{new}} - u_j^{\text{old}}\right)\left(-\frac{1}{2}\sum_{\substack{i \\ i \neq j}} T_{ij} u_i\right) \tag{20}$$

Now defining u_j as follows:

$$\Delta u_j = u_j^{\text{new}} - u_j^{\text{old}} \tag{21}$$

we obtain:

$$\Delta E_j = \Delta u_j \left(-\frac{1}{2}\sum_{\substack{i \\ i \neq j}} T_{ij} u_i\right) \tag{22}$$

and by substituting into equation 14 we obtain:

$$\Delta E_j = -\frac{1}{2} \Delta u_j S_j \tag{23}$$

Now we must consider three cases:

1. If the jth neuron does not change state, then $\Delta u_j = 0$; and thus by equation 23 the change in energy, $\Delta E_j = 0$.
2. If the jth neuron was initially 1 and transitions to a 0 state, then:

$$u_j^{\text{old}} = 1$$

$$u_j^{\text{new}} = 0$$

$$\therefore \Delta u_j = -1 < 0 \tag{24}$$

Note, however, that by equation 15, S_j must be less than 0 when a neuron transitions from 1 to 0. Therefore the product $\Delta u_j S_j$ must be positive. Thus we can conclude:

$$\Delta E_j < 0 \tag{25}$$

3. If the jth neuron was initially 0 and makes a transition to the 1 state:

$$u_j^{\text{old}} = 0$$

$$u_j^{\text{new}} = 1$$

$$\therefore \Delta u_j = 1 > 0 \tag{26}$$

Again referring to equation 15 we see that in this case:

$$S_j \geq 0 \tag{27}$$

The product $\Delta u_j S_j$ is then greater than or equal to zero from which we conclude that:

$$\Delta E_j \leq 0 \tag{28}$$

Thus we can conclude that for any possible transition of an arbitrary Hopfield neuron the energy must either decrease or remain the same. Further, from this behavior we can see that when all neurons can no longer change, a stable energy minimum has been achieved.

To evaluate the network, neurons are selected at random. The network is said to have converged when all neurons in the network have been visited and no neuron has changed state (see equations 14 and 15).

We must now address the issue of training the network so that the minima of the network correspond to a set of exemplar vectors that we wish the network to memorize. Given a set of m training vectors, $\mathbf{A}^p$, to be memorized by the network, the following procedure is used to create weights such that each stored vector will lie at an energy minimum and the network will contain a stable state for each of these stored values:

$$T_{ji} = \sum_{p=1}^{m} (2a_{pi} - 1)(2a_{pj} - 1) \quad (i \neq j)$$

$$T_{ii} = 0, \quad T_{ij} = T_{ji} \tag{29}$$

where a_{pi} is the ith element of the pth training vector. T_{ij} represents the weight between the ith and the jth neuron.

9.4 A HOPFIELD EXAMPLE

We now consider a simple example. We will discuss the results of a more elaborate example later in this section. Suppose that we wish to store two patterns into the Hopfield network illustrated in Figure 9.9. The patterns to be stored as follows:

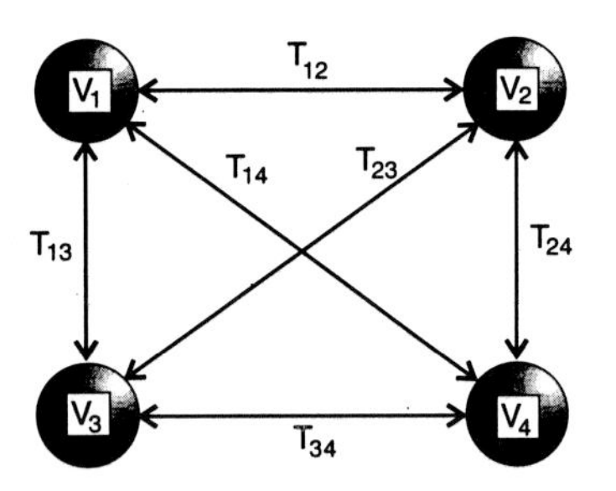

FIGURE 9.9 A 4-neuron Hopfield network.

$$a^{(1)} = \begin{pmatrix} 1 \\ 0 \\ 1 \\ 0 \end{pmatrix} \quad a^{(1)} = \begin{pmatrix} 0 \\ 1 \\ 0 \\ 1 \end{pmatrix} \tag{30}$$

Thus the training equations become:

$$\begin{aligned} T_{ji} &= \sum_{p=1}^{2} \left(2a_i^{(p)} - 1\right)\left(2a_j^{(p)} - 1\right) \quad (i \neq j) \\ T_{ii} &= 0, \quad T_{ij} = T_{ji} \end{aligned} \tag{31}$$

from which we may calculate the individual elements of the weight matrix as follows:

$$\begin{aligned} T_{12} &= \left(2a_1^{(1)} - 1\right)\left(2a_2^{(1)} - 1\right) + \left(2a_1^{(2)} - 1\right)\left(2a_2^{(2)} - 1\right) \\ &= 1(-1) + (-1)1 = -2 \end{aligned} \tag{32}$$

$$\begin{aligned} T_{13} &= \left(2a_1^{(1)} - 1\right)\left(2a_3^{(1)} - 1\right) + \left(2a_1^{(2)} - 1\right)\left(2a_3^{(2)} - 1\right) \\ &= 1(1) + (-1)(-1) = 2 \end{aligned} \tag{33}$$

$$\begin{aligned} T_{14} &= \left(2a_1^{(1)} - 1\right)\left(2a_4^{(1)} - 1\right) + \left(2a_1^{(2)} - 1\right)\left(2a_4^{(2)} - 1\right) \\ &= (1)(-1) + (-1)(1) = -2 \end{aligned} \tag{34}$$

$$\begin{aligned} T_{23} &= \left(2a_2^{(1)} - 1\right)\left(2a_3^{(1)} - 1\right) + \left(2a_2^{(2)} - 1\right)\left(2a_3^{(2)} - 1\right) \\ &= (-1)(1) + (1)(-1) = -2 \end{aligned} \tag{35}$$

$$T_{24} = \left(2a_2^{(1)} - 1\right)\left(2a_4^{(1)} - 1\right) + \left(2a_2^{(2)} - 1\right)\left(2a_4^{(2)} - 1\right) \tag{36}$$

$$= (-1)(-1) + (1)(1) = 2$$

$$T_{34} = \left(2a_3^{(1)} - 1\right)\left(2a_4^{(1)} - 1\right) + \left(2a_3^{(2)} - 1\right)\left(2a_4^{(2)} - 1\right) \tag{37}$$

$$= (1)(-1) + (-1)(1) = -2$$

The completed weight matrix is therefore given in equation 38 below:

$$T = \begin{pmatrix} 0 & -2 & 2 & -2 \\ -2 & 0 & -2 & 2 \\ 2 & -2 & 0 & -2 \\ -2 & 2 & -2 & 0 \end{pmatrix} \tag{38}$$

Now suppose that given the above trained network an input vector $[1110]^T$ is presented to the network. This means that initially the state of the network is

$$\begin{aligned} v_1 &= 1 \\ v_2 &= 1 \\ v_3 &= 1 \\ v_4 &= 0 \end{aligned} \tag{39}$$

Please note that while "a's" are used to represent input training patterns, "v's" are used to represent the state of the network.
If neuron v_2 is the first to update we obtain:

$$S_2 = \sum_{j=1}^{4} T_{2j} v_j \tag{40}$$

$$= (-2)(1) + (0)(1) + (-2)(1) + (2)(0)$$

$$= -4 < 0 \quad \therefore v_2 = 0$$

Notice that v_2 has changed state. Also note that as a result the network state matches that of the stored pattern, $[1010]^T$. It therefore only remains to demonstrate that in traversing each of the others, no state change will occur. This is easily accomplished as follows:

$$S_4 = \sum_{j=1}^{4} T_{4j} v_j \tag{41}$$

$$= (-2)(1) + (2)(0) + (-2)(1) + (0)(0)$$

$$= -4 < 0 \quad \therefore v_4 = 0$$

$$S_1 = \sum_{j=1}^{4} T_{1j} v_j \tag{42}$$

$$= (0)(1) + (-2)(0) + (2)(1) + (-2)(0)$$

$$= 2 > 0 \quad \therefore v_1 = 1$$

$$S_3 = \sum_{j=1}^{4} T_{3j} v_j \tag{43}$$

$$= (2)(1) + (-2)(0) + (0)(1) + (-2)(0)$$

$$= 2 > 0 \quad \therefore v_3 = 1$$

$$S_2 = \sum_{j=1}^{4} T_{2j} v_j \tag{44}$$

$$= (-2)(1) + (0)(0) + (-2)(1) + (2)(0)$$

$$= -4 < 0 \quad \therefore v_2 = 0$$

Notice that in equations 41 through 44 all neurons have been evaluated without altering the state of the network, $[1010]^T$. Therefore the network has converged.

9.5 DISCUSSION

As we have seen, Hopfield networks are easily trained in a single pass. Thus, the weights are prestored and it is classified as a fixed-weight neural network model as opposed to a supervised or unsupervised model, where the nets are trained adaptively using a learning algorithm.

After training, the network units are instantiated to an input vector to be recognized. Neurons are selected at random to be updated using equations 14 and 15. Convergence occurs when all neurons have been covered with no change having occurred. A cautionary not is that the direction of the first update can affect the state to which the network will ultimately converge. A parallel method for updating the net does exist. However, this method is even more subject to the problem of spurious states — discussed below — than the sequential method. For a detailed discussion on the synchronous update method for the Hopfield networks see Kung [1993]).

In the Hopfield model every local minimum of the energy function must be an attractor. An attractor is a state of equilibrium such that having achieved such a state the network will remain there forever. The Hopfield network can be fully described in the state space, but its behavior can be characterized in terms of a single scalar entity, called the energy (i.e., the value of a suitable Liapunov function). It is important to know how a content addressable memory, like the Hopfield model, responds to an arbitrary input pattern. Given an input pattern, what is the guarantee that it will evolve toward a stored pattern closely resembling the input pattern? What is the information storage capacity of such a network (i.e., how many stable equilibrium states [stored patterns] can simultaneously exist)? Hopfield and Tank [1985[have empirically shown that in the Hopfield networks containing N neurons about 0.15 N states can be stored before such errors in recall become severe.

One difficulty is that nonorthogonal members of the training set can couple to produce spurious attractors, stable minimum energy states which do not correspond to a training vector. The situation is further aggravated by the property that the negative of an input pattern is also an attractor. This characteristic, which is also pointed out by Kung [1993], is evident from the symmetry found in the equation. T_{ji} is increased for $a_{pi} = a_{pj}$ and decreased for $a_{pi} \neq a_{pj}$. It follows that the T_{ji} weights are sensitive only to the comparison (and that the individual values are lost). This necessarily results in the storage of the negative or inverted pattern. That is, for each pattern stored, another incorrect, inverted pattern is also stored. It might be tempting to accept this or even welcome it as a feature, an invariance with respect to negative images. Unfortunately these negative patterns are also capable of coupling with other patterns to form spurious minima further hampering attempts to orthogonalize the input. These problems taken together with the relatively low storage capacity of Hopfield networks would seem to make the Hopfield network an unconvincing choice for a recognition engine. A study of networks [Xuan-Jing, 1992] for character recognition that compared Hopfield networks, Hamming networks, and neocognitron concluded that Hopfield networks were unsuited to the task.

Hopfield networks appear relatively infrequently in the literature compared to many of the other neural methods for pattern recognition discussed in this book. One Hopfield net-based implementation [Gee et al., 1991], for recognition of rotated handwritten digits represents the character image as a series of n points spaced along the character outline. A modified version of the continuous Hopfield model [Tank and Hopfield 1987] was used to solve an optimization problem involving mapping these feature points in a test or unknown image to those of an exemplar pattern or template. Samples for which the network recognizer succeeded were shown but no quantitative data on accuracy were provided.

A modified version of the binary Hopfield network has been applied to the problem of digit recognition [Chen et al., 1991]. Here a single-error-detection, single-error-correction code (SEDSEC) is used to improve both the learning and recognition of the Hopfield network. SEDSEC encoding is frequently used in hardware memory systems for reliability. The results reported for this implementation are dramatic and startling. Over like data sets, where the conventional Hopfield model yielded a recognition accuracy rate of only 5%, the modified network provided an accuracy of over 90%. SEDSEC codes are formed by adding check bits to information bits present in the original image. These check bits are partial but overlapping parity operators spanning both the information bits and the check bits themselves. The number of check bits k for a binary sequence of information bits of length n can be obtained as: $2k \geq n + k + 1$. For example, an 8-bit byte of information bits would require 4 check bits. Calculation of the check bits is fast, requiring only the use of the exclusive-or (XOR) operator. The network is given an alveolate (or hexagonal) structure. The network is no longer fully connected. Neurons are only connected to their three nearest neighbors. During learning the check bits are created XOR operations. Recall locates and corrects erroneous input information.

9.6 BIT MAP EXAMPLE

Suppose we wish to create an associate memory which will recognize the four digits shown in Figure 9.10 below. Even given the relatively Spartan representation of the digits to be recognized, a 56-neuron network is required and an automated facility to perform the calculations required is desirable. Listing 9.2 below contains the necessary code to implement the binary Hopfield model.

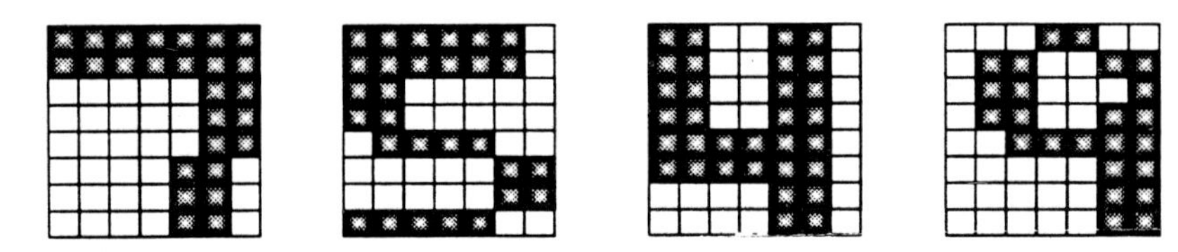

FIGURE 9.10 Example of Hopfield network input patterns.

Listing 9.2

```
/****************************************************************************
*                                                                           *
* Binary Hopfield Network                                                   *
*                                                                           *
****************************************************************************/

#include <conio.h>
#include <stdio.h>
#include <stdlib.h>
#include <string.h>
#include <math.h>
//#include <dos.h>

//----------------------------------------------------------------------------

// DEFINES
#define MAXNEURONS    64
```

```
#define MAXPATTERNS   10
#define MAXITER       600000
#define TRUE          1
#define FALSE         0

//------------------------------------------------------------------

// FUNCTION PROTOTYPES
// network fns
void InitNetRand(void);               // Scramble net state
void LoadTrainingSet(char *Fname);    // Get training set from file
void TrainNet();                      // Train net to recognize patterns
void RunNet(void);                    // Update net til convergence or MAXITER
int UpdateNeuron(int j);              // Update jTH neuron
                                      // Return TRUE if state changed

void LoadUserPattern(char *Fname);
int QueryAllClean(void);              // Return TRUE if all neurons were visited
void SetAllToDirty(void);             // Set all neurons to NOT visited
// utility fns
void Initialize(int cnt, char *Name);           // housekeeping
void DisplayPatVect(int V[MAXNEURONS]);         // show Net/Pattern (human-eye view)
void SavePatVect(int V[MAXNEURONS],             // store Net/Pattern (human-eye view)
             char *txt, int i);                 // to the archive file
void DisplayWeights(void);                      // show the weight matrix
void DisplayPatterns(void);                     // Display all trained patterns
int  QueryUserInt(char *msg);
void KeyWait(void);
void KeyWait2(char *txt);
int  random(int N);
//------------------------------------------------------------------

// GLOBALS
int PatVec[MAXNEURONS];                        // Pattern Vector
int PatMatrix[MAXPATTERNS][MAXNEURONS];        // Pattern Vector
int NEURON[MAXNEURONS];                        // Network
int T[MAXNEURONS][MAXNEURONS];                 // Weight matrix
int NumNeurons;                                // Actual size of net
int NumPatterns;                               // Actual number of patterns
int PatternX;                                  // X-dimension for human viewing
int PatternY;                                  // Y-dimension for human viewing
int Dirty[MAXNEURONS];                         // TRUE if neuron has not been updated
                                               // FALSE otherwise

FILE *ARCHIVE;

//------------------------------------------------------------------

int random(int N){
int x;
x=N*rand();
return (x/RAND_MAX);
}

void DisplayPatVect(int V[MAXNEURONS]){
  int x,y,indx;
```

```
  indx=0;
  for (y=0; y<PatternY; y++) {
    for (x=0; x<PatternX; x++) {
       if (V[indx]==1) {
          printf("X");
       } else {
            printf(".");
          } /* endif */
         indx++;
         } /* endfor */
      printf("\n");
      } /* endfor */
    printf("\n");
}

void SavePatVect(int V[MAXNEURONS], char *txt, int i) {
  int x,y,indx;
  indx=0;
  fprintf(ARCHIVE,"\n");
  for (y=0; y<PatternY; y++) {
    for (x=0; x<PatternX; x++) {
       if (V[indx]==1) {
          fprintf(ARCHIVE,"X");
       } else {
          fprintf(ARCHIVE,".");
       } /* endif */
       indx++;
       } /* endfor */
    fprintf(ARCHIVE,"\n");
    } /* endfor */
  fprintf(ARCHIVE,"\n%s ",txt);
  if (i>=0) fprintf(ARCHIVE,"%d ",i);
  fprintf(ARCHIVE,"\n\n ");
}

void DisplayWeights(){
  int i,j;
fprintf(ARCHIVE,"WEIGHTS:\n");
for (i=0; i<NumNeurons; i++) {
  fprintf(ARCHIVE,"[");
  for (j=0; j<NumNeurons; j++) {
    fprintf(ARCHIVE, " %d",T[j][i]);
    } /* endfor */
  fprintf(ARCHIVE,"]\n");
  } /* endfor */
}

void DisplayPatterns() {
  int i,p;
for (p=0; p<NumPatterns; p++) {
  for (i=0; i<NumNeurons; i++) {
    PatVec[i] =PatMatrix[p][i];
    } /* endfor */
  DisplayPatVect(PatVec);                    // show 1st training pattern
  SavePatVect(PatVec, "Training Pattern", p+1);
  printf("\n\nTraining Pattern %d of %d\n\n",p+1,NumPatterns);
```

```
  KeyWait();
  } /* endfor */
}

int QueryUserInt(char *msg){
int rv;
printf("Enter %s ==> ",msg);
scanf("%d",&rv);
return rv;
}

void KeyWait(void){
printf("Press any key to continue.\n");
while (!kbhit()) { } /* endwhile */
getch();
}

void KeyWait2(char *txt){
printf("\n\n%s\n",txt);
KeyWait();
}

void InitNetRand() {
  int i,r;

fprintf(ARCHIVE,"Creating test pattern\n");
srand(5);
//randomize();

for (i=0; i<NumNeurons; i++) {
  r=random(100);
  if (r >= 50) {
     NEURON[i]=0;
     }
   else {
     NEURON[i]=1;
     } /* endif */
  } /* endfor */

}

void LoadTrainingSet(char *Fname) {
  int pat,j, InVal;
  FILE *PATTERNFILE;

printf("Loading training set from default file: %s\n",Fname);
fprintf(ARCHIVE,"Loading training set from default file: %s\n",Fname);
PATTERNFILE = fopen(Fname,"r");
if (PATTERNFILE==NULL){
  printf("Unable to open default training Set file: %s",Fname);
  exit(0);
  }

//printf("\n");
fscanf(PATTERNFILE,"%d",&NumNeurons);            // Get number of neurons
fscanf(PATTERNFILE,"%d",&NumPatterns);           // Get number of patterns
```

```
fscanf(PATTERNFILE,"%d",&PatternX);                 // X-dimension for human viewing
fscanf(PATTERNFILE,"%d",&PatternY);                 // Y-dimension for human viewing
printf("%d Patterns Loaded\n",NumPatterns);
fprintf(ARCHIVE,"%d Patterns Loaded\n",NumPatterns);
for (pat=0; pat<NumPatterns; pat++) {
  for (j=0; j<NumNeurons; j++) {
    fscanf(PATTERNFILE,"%d",&InVal);
    PatMatrix[pat][j] =InVal;
    } // endfor
  } // endfor
fclose(PATTERNFILE);
}


void LoadUserPattern(char *Fname) {
  int j, InVal;
FILE *PATTERNFILE;

printf("Loading pattern from file: %s\n", Fname);
fprintf(ARCHIVE,"Loading pattern from file: %s\n", Fname);
PATTERNFILE = fopen(Fname,"r");
if (PATTERNFILE==NULL){
  printf("Unable to open file: %s",Fname);
  exit(0);
  }

printf("\n");
for (j=0; j<NumNeurons; j++) {
  fscanf(PATTERNFILE,"%d",&InVal);
  NEURON[j] =InVal;
  } // endfor
fclose(PATTERNFILE);
}


void TrainNet(){
  int i,j,pat;
  int Sum;
for (i=0; i<NumNeurons; i++) {
  for (j=0; j<NumNeurons; j++) {
    Sum=0;
    for (pat=0; pat<NumPatterns; pat++) {
      Sum += (2*PatMatrix[pat][i]-1) *  (2*PatMatrix[pat][j]-1);
      //Sum += PatMatrix[pat][i] *  PatMatrix[pat][j];
      } /* endfor */
    T[j][i] = T[i][j] = Sum;
    } /* endfor */
  } /* endfor */
for (i=0; i<NumNeurons; i++) {                 // Get rid of the diagonal...
  T[i][i]=0;                                   //  ...so it doesn't cause trouble later
  } /* endfor */
}
```

```
int QueryAllClean() {
  int i;
for (i=0; i<NumNeurons; i++) {
  if (Dirty[i]==TRUE) return FALSE;
  } // endfor
return TRUE;
}


void SetAllToDirty() {
  int i;
for (i=0; i<NumNeurons; i++) {
  Dirty[i]=TRUE;
  } // endfor
}


int UpdateNeuron(int j) {
  int i;
  int Sum = 0;
  int OldState = NEURON[j];
for (i=0; i<NumNeurons; i++) {          // accumulate all contributions
  Sum += T[j][i] * NEURON[i];           // remember we set diagonal of matrix T to 0 ..
                                        //      .. so no need to test for i==j
  } /* endfor */

if (Sum < 0) {
  NEURON[j] = 0;
  }
 else {
  if (Sum>0)
    NEURON[j] = 1;
  } /* endif */

if (NEURON[j] == OldState) {
  return 0;
  }
 else {
  return 1;
  } /* endif */
}


void RunNet(void) {
  int j;
  int Converged = FALSE;
  int ChngCount = 0;
  unsigned long int IterCount = 0;
  int ArchCnt=0;
 SetAllToDirty();
 while ( (!Converged) && (IterCount < MAXITER) ) {
  j = random(NumNeurons);               //next updating neuron j);
  ChngCount += UpdateNeuron(j);         // increment if neuron changed state
  DisplayPatVect(NEURON);
    printf("RUNNING...  Iteration=%d \n",IterCount);
```

```
 if (ArchCnt>=9) {
  SavePatVect(NEURON, "Net output at iteration =", IterCount+1);
   ArchCnt=0;
    } else {
    ArchCnt++;
    } /* endif */
  Dirty[j] = FALSE;                              // Record that we've  covered this neuron


  if (QueryAllClean()) {                         // Check if we hit all neurons at least once
    // here if we have hit all at least once
    //DisplayPatVect(NEURON);
     if (ChngCount == 0) {                       // Check if any neurons changed this pass
      // if we're here then were converged
       Converged = TRUE;
       printf("\nCONVERGED");
      SavePatVect(NEURON, "Net after convergence at iteration=", IterCount);
       }
      else {
      // if here then NOT converged so reinit for another pass
       SetAllToDirty();
       ChngCount=0;
       } /* endif */
    } /* endif */
  IterCount++;                                   // Increment iteration counter
  } /* endwhile */
}


void Initialize(int cnt, char *Name) {           // housekeeping
  char TrnName[50];
  char TstName[50];
ARCHIVE = fopen("ARCHIVE.LST","w");
if (ARCHIVE==NULL){
  printf("Unable to open default Training Set file: ARCHIVE.LST");
  exit(0);
   }
if (cnt>1) {
  // Get test pattern from file specified by command line arg
  strcpy(TrnName,Name);
  strcpy(TstName,Name); strcat(TstName,".tst");
  TrnName[4]=0; strcat(TrnName,"n4.trn");
  LoadTrainingSet(TrnName);
  if (cnt==2) LoadUserPattern(TstName);
   }
 else {
  // Initialize net with random test pattern
  // no command line parms
  LoadTrainingSet("HOPNET1.TRN");                // Use default training set
  InitNetRand();                             // use random pattern
   } /* endif */

if (cnt>2) InitNetRand();                    // parm count >2 —ignore pat file & randomize
KeyWait();
}
```

```
int main(int argc, char *argv[]) {
  int i;
Initialize(argc, argv[1]);
DisplayPatVect(NEURON);                // show net is set to test pattern
SavePatVect(NEURON, "Test Pattern", -1);
KeyWait2("TEST PATTERN");
DisplayPatterns();
TrainNet();
DisplayWeights();
RunNet();
fclose(ARCHIVE);
}
```

Figure 9.11 illustrates the progress of the network (trained as shown in Figure 9.10) in recognizing a corrupted version of the pattern for the digit seven. The state of the network is shown at intervals of 30 neuron updates. Notice that by iteration 150 the network has achieved a stable state corresponding to the learned "7" pattern.

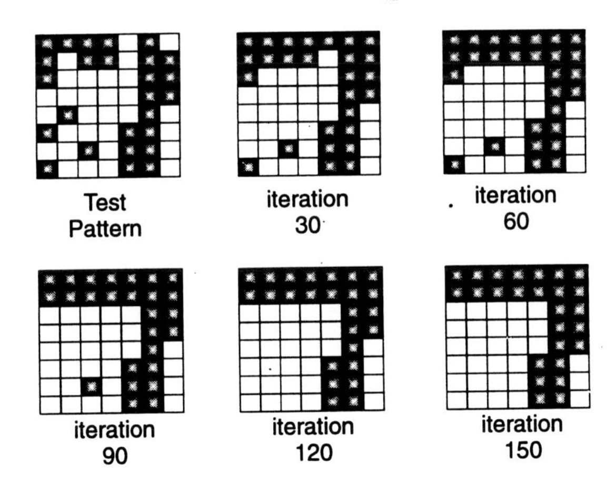

FIGURE 9.11 Convergence of a Hopfield network illustrating a successful retrieval.

Notice that the retrieval (see Figure 9.11) was successful in that one of the intentionally saved patterns was retrieved. By contrast, Figure 9.12 illustrates the situation in which the network converges to one of the inverted attractors.

We observe that the test pattern has converged to a stable state after iteration 300. Further, we may observe that this stable state (attractor) is the inverted version of the training pattern for the digit "9". Clearly our test pattern was challenging in that it is quite distant from the training inputs. Nevertheless, it serves to demonstrate this characteristic of the training of Hopfield networks. It is not difficult to discover other test patterns which converge to spurious attractors.

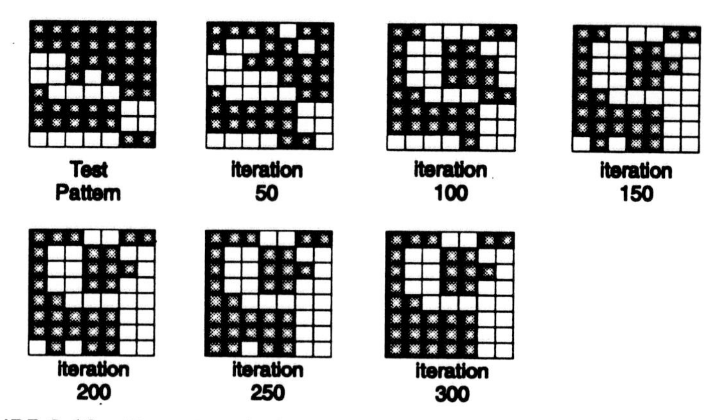

FIGURE 9.12 Illustration of a Hopfield network converging to an inverted attractor.

9.7 BAM NETWORKS

Biodirectional associative memories (BAM) store a set of pattern associations [Kosko, 1987; 1988]. Like the Hopfield network, they are recurrent in that they have feedback connections and will require some number of iterations to settle into an equilibrium state. Unlike the Hopfield networks they may be heteroassociative. Figure 9.13 illustrates the architecture of a BAM network.

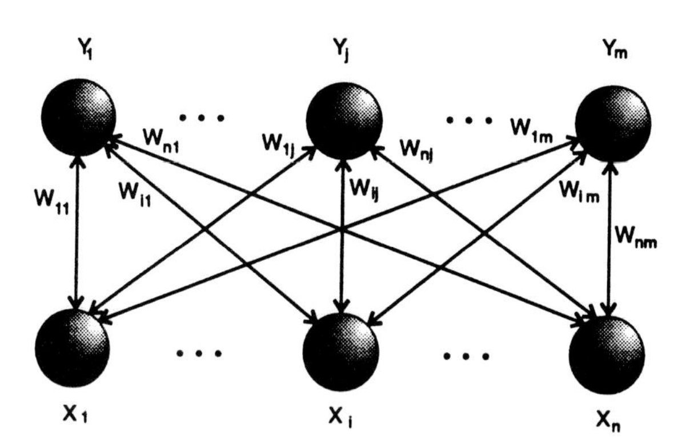

FIGURE 9.13 Architecture of a BAM network.

Notice that the network consists of two layers which may be of differing dimension. In a BAM network neither layer can be intrinsically termed input or output. A pattern incident on either layer should, in principle, ultimately result in an appropriate and corresponding pattern at the other, once the network has achieved equilibrium. We have therefore termed these layers X-layer and Y-layer. Note that the X- and Y-layers are fully connected. Note also that there are no lateral connections between neurons within the same layer. The connections between the layer are bidirectional. If the weight matrix from the X- to the Y-layer is designated **W**, then the weight matrix from Y to X will be the transpose of **W**, $\mathbf{W}^T$. The network functions by iteratively sending signals back and forth between the layer until an equilibrium has been achieved. That is, the network alternates between forward passes ($X \rightarrow Y$) and backward passes ($Y \rightarrow X$) until no further changes occur.

During a forward pass, the input to the jth Y-layer neuron is calculated as follows:

$$net_j^{(y)} = \sum_{i=1}^{n} w_{ij} x_i \tag{45}$$

and the activation of the Y-layer neurons is computed as:

$$y_j = \begin{cases} 1 & net_j^{(y)} > \theta_j \\ y_j & net_j^{(y)} = \theta_j \\ -1 & net_j^{(y)} < \theta_j \end{cases} \tag{46}$$

During the backward pass, the input to the X-layer neuron is calculated as follows:

$$net_j^{(x)} = \sum_{j=1}^{m} w_{ij} x_i \tag{47}$$

and the activation of the X-layer neurons is computed as:

$$x_i = \begin{cases} 1 & net_i^{(x)} > \theta_i \\ x_i & net_i^{(x)} = \theta_i \\ -1 & net_i^{(x)} < \theta_i \end{cases} \tag{48}$$

The overall mechanism by which the BAM network updates until convergence occurs is illustrated in Figure 9.14.

Training of the BAM network requires a training set consisting of vector association pairs as follows:

$$\left\{\left(\mathbf{a}^{(1)}, \mathbf{b}^{(1)}\right), \left(\mathbf{a}^{(2)}, \mathbf{b}^{(2)}\right), \; \ldots, \; \left(\mathbf{a}^{(p)}, \mathbf{b}^{(p)}\right)\right\} \tag{49}$$

where

$$\mathbf{a}^{(p)} = \mathrm{a}_1^{(p)}, \mathrm{a}_2^{(p)}, \; \ldots, \; \mathrm{a}_n^{(p)} \tag{50}$$

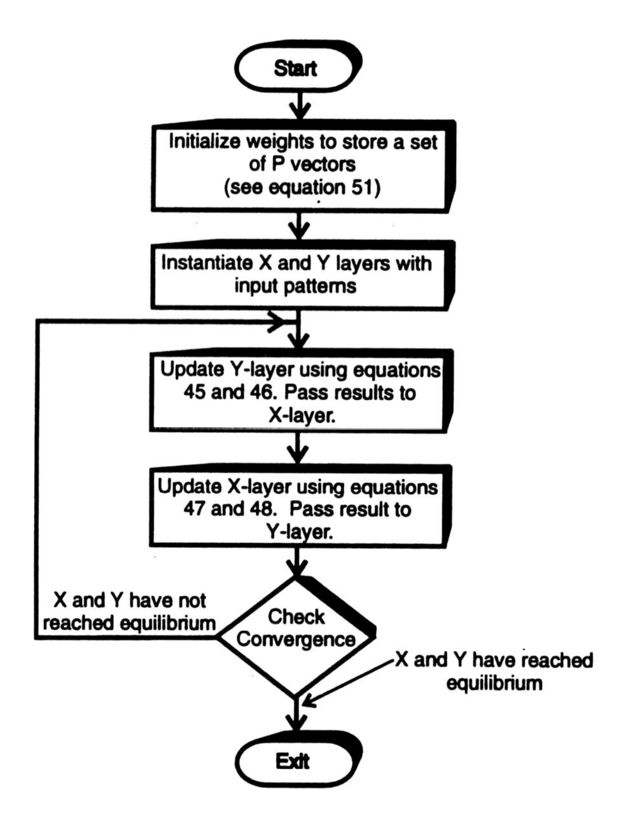

FIGURE 9.14 Flow chart for a BAM network.

and

$$\mathbf{b}^{(p)} = b_1^{(p)}, b_2^{(p)}, \ \ldots, \ b_m^{(p)} \tag{51}$$

For bipolar data the weights may be calculated as follows:

$$W_{ij} = \sum_p a_i^{(p)} b_j^{(p)} \tag{52}$$

If the data were binary, the weights would be calculated as:

$$W_{ij} = \sum_p \left(2a_i^{(p)} - 1\right)\left(2b_j^{(p)} - 1\right) \tag{53}$$

We now present a simple example illustrating the operation of the BAM network.

9.8 A BAM EXAMPLE

Figure 9.15 illustrates two bit map patterns together with a bipolar code to which we wish to associate the respective patterns. The corresponding BAM network is illustrated in Figure 9.16.

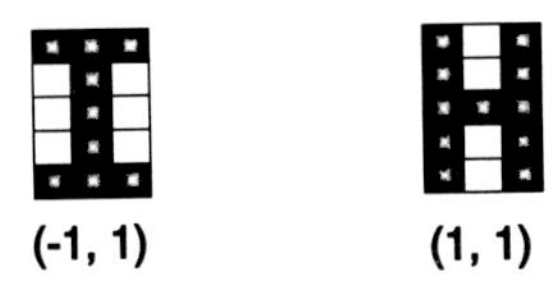

FIGURE 9.15 Example of BAM input patterns.

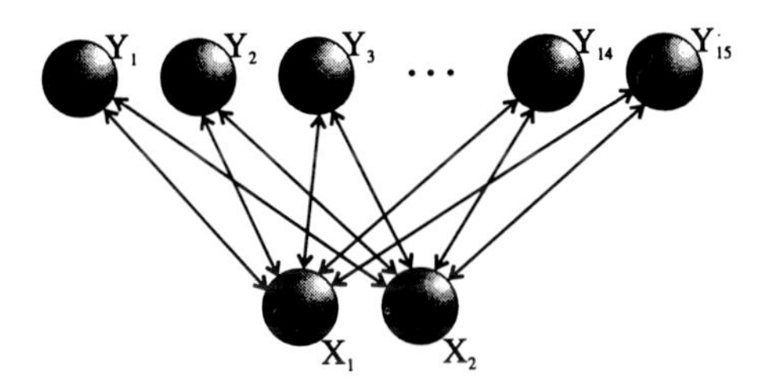

FIGURE 9.16 Example of BAM network with 2 X-layer neurons and 15 Y-layer neurons.

Thus we have two associations pairs, $\{(\mathbf{a}^{(1)}, \mathbf{b}^{(1)}),(\mathbf{a}^{(2)}, \mathbf{b}^{(2)})\}$, with the following values:

$$\mathbf{a}^{(1)} = \begin{pmatrix} 1 \\ 1 \\ 1 \\ -1 \\ 1 \\ -1 \\ -1 \\ 1 \\ -1 \\ -1 \\ 1 \\ -1 \\ 1 \\ 1 \\ 1 \end{pmatrix}^T \quad \mathbf{b}^{(1)} = \begin{pmatrix} -1 \\ 1 \end{pmatrix}^T \quad \mathbf{a}^{(2)} = \begin{pmatrix} 1 \\ -1 \\ 1 \\ 1 \\ -1 \\ 1 \\ 1 \\ 1 \\ 1 \\ 1 \\ -1 \\ 1 \\ 1 \\ -1 \\ 1 \end{pmatrix}^T \quad \mathbf{b}^{(2)} = \begin{pmatrix} 1 \\ 1 \end{pmatrix}^T \tag{54}$$

The corresponding weight matrix is calculated using equation 53 as follows:

$$\mathbf{W} = \begin{pmatrix} -1 & 1 \\ -1 & 1 \\ -1 & 1 \\ 1 & -1 \\ -1 & 1 \\ 1 & -1 \\ 1 & -1 \\ -1 & 1 \\ 1 & -1 \\ 1 & -1 \\ -1 & 1 \\ 1 & -1 \\ -1 & 1 \\ -1 & 1 \\ -1 & 1 \end{pmatrix} + \begin{pmatrix} 1 & 1 \\ -1 & -1 \\ 1 & 1 \\ 1 & 1 \\ -1 & -1 \\ 1 & 1 \\ 1 & 1 \\ 1 & 1 \\ 1 & 1 \\ 1 & 1 \\ -1 & -1 \\ 1 & 1 \\ 1 & 1 \\ -1 & -1 \\ 1 & 1 \end{pmatrix} = \begin{pmatrix} 0 & 2 \\ -2 & 0 \\ 0 & 2 \\ 2 & 0 \\ -2 & 0 \\ 2 & 0 \\ 2 & 0 \\ 0 & 2 \\ 2 & 0 \\ 2 & 0 \\ -2 & 0 \\ 2 & 0 \\ 0 & 2 \\ -2 & 0 \\ 0 & 2 \end{pmatrix} \tag{55}$$

Let us now verify that when one of the training patterns is input to the network that the corresponding associated training pattern is properly reproduced by the BAM. To this end we place $\mathbf{b}^{(1)}$ in our X-layer and set the Y-layer to zeros. (Setting a layer of a BAM network to zeros in effect abstains from giving the BAM any information at that layer about the association pair. It is possible to provide partial information about the vector at both levels simultaneously if desired.) The calculations are as follows:

$$net^{(x)} = \begin{pmatrix} -1 & 1 \end{pmatrix} \begin{pmatrix} 0 & -2 & 0 & 2 & -2 & 2 & 2 & 0 & 2 & 2 & -2 & 2 & 0 & -2 & 0 \\ 2 & 0 & 2 & 0 & 0 & 0 & 0 & 2 & 0 & 0 & 0 & 0 & 2 & 0 & 2 \end{pmatrix} \tag{56}$$

$$= \begin{pmatrix} 2 & 2 & 2 & -2 & 2 & -2 & -2 & 2 & -2 & -2 & 2 & -2 & 2 & 2 & 2 \end{pmatrix}$$

$$f\left(net^{(x)}\right) = \begin{pmatrix} 1 & 1 & 1 & -1 & 1 & -1 & -1 & 1 & -1 & -1 & 1 & -1 & 1 & 1 & 1 \end{pmatrix} \tag{57}$$

Note that the vector $\mathbf{a}^{(1)}$ has now appeared at the Y-layer as desired.

Now, to illustrate the bidirectional property of the BAM network, as well as its tolerance to noise, we will place a slightly modified version of the $\mathbf{a}^{(2)}$ pattern on the Y-layer. The X-layer is again set to zeros. The vector, **c**, incident on the x is as follows:

$$\mathbf{c} = \begin{pmatrix} 1 & \mathbf{0} & 1 & 1 & -1 & 1 & 1 & 1 & 1 & 1 & -1 & 1 & 1 & \mathbf{1} & 1 \end{pmatrix} \tag{58}$$

Now we may calculate the $net^{(y)}$ as follows:

$$net^{(y)} = \mathbf{cW} \tag{59}$$

$$net^{(y)} = \begin{pmatrix}1 & \mathbf{0} & 1 & 1 & -1 & 1 & 1 & 1 & 1 & 1 & -1 & 1 & 1 & \mathbf{1} & 1\end{pmatrix} \begin{pmatrix} 0 & 2 \\ -2 & 0 \\ 0 & 2 \\ 2 & 0 \\ -2 & 0 \\ 2 & 0 \\ 2 & 0 \\ 0 & 2 \\ 2 & 0 \\ 2 & 0 \\ -2 & 0 \\ 2 & 0 \\ 0 & 2 \\ -2 & 0 \\ 0 & 2 \end{pmatrix} \tag{60}$$

$$net^{(y)} = \begin{pmatrix}18 & 10\end{pmatrix} \tag{61}$$

Applying the activation function, we then obtain:

$$\mathrm{b}_{ij} < \frac{L}{(L-1+m)} \tag{62}$$

Notice that the network has converged properly to provide us with the $\mathbf{b}^{(2)}$ pattern at the X-layer.

REFERENCES AND BIBLIOGRAPHY

Amari, S. I., "Learning patterns and pattern sequences by self-organizing nets," *IEEE Trans. Comput.*, vol. 21, pp. 1197–1206, 1972.

Anderson, J. A., "A simple neural network generating an interactive memory," *Math. Biosci.*, vol. 14, pp. 197–220, 1972.

Anderson, J. A. and Bower, G. H., *Human Associative Memory,* V. H. Vincent, Washington, D.C., 1973.

Carpenter, G. A., "Neural network models for pattern recognition and associative memory," *Neural Networks,* vol. 2, pp. 243–257, 1989.

Chen, L. C., Fan, J. L., and Chen, Y. S., "A high speed modified Hopfield neural network and a design of character recognition system," *Proc. 1991 25th Annual 1991 IEEE Int. Carnahan Conf. Security Technology,* pp. 308–314, 1991.

Fausett, L., *Fundamentals of Neural Networks,* Prentice-Hall, Englewood Cliffs, NJ, 1994.

Freeman, J. A. and Skapura, D. M., *Neural Networks: Algorithms, Applications, and Programming Techniques,* Addison-Wesley, Reading, MA, 1991.

Gee, A. H., Aiyer, S. V. B., and Prager, R. W., "A Subspace Approach to Invariant Pattern Recognition using Hopfield Networks," Cambridge Univ. Eng. Dept. Tech. Report No. CUED/F-INFENG/TR62, 1991.

Grossberg, S., "Nonlinear neural networks: principles, mechanisms and architectures," *Neural Networks,* Vol. 1, pp. 17–61, 1988.

Hopfield, J. J., "Neural networks and physical systems with emergent collective computational abilities," *Proc. Natl. Sci.,* vol. 79, pp. 2554–2558, 1982.

Hopfield, J. J., "Neurons with graded response have collective computational properties like those of two-state neurons," *Proc. Natl. Sci.,* vol. 81, pp. 3088–3092, 1984.

Hopfield, J. J., "Neuron computation decisions in optimizing problems," *Biol. Cybern.,* vol. 52, pp. 141–152, 1985.

Hopfield, J. J. and Tank, D. W., "Neural computation of decisions in optimization problems," *Biol. Cybern.,* vol. 52, pp. 141–152, 1985.

Kohonen, T., "Correlation matrix memories," *IEEE Trans. Comput.,* vol. C-21, no. 4, pp. 353–359, 1972.

Kohonen, T., *Associative Memory: A System-Theoretical Approach,* Springer-Verlag, Berlin, 1977.

Kohonen, T., *Content-Addressable Memories,* Springer-Verlag, Berlin, 1980.

Kosko, B., "Adaptive bidirectional associative memories," *Appl. Optics,* vol. 26, no. 23, pp. 4947–4959, 1987.

Kosko, B., "Bidirectional associative memories," *IEEE Trans. Syst. Man Cyber.,* vol. 18, no. 1, pp. 49–60, 1988.

Kung, S. Y., *Digital Neural Networks,* Prentice-Hall, Englewood Cliffs, NJ, 1993.

Levine, D. S., *Introduction to Neural and Cognitive Modeling,* LEA Publishers, Hillside, NJ, 1990.

Nakano, K., "Associatron — A model of associative memory," *IEEE Trans. Syst. Man Cybern.,* pp. 380–388, 1972.

Pao, Y. H. and Merat, F. L., "Distributed associative memory for patterns," *IEEE Trans. Syst. Man Cybern.,* vol. 5, pp. 620–625, 1975.

Tank, D. W. and Hopfield, J. J., "Neural computation by concentrating information in time," *Proc. Natl. Acad. Sci.,* vol. 84, pp. 1896–1990, 1987.

Xuan-Jing, S., "A study of neural network application in handwritten digit recognition," *SPIE,* vol. 1766, pp. 684–689, 1992.

Zurada, J. M., *Introduction to Artificial Neural Systems,* West Publ., New York, 1992.

10 Adaptive Resonance Theory (ART)

10.1 GENERAL

One of the main goals of computer science is to develop an intelligent machine that can perform satisfactorily in an unaided fashion in a complex environment. Paradigms such as adaptive resonance theory (ART) that learn in an unsupervised fashion represent attempts to fulfill this goal.

The architecture of ART is based on the idea of adaptive resonant feedback between two layers of nodes as developed by Grossberg [1976]. The ART1 model described in Carpenter and Grossberg [1987a] was designed to cluster binary input patterns. The ART2 model discussed in Carpenter and Grossberg [1987b] was developed for clustering analog input patterns. Further refinements of these models, known as ART3, are developed in Carpenter and Grossberg [1990].

In case of the ART paradigm, autonomous learning and pattern recognition proceed in a stable fashion in response to an arbitrary sequence of input patterns. In this paradigm a self-regulating control structure is embedded into a competitive learning model.

10.2 DISCOVERING THE CLUSTER STRUCTURE

Patterns can be viewed as points in an N-dimensional feature space; and we expect that patterns that are similar in some respect, on the basis of class membership or other attribute values, would be close to each other in the pattern space. Thus patterns belonging to class C_i would cluster more closely to one another than any pattern belonging to class C_j. Of course, in many practical applications these clusters overlap. Figure 10.1 shows pattern space with various clusters formed based on a distance measure.

Unsupervised learning algorithms try to identify several prototypes or exemplars that can serve as cluster centers. A prototype can be either one of the actual patterns or a synthesized pattern vector centrally located in the respective cluster. K-means algorithm (see chapter 8), ISODATA algorithm, vector quantization (VQ) techniques (see next section) (see, for example Pao [1989] or Tou and Gonzales [1974] for a detailed discussion on these algorithms) are examples of decision-theoretical approaches for cluster formation. ART structure is a neural network approach for cluster formation in an unsupervised learning domain.

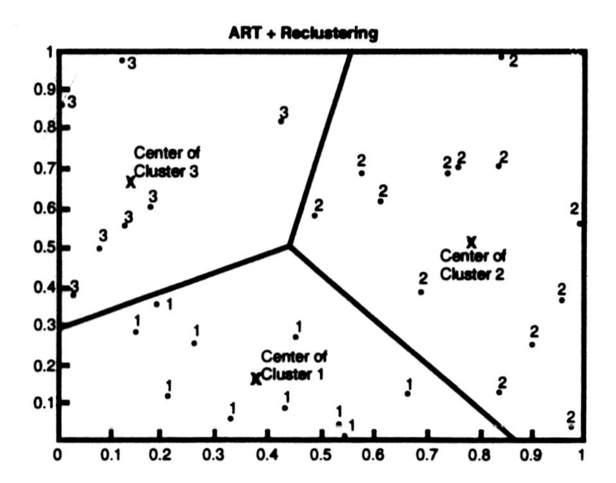

FIGURE 10.1 Clusters based on distance measure. (Adapted from Kung [1993].)

In many practical applications involving unsupervised learning, the number of classes may not be known *a priori*. Thus while determining the network architecture, the number of output nodes cannot be accurately determined in advance.

10.3 VECTOR QUANTIZATION

The objective of unsupervised learning is to separate the input data into some number of meaningful classes. The most obvious and intuitive approach to this problem is to cluster input vectors based on distance functions within a Euclidean space. We have already encountered the K-means algorithm, which is just such a clustering method based on Euclidian distance in chapter 8. The K-means algorithm requires that the number of cluster centers be known in advance. In real-world unsupervised learning, however, *a priori* knowledge of the number of meaningful classes required to describe the input data may not be available. For this reason a means by which to adaptively adjust the number of cluster centers becomes highly desirable.

VQ, presented in this section, and ART, presented in following sections of this chapter, represent two distinct approaches to the dynamic allocation of cluster centers. Note that vector quantization is a nonneural method whereas ART is a neural approach.

The vector quantization method begins with no clusters allocated. (The first pattern will force the creation of a cluster to hold it.) Thereafter, as each new input pattern is encountered, the Euclidean distance between it and any allocated clusters is calculated. If we designate the pth input vector as $\mathbf{X}^{(p)}$ and the jth cluster center by vector $\mathbf{C}_j$, then the Euclidean distance, d, is calculated as shown below in equation 1:

$$d = \left\| \mathbf{X}^{(p)} - \mathbf{C}_j \right\| = \left[\sum_{i=1}^{N} \left(x_i^{(p)} - c_{ji} \right)^2 \right]^{1/2} \tag{1}$$

where N is the dimension of the **X** and **C** are vectors.

Once the distance between the current pattern and all allocated clusters is known, the cluster closest to the input pattern $\mathbf{C}_k$ may be chosen such that:

$$\left\|\mathbf{X}^{(p)} - \mathbf{C}_k\right\| < \left\|\mathbf{X}^{(p)} - \mathbf{C}_j\right\| = \begin{cases} j = 1, \ \ldots, \ M \\ j \neq k \end{cases} \tag{2}$$

where M is the number of allocated clusters.

After the closest cluster, k has been determined, the distance $\|\mathbf{X}^{(p)} - \mathbf{C}_k\|$ must be tested against the distance threshold, ρ. On the basis of this comparison, one of two possible actions will be taken:

1. When $\|\mathbf{X}^{(p)} - \mathbf{C}_k\| < \rho$, the vector $\mathbf{X}^{(p)}$ is within tolerance. In this case the pattern is assigned to the kth cluster. That is, if we let S_k designate the set of patterns associated with the kth cluster, then $\mathbf{X}^{(p)} \in S_k$.
2. When $\|\mathbf{X}^{(p)} - \mathbf{C}_k\| > \rho$, the vector $\mathbf{X}^{(p)}$, is outside of the tolerance limit and therefore may not be assigned to the cluster (even though this was the closest existing cluster). In this case a new cluster is allocated for $\mathbf{X}^{(p)}$.

Once a pattern has been attached to a cluster, the cluster center of the newly modified cluster must be adjusted. The new cluster center is formed by taking the average of all member vectors (in a manner reminiscent of K-means). The procedure for updating the cluster center is given in equation 3 below:

$$\mathbf{C}_k = \frac{1}{N_x}\sum_{x \in S_k} \mathbf{x} \tag{3}$$

Note that in the above equation, k indicates the cluster center to which the current input pattern has been attached even if it has been newly created for this pattern.

The overall vector quantization process is illustrated in Figure 10.2. Listing 10.1 provides complete source code for a vector quantization implementation.

Listing 10.1

```
/*****************************************************************************
*                                                                           *
* VECTOR QUANTIZATION                                                       *
*                                                                           *
*****************************************************************************/

#include <stdio.h>
#include <stdlib.h>
#include <string.h>
#include <conio.h>
#include <math.h>
```

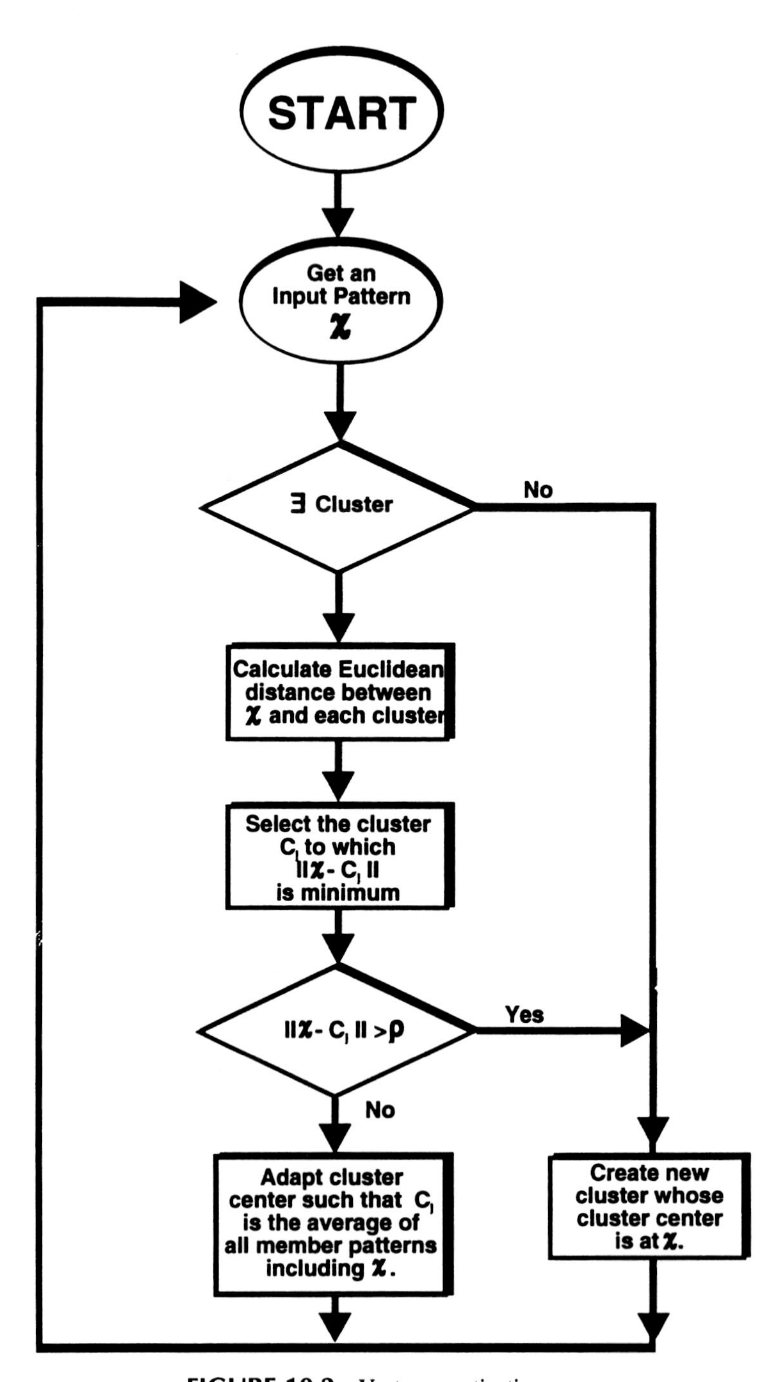

FIGURE 10.2 Vector quantization process.

```
// FUNCTION PROTOTYPES

// DEFINES
#define      SUCCESS      1
#define      FAILURE      0
#define      TRUE         1
#define      FALSE        0
#define      MAXVECTDIM   20
#define      MAXPATTERN   20
#define      MAXCLUSTER   10

// ***** Defined structures & classes *****
struct aCluster {
  double     Center[MAXVECTDIM];
  int        Member[MAXPATTERN];          //Index of Vectors belonging to this cluster
  int        NumMembers;
};

struct aVector {
  double     Center[MAXVECTDIM];
  int        Size;
};

class VQsyst {
private:
  double     Pattern[MAXPATTERN][MAXVECTDIM+1];
  aCluster   Cluster[MAXCLUSTER];
  int        NumPatterns;                 // Number of patterns
  int        SizeVector;                  // Number of dimensions in vector
  int        NumClusters;                 // Curr number of clusters
  double     Threshold;
  void       Attach(int,int);             // Add spec'd pattern to spec'd
                                          // Clusters membership list.
  int        AllocateCluster();           //
  void       CalcNewClustCenter(int);     // find cluster center for modified
                                          // clusters
  double     EucNorm(int, int);           // Calc Euclidean norm vector
  int        FindClosestCluster(int);     //ret indx of clust closest to pattern
                                          //whose index is arg
 public:
  VQsyst();
  int LoadPatterns(char *fname);          // Get pattern data to be clustered
  void RunVQ();                           // Overall control Vector Quant process
  void ShowClusters();                    // Show results on screen
  void SaveClusters(char *fname);         // Save results to file
 };

//================IMPLENTATION OF VQ METHODS=====================

//***************************************************************************
// Constructor
//***************************************************************************

VQsyst::VQsyst(){
  NumClusters=0;
  }
```

```
//*****************************************************************************
// LoadPatterns                                                               *
//  Loads pattern information from disk.                                      *
//      1) Number of patterns to be processed                                 *
//      2) Size of the vectors to be processed                                *
//      3) Max dist permitted between cluster centers                         *
//      2) Pattern definitions                                                *
//*****************************************************************************

int VQsyst::LoadPatterns(char *fname){
  FILE *InFilePtr;
  int    i,j;
  double x;
if((InFilePtr = fopen(fname, "r")) == NULL)
   return FAILURE;
fscanf(InFilePtr, "%d", &NumPatterns);           // Read # of patterns
fscanf(InFilePtr, "%d", &SizeVector);            // Read dimension of vector
fscanf(InFilePtr, "%lg", &Threshold);            // Read Euc dist Threshold.
for (i=0; i<NumPatterns; i++) {                  // For each vector
  for (j=0; j<SizeVector; j++) {                 // create a pattern
    fscanf(InFilePtr,"%lg",&x);                  // consisting of all elements
    Pattern[i][j]=x;
   printf("Pattern[%d][%d]=%f\n",i,j,Pattern[i][j]);
    } /* endfor */
  } /* endfor */
printf("\n");
return SUCCESS;
}

//*****************************************************************************
// RunVQ                                                                      *
//  Provides overall control of VQ algorithm.  K clusters                     *
//  We choose the first K vectors to do this                                  *
//*****************************************************************************

void VQsyst::RunVQ(){
 int pat,Winner;
 double dist;
for (pat=0; pat<NumPatterns; pat++) {
  Winner = FindClosestCluster(pat);              // Attach the curr input to closest
                                                 // cluster
  if (Winner == -1) {
   Winner=AllocateCluster();
    }
   else {
    dist= EucNorm(pat,Winner);
    dist= sqrt(dist);                            // because eucnorm doesn't do this
    if (dist>Threshold) {
      Winner=AllocateCluster();                  // Above threshold so allocate new
      printf("Creating NEW cluster number:%d\n",Winner);
       } /* endif */                             // cluster
    } /* endif */
 printf("patern %d assigned to cluster %d\n",pat,Winner);
 Attach(Winner,pat);                             // Attach pattern to winner
 CalcNewClustCenter(Winner);                     // Adapt clust center
  } /* endfor */
}
```

```
//*****************************************************************************
// AllocateCluster                                                           *
//   Designate the next free cluster as active                               *
//*****************************************************************************

int VQsyst::AllocateCluster(){
  int n;
n=NumClusters;
NumClusters++;
return n;
}

//*****************************************************************************
// EucNorm                                                                   *
//  Returns the Euclidian norm between a pattern, p, and a cluster          *
//  center,c-1 he first K vectors to do this                                 *
//*****************************************************************************

double VQsyst::EucNorm(int p, int c){          // Calc Euclidean norm of vector difference
double dist;                                   // between pattern vector, p, and cluster
int i;                                         // center, c.
dist=0;
for (i=0; i<SizeVector ;i++){
  dist += (Cluster[c].Center[i]-Pattern[p][i])*(Cluster[c].Center[i]-Pattern[p][i]);
  } /* endfor */
return dist;
}

//*****************************************************************************
// Find ClosestCluster                                                       *
//  Returns the index of the cluster to which the pattern, pat, has the     *
//  closest Euclidean distance.                                              *
//*****************************************************************************

int VQsyst::FindClosestCluster(int pat){
  int i, ClustID;
  double MinDist, d;
MinDist =9.9e+99;
ClustID=-1;
for (i=0; i<NumClusters; i++) {
  d=EucNorm(pat,i);
  if (d<MinDist) {
    MinDist=d;
    ClustID=i;
    } /* endif */
  } /* endfor */
if (ClustID<0) {
  printf("No Clusters exist\n");
  } /* endif */
return ClustID;
}
```

```
//*******************************************************************************
// Attach                                                                       *
//  Adds the pattern (whose index is p) to the cluster center whose index *
//  is p.                                                                       *
//*******************************************************************************

void VQsyst::Attach(int c,int p){
int MemberIndex;

  MemberIndex=Cluster[c].NumMembers;
  Cluster[c].Member[MemberIndex]=p;
  Cluster[c].NumMembers++;
}

//****************************************************************************
// CalcNewClustCenter                                                        *
//  Calculate a new cluster center for the specified cluster,c               *
//                                                                           *
//****************************************************************************

void VQsyst::CalcNewClustCenter(int c){
  int VectID,j,k;
  double tmp[MAXVECTDIM];

  for (j=0; j<SizeVector; j++) {                  // clear workspace
    tmp[j]=0.0;
    } /* endfor */
  for (j=0; j<Cluster[c].NumMembers; j++) {       //traverse member vectors
    VectID=Cluster[c].Member[j];
    for (k=0; k<SizeVector; k++) {                //traverse elements of vector
      printf("Cluster[%d]  Pattern[%d][%d]=%f,
Member_ID=%d\n",c,VectID,k,Pattern[VectID][k],VectID);
      tmp[k] += Pattern[VectID][k];               // add (member) pattern element into temp
       } /* endfor */
    } /* endfor */
  for (k=0; k<SizeVector; k++) {                  //traverse elements of vector
    tmp[k]=tmp[k]/Cluster[c].NumMembers;
    Cluster[c].Center[k]=tmp[k];
    } /* endfor */
}
//****************************************************************************
// ShowClusters                                                              *
//  Display the cluster centers for each of the allocated clusters          *
//****************************************************************************

void VQsyst::ShowClusters(){
  int cl;
for (cl=0; cl<NumClusters; cl++) {
  printf("\nCLUSTER %d ==>[%f,%f]\n", cl,Cluster[cl].Center[0],Cluster[cl].Center[1]);
  } /* endfor */
}

//****************************************************************************
// SaveClusters                                                              *
//  Store the cluster centers for each of the allocated clusters            *
//  to the designated file                                                   *
//****************************************************************************
```

```
void VQsyst::SaveClusters(char *fname){
  FILE *OutFilePtr;
  int cl;
if((OutFilePtr = fopen(fname, "r")) == NULL){
  printf("Unable to open file %s for output",fname);
   }
 else {
  for (cl=0; cl<NumClusters; cl++) {
    fprintf(OutFilePtr,"\nCLUSTER %d ==>[%f,%f]\n",
         cl,Cluster[cl].Center[0],Cluster[cl].Center[1]);
    } /* endfor */
  fclose(OutFilePtr);
  }
}

//***************************************************************************
// Main                                                                      *
//                                                                           *
//                                                                           *
//***************************************************************************
main(int argc, char *argv[]) {
  VQsyst VQ;
if (argc<3) {
  printf("USAGE: VQ PATTERN_FILE(input) CLUSTER_FILE(output)\n");
  exit(0);
  }
if (VQ.LoadPatterns(argv[1])==FAILURE ){
  printf("UNABLE TO READ PATTERN_FILE:%s\n",argv[1]);
  exit(0);
  }
VQ.RunVQ();
VQ.ShowClusters();
}
```

Note that vector quantization suffers from two drawbacks in that:

1. It will clearly be sensitive to the sequence of presentation of the input vectors. (We take note that K-means is not sensitive to the order of presentation. We may further observe that it is not at all difficult to introduce the idea of a distance threshold into the K-means algorithm.
2. The threshold distance at which new clusters will be created must be arbitrarily selected (in advance). Unfortunately an inauspicious choice of this parameter can have undesirable results. We will demonstrate this in the examples in sections 10.3.1 and 10.3.2 below.

Before moving on to present examples of VQ, we will note that the implementation we have shown requires that all individual input pattern vectors must be retained to be used in the calculation of new cluster centers. As an alternative, where memory or speed is a priority, a sliding window approach could be taken with respect to the calculation of new cluster centers.

10.3.1 VQ Example 1

Figure 10.3 shows 12 points in a two-dimensional Euclidean space. Our task for this example will be to cluster these utilizing the vector quantization algorithm.

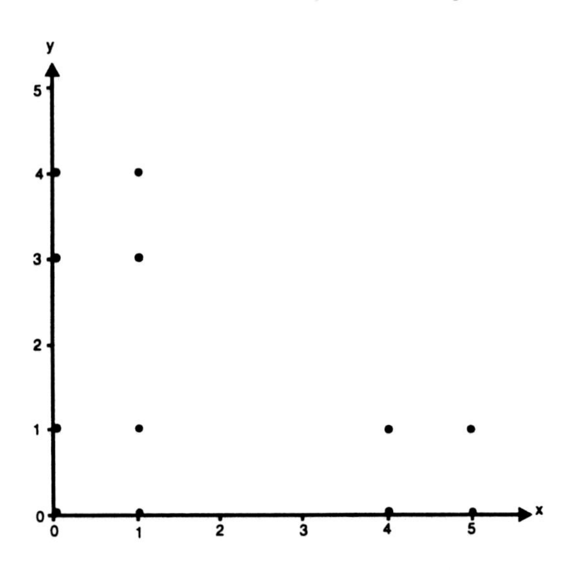

FIGURE 10.3 Input pattern for VQ example 1.

The calculations are as follows given a distance threshold of $\rho = 2.0$.
The input patterns are:

Pattern[0][0]=0.000000
Pattern[0][1]=0.000000
Pattern[1][0]=1.000000
Pattern[1][1]=0.000000
Pattern[2][0]=0.000000
Pattern[2][1]=1.000000
Pattern[3][0]=1.000000
Pattern[3][1]=1.000000
Pattern[4][0]=0.000000
Pattern[4][1]=3.000000
Pattern[5][0]=1.000000
Pattern[5][1]=3.000000
Pattern[6][0]=0.000000
Pattern[6][1]=4.000000
Pattern[7][0]=1.000000
Pattern[7][1]=4.000000
Pattern[8][0]=4.000000
Pattern[8][1]=0.000000
Pattern[9][0]=5.000000
Pattern[9][1]=0.000000
Pattern[10][0]=4.000000
Pattern[10][1]=1.000000
Pattern[11][0]=5.000000
Pattern[11][1]=1.00000

PATTERN 0:
No Clusters exist so allocate a new cluster 0
pattern 0 assigned to cluster 0

The new cluster centers are:
CLUSTER 0 ==>[0.000000,0.000000]
CLUSTER Membership
Cluster 0 ==>{0 }

PATTERN 1:
The closest cluster is: 0
Distance 1.000000 < 2.000000
Therefore cluster 0 passed the distance test.
pattern 1 assigned to cluster 0
The new cluster centers are:
CLUSTER 0 ==>[0.500000,0.000000]
CLUSTER Membership
Cluster 0 ==>{0 1 }

PATTERN 4:
The closest cluster is: 0
distance 2.549510 > 2.000000
Therefore cluster 0 failed the distance test.
so create NEW cluster number: 1
pattern 4 assigned to cluster 1

The new cluster centers are:
CLUSTER 0 ==>[0.500000,0.500000]
CLUSTER 1 ==>[0.000000,3.000000]
CLUSTER Membership
Cluster 0 ==>{0 1 2 3 }

Cluster 1 ==>{4 }

PATTERN 5:
The closest cluster is: 1
Distance 1.000000 < 2.000000
Therefore cluster 1 passed the distance test
pattern 5 assigned to cluster 1

The new cluster centers are:
CLUSTER 0 ==>[0.500000,0.500000]
CLUSTER 1 ==>[0.500000,3.000000]
CLUSTER Membership
Cluster 0 ==>{0 1 2 3 }

Cluster 1 ==>{4 5 }

PATTERN 2:
The closest cluster is: 0
Distance 1.118034 < 2.000000
Therefore cluster 0 passed the distance test.
pattern 2 assigned to cluster 0

The new cluster centers are:
CLUSTER 0 ==>[0.333333,0.333333]
CLUSTER Membership
Cluster 0 ==>{0 1 2 }

PATTERN 3:
The closest cluster is: 0
Distance 0.942809 < 2.000000
Therefore cluster 0 passed the distance test.
pattern 3 assigned to cluster 0

The new cluster centers are:
CLUSTER 0 ==>[0.500000,0.500000]
CLUSTER Membership
Cluster 0 ==>{0 1 2 3 }

PATTERN 7:
The closest cluster is: 1
Distance 0.942809 < 2.000000
Therefore cluster 1 passed the distance test
pattern 7 assigned to cluster 1
The new cluster centers are:
CLUSTER 0 ==>[0.500000,0.500000]
CLUSTER 1 ==>[0.500000,3.500000]
CLUSTER Membership
Cluster 0 ==>{0 1 2 3 }

Cluster 1 ==>{4 5 6 7 }

PATTERN 8:
The closest cluster is: 0
distance 3.535534 > 2.000000
Therefore cluster 0 failed the distance test
so create NEW cluster number:2
pattern 8 assigned to cluster 2

The new cluster centers are:
CLUSTER 0 ==>[0.500000,0.500000]
CLUSTER 1 ==>[0.500000,3.500000]
CLUSTER 2 ==>[4.000000,0.000000]
CLUSTER Membership
Cluster 0 ==>{0 1 2 3 }

Cluster 1 ==>{4 5 6 7 }

```
PATTERN 6:
The closest cluster is: 1
Distance 1.118034 < 2.000000
Therefore cluster 1 passed the distance test.
patern 6 assigned to cluster 1

The new cluster centers are:
CLUSTER 0 ==>[0.500000,0.500000]
CLUSTER 1 ==>[0.333333,3.333333]
CLUSTER Membership
 Cluster 0 ==>{0 1 2 3 }

 Cluster 1 ==>{4 5 6 }
```

```
 Cluster 2 ==>{8 }


PATTERN 9:
The closest cluster is: 2
Distance 1.000000 < 2.000000
Therefore cluster 2 passed the distance test.
pattern 9 assigned to cluster 2

The new cluster centers are:
CLUSTER 0 ==>[0.500000,0.500000]
CLUSTER 1 ==>[0.500000,3.500000]
CLUSTER 2 ==>[4.500000,0.000000]
CLUSTER Membership
 Cluster 0 ==>{0 1 2 3 }

 Cluster 1 ==>{4 5 6 7 }

 Cluster 2 ==>{8 9 }
```

```
PATTERN 10:
The closest cluster is: 2
Distance 1.118034 < 2.000000
Therefore cluster 2 passed the distance test
pattern 10 assigned to cluster 2

The new cluster centers are:
CLUSTER 0 ==>[0.500000,0.500000]
CLUSTER 1 ==>[0.500000,3.500000]
CLUSTER 2 ==>[4.333333,0.333333]
CLUSTER Membership
 Cluster 0 ==>{0 1 2 3 }

 Cluster 1 ==>{4 5 6 7 }

 Cluster 2 ==>{8 9 10 }
```

```
PATTERN 11:
The closest cluster is: 2
Distance 0.942809 < 2.000000
Therefore cluster 2 passed the distance test.
pattern 11 assigned to cluster 2

The new cluster centers are:
CLUSTER 0 ==>[0.500000,0.500000]
CLUSTER 1 ==>[0.500000,3.500000]
CLUSTER 2 ==>[4.500000,0.500000]
CLUSTER Membership
 Cluster 0 ==>{0 1 2 3 }

 Cluster 1 ==>{4 5 6 7 }

 Cluster 2 ==>{8 9 10 11 }
```

Note that we have concluded that there are three clusters with cluster centers at $\mathbf{C}_1 = (0.5, 0.5)$, $\mathbf{C}_2 = (0.5, 3.5)$, $\mathbf{C}_3 = (4.5, 0.5)$. The cluster membership list can also be seen to be as follows $S(1)=\{0, 1, 2, 3\}$, $S(2)=\{4, 5, 6, 7\}$, $S(3)=\{8, 9, 10, 11\}$. These results are shown graphically in Figure 10.4. Also note that the clusters we identified are very much what we as humans might expect from looking at the pattern. However, we will now repeat the example using a different distance threshold.

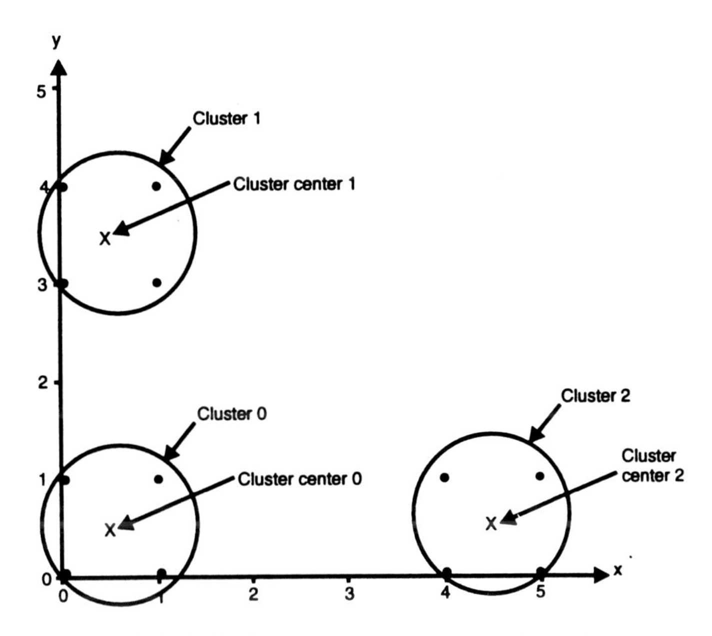

FIGURE 10.4 VQ cluster formation with ρ = 2.0.

10.3.2 VQ Example 2

We will now explore how VQ causes the same pattern used in example 1 (section 10.3.1) when the distance threshold is elevated to ρ = 3.5. Not surprisingly, we achieve a different result. The calculations follow.

The input patterns are:

```
Pattern[0][0]=0.000000
Pattern[0][1]=0.000000
Pattern[1][0]=1.000000
Pattern[1][1]=0.000000
Pattern[2][0]=0.000000
Pattern[2][1]=1.000000
Pattern[3][0]=1.000000
Pattern[3][1]=1.000000
Pattern[4][0]=0.000000
Pattern[4][1]=3.000000
Pattern[5][0]=1.000000
Pattern[5][1]=3.000000
Pattern[6][0]=0.000000
Pattern[6][1]=4.000000
Pattern[7][0]=1.000000
Pattern[7][1]=4.000000
Pattern[8][0]=4.000000
Pattern[8][1]=0.000000
Pattern[9][0]=5.000000
Pattern[9][1]=0.000000
Pattern[10][0]=4.000000
Pattern[10][1]=1.000000
Pattern[11][0]=5.000000
Pattern[11][1]=1.000000
```

PATTERN 0:
No Clusters exist so allocate a new cluster 0
pattern 0 assigned to cluster 0

The new cluster centers are:
CLUSTER 0 ==>[0.000000,0.000000]
CLUSTER Membership
Cluster 0 ==>{0 }

PATTERN 1:
The closest cluster is: 0
Distance 1.000000 < 3.500000
Therefore cluster 0 passed the distance test
pattern 1 assigned to cluster 0

The new cluster centers are:
CLUSTER 0 ==>[0.500000,0.000000]
CLUSTER Membership
Cluster 0 ==>{0 1 }

Cluster 0 ==>{0 1 2 3 }
PATTERN 4:
The closest cluster is: 0
Distance 2.549510 < 3.500000
Therefore cluster 0 passed the distance test
pattern 4 assigned to cluster 0

The new cluster centers are:
CLUSTER 0 ==>[0.400000,1.000000]
CLUSTER Membership
Cluster 0 ==>{0 1 2 3 4 }

PATTERN 5:
The closest cluster is: 0
Distance 2.088061 < 3.500000
Therefore cluster 0 passed the distance test
pattern 5 assigned to cluster 0

The new cluster centers are:
CLUSTER 0 ==>[0.500000,1.333333]
CLUSTER Membership
Cluster 0 ==>{0 1 2 3 4 5 }

PATTERN 6:
The closest cluster is: 0
Distance 2.713137 < 3.500000
Therefore cluster 0 passed the distance test
pattern 6 assigned to cluster 0

PATTERN 2:
The closest cluster is: 0
Distance 1.118034 < 3.500000
Therefore cluster 0 passed the distance test
pattern 2 assigned to cluster 0

The new cluster centers are:
CLUSTER 0 ==>[0.333333,0.333333]
CLUSTER Membership
Cluster 0 ==>{0 1 2 }

PATTERN 3:
The closest cluster is: 0
Distance 0.942809 < 3.500000
Therefore cluster 0 passed the distance test
pattern 3 assigned to cluster 0

The new cluster centers are:
CLUSTER 0 ==>[0.500000,0.500000]
CLUSTER Membership

The new cluster centers are:
CLUSTER 0 ==>[0.500000,2.000000]
CLUSTER 1 ==>[4.000000,0.000000]
CLUSTER Membership
Cluster 0 ==>{0 1 2 3 4 5 6 7 }

Cluster 1 ==>{8 }

PATTERN 9:
The closest cluster is: 1
Distance 1.000000 < 3.500000
Therefore cluster 1 passed the distance test
pattern 9 assigned to cluster 1

The new cluster centers are:
CLUSTER 0 ==>[0.500000,2.000000]
CLUSTER 1 ==>[4.500000,0.000000]
CLUSTER Membership
Cluster 0 ==>{0 1 2 3 4 5 6 7 }

Cluster 1 ==>{8 9 }

PATTERN 10:
The closest cluster is: 1
Distance 1.118034 < 3.500000
Therefore cluster 1 passed the distance test
pattern 10 assigned to cluster 1

```
The new cluster centers are:
CLUSTER 0 ==>[0.428571,1.714286]
CLUSTER Membership
 Cluster 0 ==>{0 1 2 3 4 5 6 }


PATTERN 7:
The closest cluster is: 0
Distance 2.356060 < 3.500000
Therefore cluster 0 passed the distance test
pattern 7 assigned to cluster 0

The new cluster centers are:
CLUSTER 0 ==>[0.500000,2.000000]
CLUSTER Membership
 Cluster 0 ==>{0 1 2 3 4 5 6 7 }

PATTERN 8:
The closest cluster is: 0
Distance 4.031129 > 3.500000
Therefore cluster 0 failed the distance test
so create NEW cluster number:1
pattern 8 assigned to cluster 1
```

```
The new cluster centers are:
CLUSTER 0 ==>[0.500000,2.000000]
CLUSTER 1 ==>[4.333333,0.333333]
CLUSTER Membership
 Cluster 0 ==>{0 1 2 3 4 5 6 7 }

 Cluster 1 ==>{8 9 10 }


PATTERN 11:
The closest cluster is: 1
Distance 0.942809 < 3.500000
Therefore cluster 1 passed the distance test
pattern 11 assigned to cluster 1

The new cluster centers are:
CLUSTER 0 ==>[0.500000,2.000000]
CLUSTER 1 ==>[4.500000,0.500000]
CLUSTER Membership
 Cluster 0 ==>{0 1 2 3 4 5 6 7 }

 Cluster 1 ==>{8 9 10 11 }
```

Note that we have concluded that there are only two clusters with cluster centers at $\mathbf{C}_1 = (0.5, 2.0)$, $\mathbf{C}_2 = (0.5, 3.5)$. The cluster membership list can also be seen to be as follows $S(1) = \{0, 1, 2, 3, 4, 5, 6, 7\}$, $S(2) = \{8, 9, 10, 11\}$. These results are shown graphically in Figure 10.5. Note that the clusters we identified are no longer quite what we might expect. We may reasonably observe that choosing a threshold distance too high may easily obscure meaningful categories. Conversely, choosing the threshold too low may lead to a proliferation of many nonuseful categories. (In the extreme case each input pattern vector would get its own cluster.)

10.3.3 VQ Example 3

As a final example of the vector quantization process, we demonstrate the dependency on the order in which vectors are presented. For this example we will retain the distance threshold of $\rho = 3.5$ while reordering the input data. Note that although the same number of clusters are generated, the population within the clusters and the position of the cluster centers change greatly. Figure 10.6 shows the grouping of patterns (clusters) formed by applying the VQ algorithm in this case. The calculations for this case follow.

The input patterns are:

```
Pattern[0][0]=1.000000
Pattern[0][1]=0.000000
Pattern[1][0]=4.000000
Pattern[1][1]=0.000000
Pattern[2][0]=0.000000
Pattern[2][1]=1.000000
Pattern[3][0]=0.000000
Pattern[3][1]=3.000000
Pattern[4][0]=1.000000
Pattern[4][1]=1.000000
```

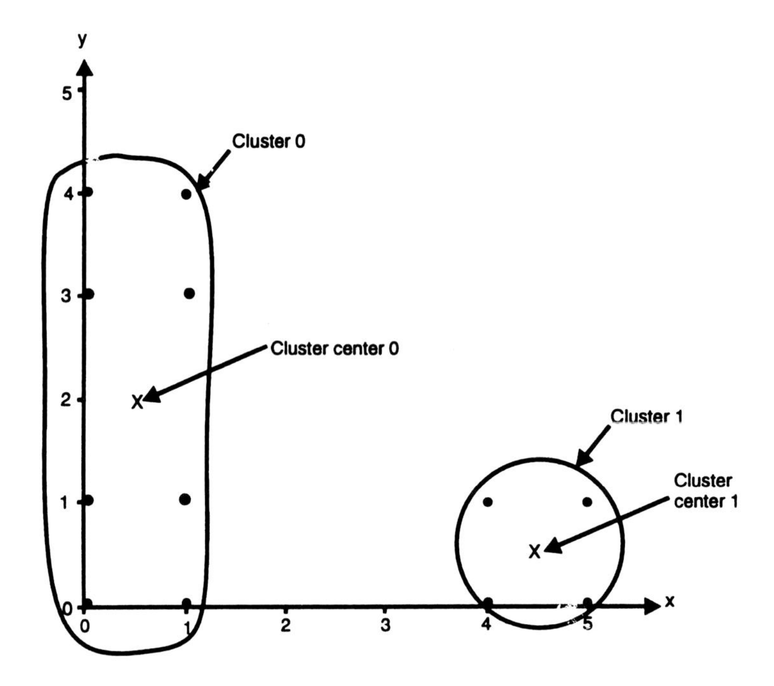

FIGURE 10.5 VQ cluster formation with ρ = 3.

Pattern[5][0]=4.000000
Pattern[5][1]=1.000000
Pattern[6][0]=5.000000
Pattern[6][1]=1.000000
Pattern[7][0]=0.000000
Pattern[7][1]=0.000000
Pattern[8][0]=1.000000
Pattern[8][1]=3.000000
Pattern[9][0]=0.000000
Pattern[9][1]=4.000000
Pattern[10][0]=1.000000
Pattern[10][1]=4.000000
Pattern[11][0]=5.000000
Pattern[11][1]=0.000000

PATTERN 0:
No Clusters exist so allocate a new cluster 0
pattern 0 assigned to cluster 0

The new cluster centers are:
CLUSTER 0 ==>[1.000000,0.000000]
CLUSTER Membership
Cluster 0 ==>{0 }

PATTERN 2:
The closest cluster is: 0
Distance 2.692582 < 3.500000
Therefore cluster 0 passed the distance test
pattern 2 assigned to cluster 0

The new cluster centers are:
CLUSTER 0 ==>[1.666667,0.333333]
CLUSTER Membership
Cluster 0 ==>{0 1 2 }

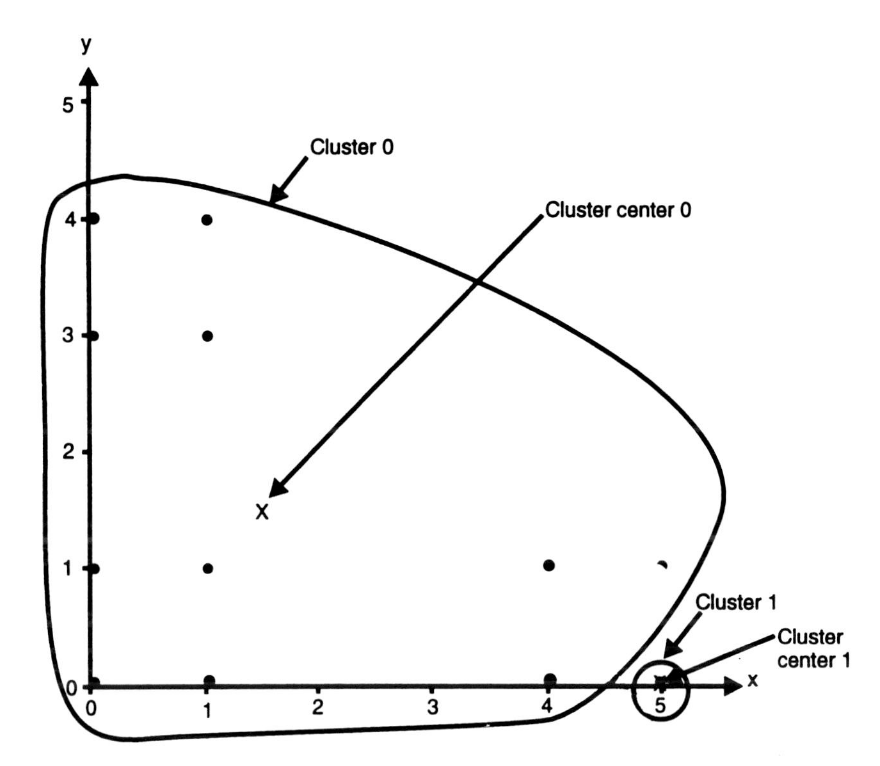

FIGURE 10.6 VQ cluster formation for ρ = 3.5 when order of presentation is altered.

PATTERN 1:
The closest cluster is: 0
Distance 3.000000 < 3.500000
Therefore cluster 0 passed the distance test
pattern 1 assigned to cluster 0

The new cluster centers are:
CLUSTER 0 ==>[2.500000,0.000000]
CLUSTER Membership
Cluster 0 ==>{0 1 }

PATTERN 3:
The closest cluster is: 0
Distance 3.144660 < 3.500000
Therefore cluster 0 passed the distance test
pattern 3 assigned to cluster 0

The new cluster centers are:
CLUSTER 0 ==>[1.250000,1.000000]
CLUSTER Membership
Cluster 0 ==>{0 1 2 3 }

PATTERN 4:
The closest cluster is: 0
Distance 0.250000 < 3.500000
Therefore cluster 0 passed the distance test
pattern 4 assigned to cluster 0

The new cluster centers are:
CLUSTER 0 ==>[1.200000,1.000000]
CLUSTER Membership
Cluster 0 ==>{0 1 2 3 4 }

PATTERN 8:
The closest cluster is: 0
Distance 2.298097 < 3.500000
Therefore cluster 0 passed the distance test
pattern 8 assigned to cluster 0

The new cluster centers are:
CLUSTER 0 ==>[1.777778,1.111111]
CLUSTER Membership
Cluster 0 ==>{0 1 2 3 4 5 6 7 8 }

```
PATTERN 5:
The closest cluster is: 0
Distance 2.800000 < 3.500000
Therefore cluster 0 passed the distance test
pattern 5 assigned to cluster 0

The new cluster centers are:
CLUSTER 0 ==>[1.666667,1.000000]
CLUSTER Membership
 Cluster 0 ==>{0 1 2 3 4 5 }


PATTERN 6:
The closest cluster is: 0
Distance 3.333333 < 3.500000
Therefore cluster 0 passed the distance test
pattern 6 assigned to cluster 0

The new cluster centers are:
CLUSTER 0 ==>[2.142857,1.000000]
CLUSTER Membership
 Cluster 0 ==>{0 1 2 3 4 5 6 }


PATTERN 7:
The closest cluster is: 0
Distance 2.364706 < 3.500000
Therefore cluster 0 passed the distance test
pattern 7 assigned to cluster 0

The new cluster centers are:
CLUSTER 0 ==>[1.875000,0.875000]
CLUSTER Membership
 Cluster 0 ==>{0 1 2 3 4 5 6 7 }


PATTERN 9:
The closest cluster is: 0
Distance 3.392075 < 3.500000
Therefore cluster 0 passed the distance test
pattern 9 assigned to cluster 0

The new cluster centers are:
CLUSTER 0 ==>[1.600000,1.400000]
CLUSTER Membership
 Cluster 0 ==>{0 1 2 3 4 5 6 7 8 9 }


PATTERN 10:
The closest cluster is: 0
Distance 2.668333 < 3.500000
Therefore cluster 0 passed the distance test
pattern 10 assigned to cluster 0

The new cluster centers are:
CLUSTER 0 ==>[1.545455,1.636364]
CLUSTER Membership
 Cluster 0 ==>{0 1 2 3 4 5 6 7 8 9 10 }


PATTERN 11:
The closest cluster is: 0
distance 3.822508 > 3.500000
Therefore cluster 0 failed the distance test
so create NEW cluster number:1
pattern 11 assigned to cluster 1

The new cluster centers are:
CLUSTER 0 ==>[1.545455,1.636364]
CLUSTER 1 ==>[5.000000,0.000000]
CLUSTER Membership
 Cluster 0 ==>{0 1 2 3 4 5 6 7 8 9 10 }

 Cluster 1 ==>{11 }
```

10.4 ART PHILOSOPHY

The ART network is an unsupervised vector classifier that accepts input vectors that are classified according to the stored pattern they most resemble. It also provides for a mechanism allowing adaptive expansion of the output layer of neurons until an adequate size is reached based on the number of classes, inherent in the observation.

The ART network can adaptively create a new neuron corresponding to an input pattern if it is determined to be "sufficiently" different from existing clusters. This determination, called the vigilance test, is incorporated into the adaptive backward network. Thus, the ART architecture allows the user to control the degree of similarity of patterns placed in the same cluster.

Figure 10.7 shows the simplified configuration of the ART architecture.

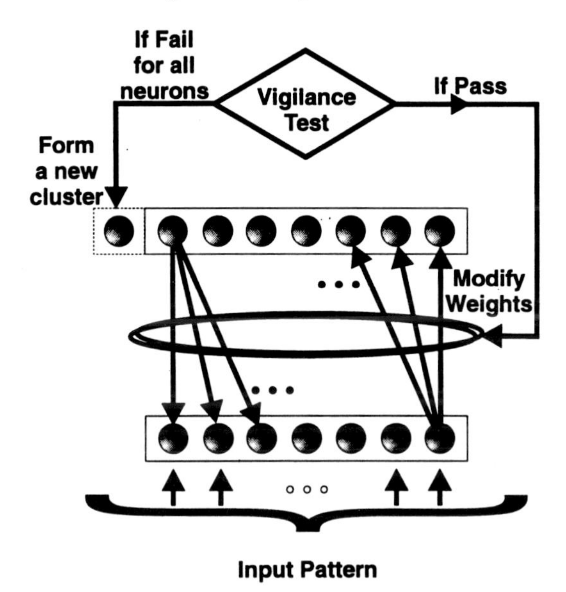

FIGURE 10.7 Simplified ART architecture.

10.5 THE STABILITY-PLASTICITY DILEMMA

The human brain has the ability to learn and memorize many new things in a fashion that does not necessarily cause the existing ones to be forgotten. In order to design a truly intelligent pattern recognition machine, compatible with the human brain, it would be highly desirable to impart this ability to our models.

Most pattern recognition paradigms discussed in earlier chapters will tend to forget old information if we attempt to store new patterns in an incremental fashion. In many of these paradigms a set of exemplars are first acquired and then used to train the system. The weight values in the system are adjusted to encode the patterns during the training. The exemplar patterns are often applied iteratively until the network has learned the entire set. Once the training is determined to be adequate, the system can perform in the operational mode (see Figures 4.4[a] and 4.4[b]) and no additional parameter (or weight) modifications are permitted.

In real-world applications, though, the network can be exposed to a constantly changing environment, such that training that does not evolve will ultimately become inaccurate. Exemplars of new classes are also encountered during the network operational phase. Consider a simple example. Suppose we are assigned the task of training a pattern classifier to recognize various human faces and place them into various categories based on different security levels in an organization. The appropriate images can be collected and used to train the network.

For most of the paradigms discussed in earlier chapters, after the model has successfully learned to recognize all of the faces of the employees within the organization, the training period is ended and no further modification of the parameters (or weights) is allowed. If at some future time the employees leave or new ones join the organization, we would have to delete or add their faces accordingly to the store of knowledge in our model. However, for these paradigms if a fully trained model must learn a new pattern (face in this case), it might disrupt the parameters so badly that complete retraining may be required. Training on only new faces could result in the network learning those faces quite well, but forgetting the previously learned faces.

The ability of a network to adapt and learn a new pattern well at any stage of operation is called *plasticity.*

Grossberg [1987] describes the stability-plasticity dilemma as follows:

> How can a learning system be designed to remain plastic, or adaptive, in response to significant events and yet remain stable in response to irrelevant events? How does the system know how to switch between its stable and its plastic modes to achieve stability without chaos? In particular how can it preserve its previously learned knowledge while continuing to learn new things? What prevents the new learning from washing away the memories of prior learning?

ART networks attempt to address the stability-plasticity dilemma. As such ART provides a mechanism by which the network can learn new patterns without forgetting (or degrading) old knowledge. For example, in the context of the character recognition problem, this could be useful in contexts such as training writer specific handwriting in an on-line system or in adding new fonts to an existing off-line system without needing to retrain the network from scratch.

The incorporation of a tolerance measure (vigilance test) allows ART architecture to resolve the stability-plasticity dilemma. New patterns from the environment can create additional classification categories, but they cannot cause an existing memory to be changed unless the two match closely.

In a physical system, when a small vibration of proper frequency causes a large-amplitude vibration, it is termed as resonance. The ART architecture gets its name due to the fact that the information in the form of a processing element output reverberates back and forth between layers. The neural network equivalent of resonance occurs when a proper pattern develops and a stable oscillation ensues. The pattern of activity that develops in the resonant state is called short-term memory (STM). The STM traces exist only in association with a single application of an input vector.

Learning (i.e., modification of weights) in the ART paradigm occurs only during the resonant period. The time required for updates in the weights between the processing elements is much longer than the time required to achieve resonance. These weights associated with the processing elements in different layers are called long-term memory (LTM) traces. The LTM traces encode information that remains a part of the network for an extended period.

10.6 ART1: BASIC OPERATION

This section discusses the ART1 network architecture and operation. ART1 inputs are binary valued as would be the case if the raw bit map of a character image were used as input to the classifier. ART2 extends the architecture to allow real-valued inputs, such as gray levels for each pixel for an image.

Figure 10.8 shows the overall architecture of the ART network. The ART1

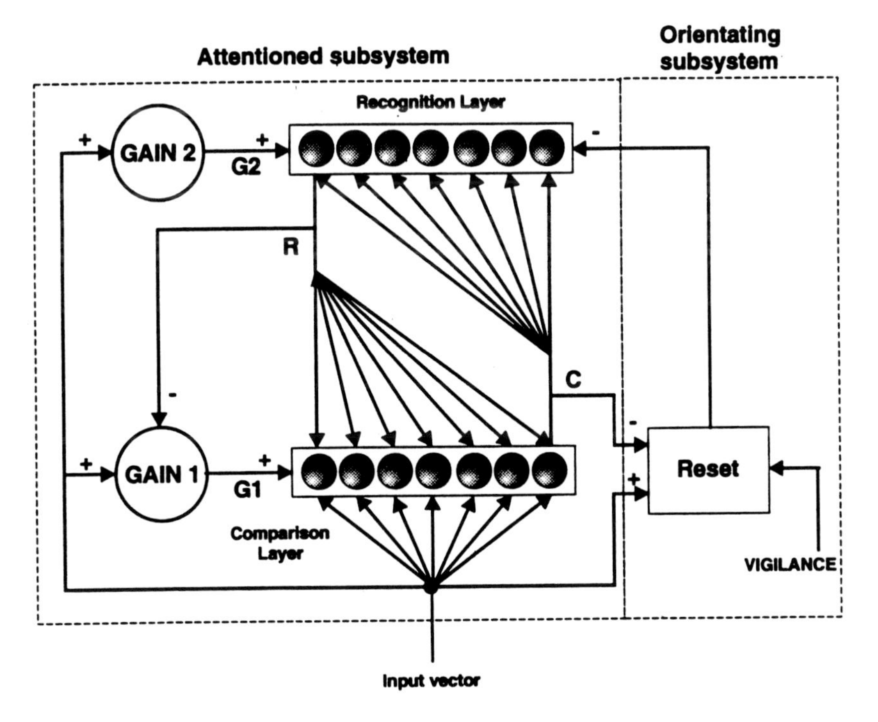

FIGURE 10.8 Overall ART architecture.

architecture consists of two layers of neurons called the *comparison layer* and the *recognition layer.* The classification decision is indicated by a single neuron in the recognition layer that fires. The neurons in the comparison layer respond to input features in the pattern, analogous to the cell groups in a sensory area of the cerebral cortex. The synaptic connections (weights) between these two layers are modifiable in both directions, according to two different learning rules. The recognition layer neurons have inhibitory connections that allow for a competition. This mechanism is common in artificial neural net architectures, inspired by the visual neurophysiology of the biological systems. The network architecture also consists of three additional modules labeled Gain 1, Gain 2, and Reset (see Figure 10.8).

The attention system consists of two layers of neurons (comparison and recognition) with feed-forward and feed-backward characteristics. This system determines whether the input pattern matches one of the prototypes stored. If a match occurs, resonance is established. The orienting subsystem is responsible for sensing mismatch between the bottom-up and top-down patterns on the recognition layer.

The recognition layer response to an input vector is compared to the original input vector through a mechanism termed vigilance. Vigilance provides a measure of the distance between the input vector and the cluster center corresponding to the firing recognition layer neuron. When vigilance falls below a preset threshold, a new category must be created and the input vector must be stored into that category. That is, a previously unallocated neuron within the recognition layer is allocated to a new cluster category associated with the new input pattern.

The recognition layer follows the winner-take-all paradigm (this behavior is sometimes referred to as MAXNET [Kung, 1993]). If the input vector passes the vigilance, the winning neuron (the one most like the input vector) is trained such that its associated cluster center in feature space is moved toward the input vector. The comparison layer is alternately termed F1 and is a bottom-up layer. The recognition layer is alternately termed F2 and is a top-down layer.

The recognition layer and the comparison layers are shown in more detail in Figures 10.9 and 10.10, respectively.

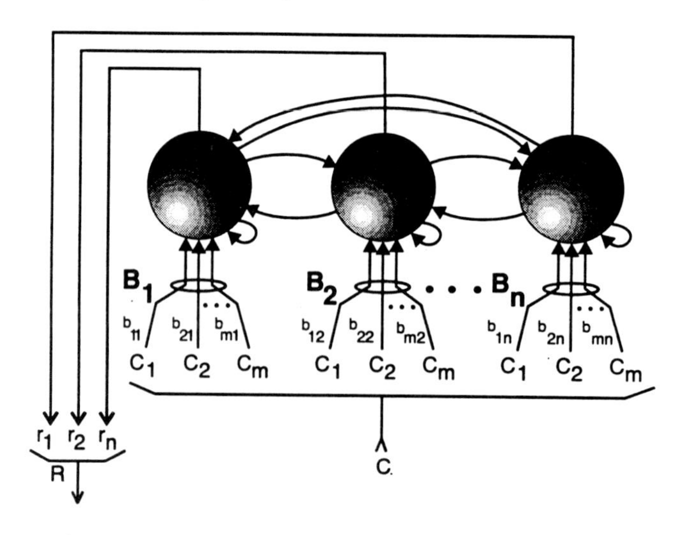

FIGURE 10.9 ART recognition layer.

Each recognition layer neuron, j, has a real-valued weight vector $\mathbf{B}_j$ associated with it (see Figure 10.9). This vector represents a stored exemplar pattern for a category of input patterns. Each neuron receives as input, the output of the comparison layer (vector **C**) through its weight vector, $\mathbf{B}_j$.

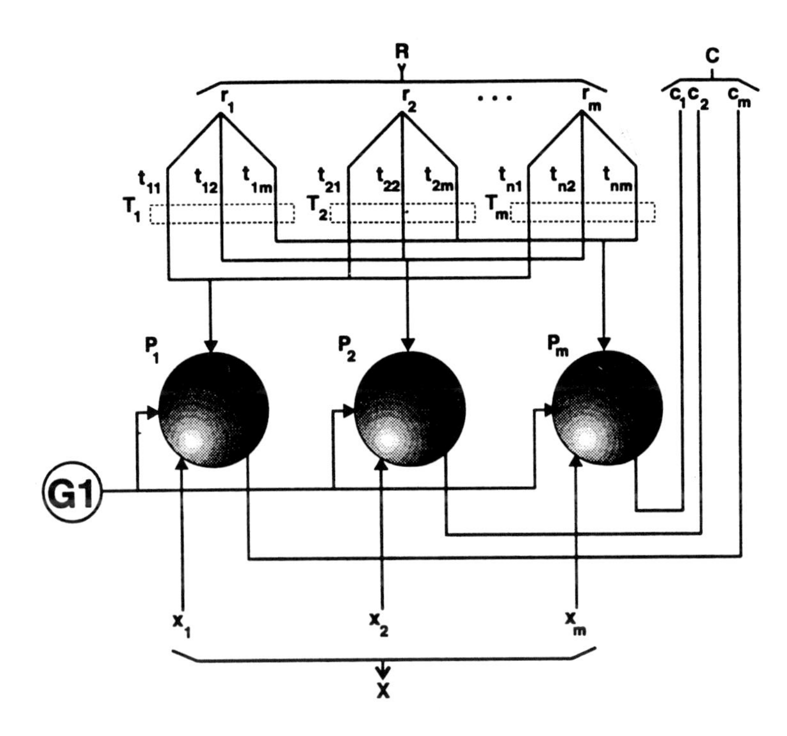

FIGURE 10.10 ART comparison layer.

The output of the recognition layer neuron, j, is given as:

$$\text{net}_j = \sum_{i=1}^{M} b_{ij} c_i$$

$$r_j = f(\text{net}_j) = \begin{cases} 1 & \text{for net}_j > \text{net}_i \text{ for all } i \neq j \\ 0 & \text{Otherwise} \end{cases} \tag{4}$$

where c_i is the output of ith comparison layer neuron; f is a step function and thus r_j results in a binary value. M is the number of neurons in the comparison layer.

As shown in Figure 10.10 each neuron, i, in the comparison layer receives the following three inputs:

1. A component of the input pattern is X, i.e., x_i.
2. The gain signal G1 is a scalar (binary value); thus the same value is input to each neuron.
3. A feedback signal from the recognition layer is a weighted sum of the recognition layer outputs

The feedback P_i through binary weights t_{ji} is given by:

$$P_i = \sum_{j=1}^{N} t_{ji} r_j \quad \text{for } i = 1, \ldots, M \tag{5}$$

where r_j is the output of the jth recognition layer neuron and N is the number of neurons in the recognition layer; in Figure 10.10, $\mathbf{T}_j$ is the weight vector associated with the recognition layer neuron j. Vector **C** represents the output of the comparison layer, with c_i representing the output of the ith neuron.

Gain 1 is one when the **R** vector is zero and the logical "OR" of the components of the input vector, **X**, is one, as seen in equation 6.

$$G_1 = \left(\overline{r_1 \mid r_2 \mid \ldots \mid r_N}\right) \bullet \left(x_1 \mid x_2 \mid \ldots \mid x_M\right) \tag{6}$$

The following C++ code segment illustrates the calculation of Gain 1:

```
int ARTNET::Gain1(){
  int i,G;
G=Gain2();
for (i=0; i<M; i++) {
  if (RVect[i]==1)
    return 0;
  } /* endfor */
return G;
}
```

Gain 2 is one when the logical OR of the components of the input vector, **X**, is 1, as seen in equation 7:

$$G_2 = \left(x_1 \mid x_2 \mid \ldots \mid x_M\right) \tag{7}$$

The following C++ code segment illustrates the calculation of Gain 2:

```
int ARTNET::Gain2(){
  int i;
for (i=0; i<M; i++) {
  if (XVect[i]==1)
    return 1;
  } /* endfor */
}
```

The comparison layer utilizes a two-thirds rule which states that if 2 of the 3 inputs are 1, then a 1 is output. Otherwise the result is zero. Equation 8 shows the two-thirds rule:

$$c_j = \begin{cases} 0 & \text{for } G_1 + x_j + P_j < 2 \\ 1 & \text{for } G_1 + x_j + P_j \geq 2 \end{cases} \tag{8}$$

The C++ code segment below illustrates the neuron output by the two-thirds rule:

```
void ARTNET::RunCompLayer(){
  int i,x;
for (i=0; i<M; i++) {
  x=XVect[i]+Gain1()+PVect[i];
  if (x>=2) {
    CVect[i]=1;
    }
   else {
    CVect[i]=0;
    } /* endif */
  } /* endfor */
}
```

The ART process occurs in stages. Initially there is no input; thus from equation 7 we can see that G2 is zero. When an input vector, shown in Figures 10.10 and 10.11 as **X**, is first presented to the network, the network enters the recognition phase. The **R** vector feedback from the recognition layer is always set to zero at the beginning of the recognition phase. Based on equations 6 and 7 we can see that presentation of **X** at this stage makes both G1 and G2 equal to 1.

As can be seen based on the initial conditions in the recognition phase, the output C of the comparison layer will be the unmodified input vector **X**. Thus the comparison layer passes **X** through to the recognition layer, as shown in Figure 10.11.

Next, each neuron in the recognition layer computes a dot product between its weight vector $\mathbf{B}_j$ (real valued) and the **C** vector (which is the output of the comparison layer). The winning vector fires, inhibiting all other neurons in the recognition layer (note the lateral inhibition in Figure 10.9). Thus a single component r_j of the **R** vector will be one and all other components of **R** will be zero. This initiates the comparison phase.

In other words, the recognition phase results in each recognition layer neuron comparing its prototype (stored in the bottom-up weights) with the input pattern (the dot product of $\mathbf{B}_j$ and **C**). The mutual inhibition mechanism causes the one with the best match to fire. See Figure 10.12.

During the comparison phase a determination must be made as to whether an input pattern is sufficiently similar to the winning stored prototype to be assimilated by that prototype. A test for this termed vigilance is performed during this phase.

In the comparison phase, the vector **R** is no longer zero so Gain 1 will be zero. By the two-thirds rule only neurons with simultaneous 1's in the **X** and **P** vectors

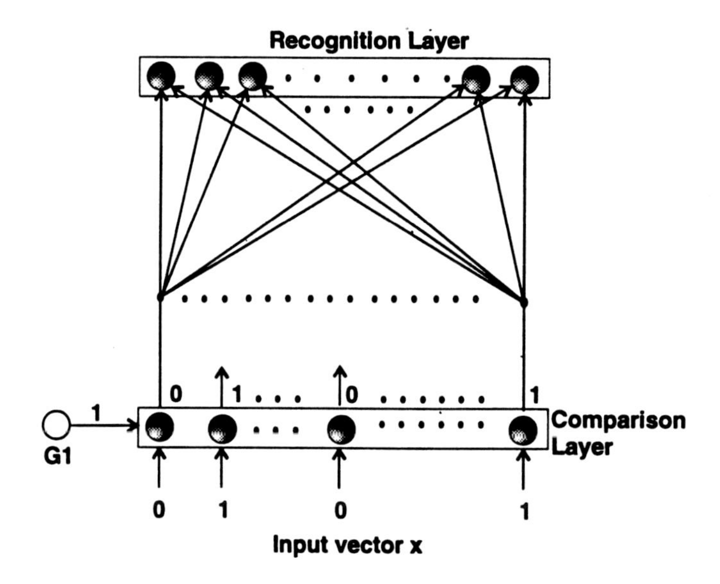

FIGURE 10.11 ART operation, step 1. G1 = 1. The input vector is passed through the comparison layer to the recognition layer.

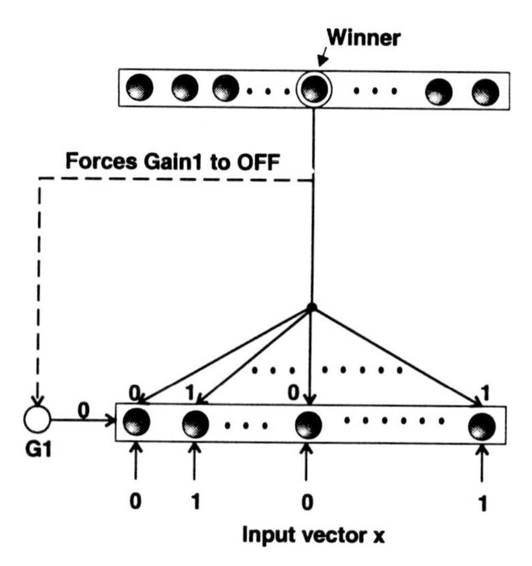

FIGURE 10.12 ART operation, step 2. The best neuron of the recognition layer has been selected as winner. The winner sends its signal back to the comparison layer through its top-down weights.

will fire. Note that the weights t_{ij} are binary valued. This top-down feedback path then forces components of **C** to zero whenever the input vector **X** fails to match the stored pattern.

Let D be the number of ones in the **x** vector and K be the number of ones in the **C** vector. Then the similarity ratio, **S**, is simply: $\mathbf{S} = K/D$.

The similarity vector, **S**, is therefore a metric for likeness between the prototype and the input pattern. Now we must establish a criterion by which to accept or reject clusters according to this metric. The test for vigilance can be represented as follows:

$$\mathbf{S} > \rho \rightarrow \text{ Vigilance test passed}$$

$$\mathbf{S} \leq \rho \rightarrow \text{ Vigilance test failed}$$

If the vigilance is passed, there is no substantial difference between the input vector and the winning prototype. Thus the required action is simply to store the input vector into the winning neuron cluster center. In this case, there is no reset signal. Therefore when the search phase is entered, the weights for this input vector are adjusted. At this point, the operation of the network is complete.

If **S** is below a preset threshold, the vigilance level, then the pattern **P** is not sufficiently similar to the winning neuron cluster center and the firing neuron should be inhibited. The inhibition is done by the reset block which resets the currently firing neuron throughout the duration of the current classification. (See Figure 10.13.) This concludes the comparison phase.

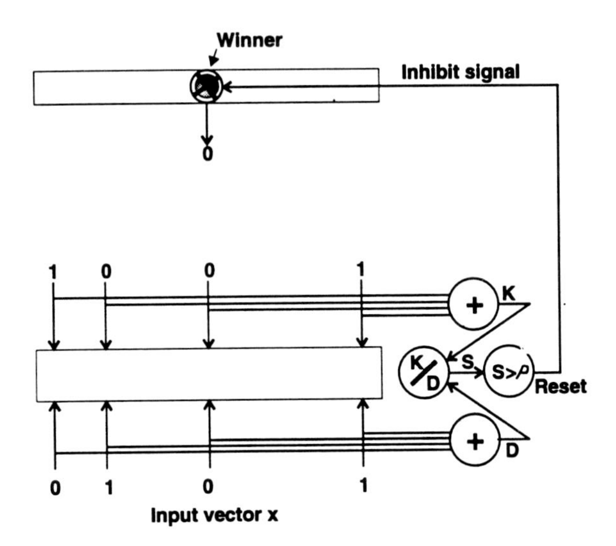

FIGURE 10.13 ART operation, step 3. The input vector, **x**, and the **P** vector from the recognition layer are compared. Vigilance has failed. Therefore the winning neuron is inhibited by the reset mechanism.

The search phase is then entered and if no reset signal has been generated, the match is considered adequate and the classification is complete. Otherwise, with the firing **R** layer neuron disabled the **R** vector is once again set to zero. As a result,

Gain 1 (G1) goes to one so that **X** once again appears on **C** and a different neuron in the recognition layer wins. (See Figure 10.14.) The new winner is checked against vigilance just as before and the process repeats until either:

1. A neuron is found that matches **X** with a similarity above the vigilance level (**S** > ρ). The weight vectors, $\mathbf{T}_j$ and $\mathbf{B}_j$ of the firing neuron, are adjusted, **or**
2. All stored patterns have been tried. Then a previously unallocated neur is associated with the pattern, and $\mathbf{T}_j$ and $\mathbf{B}_j$ are set to match the pattei

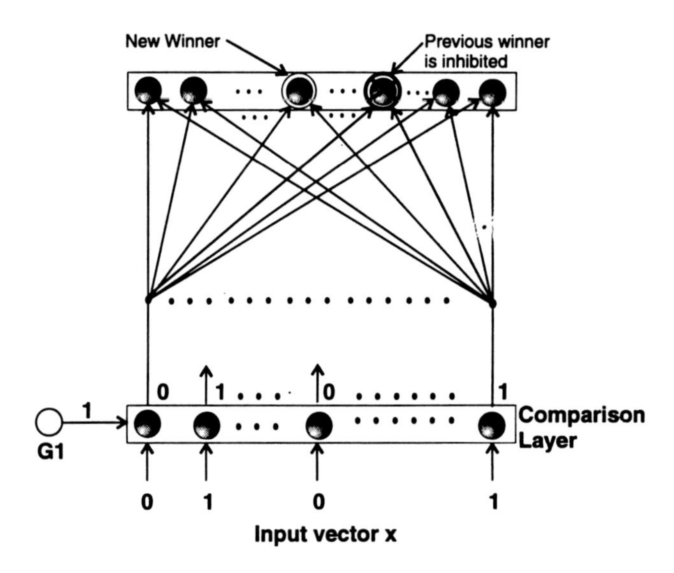

FIGURE 10.14 ART operation, step 4. The previous winning neuron remains disabled. The input vector is again presented to the recognition layer (G1 = 1) and a new winner is selected.

10.7 ART1: ALGORITHM

Initially the weights b_{ij} are initialized to the same low value which should be

$$b_{ij} < \frac{L}{(L-1+m)} \tag{9}$$

where m is the number of components in the input vector and L is a constant, typically $L = 2$.

The algorithm for the ART1 architecture is as follows:

1. When an input pattern, **X**, is presented to the network, the recognition layer selects the winner as the maximum of all the net outputs:

$$net_j = \sum_{i=1}^{N} b_{ij} c_i \tag{10}$$

where N is the number of neurons in the comparison layer.

2. Perform the vigilance test. A neuron j is declared to pass the vigilance test, if and only if,

$$\frac{net_j}{\sum_{i=1}^{N} x_i} > \rho \tag{11}$$

where ρ is the vigilance threshold.

2a. If the winner fails the test, mask the current winner and go to step 1 to select another winner.
2b. Repeat the cycle (steps 1 through 2a) until a winner is determined that passes the vigilance test; then go to step 4.
3. If no neuron passes the vigilance test, create a new neuron to accommodate the new pattern.
4. Adjust the feed-forward weights for the winner neuron. Update the feed-back weights from the winner neuron to its inputs.

The equations governing the training of the bottom-up and top-down weights are

$$b_{ij} = \frac{Lc_1}{\left(L - 1 + \Sigma c_k\right)} \tag{12}$$

$$t_{ij} = c_i$$

where c_i is the ith component of the comparison layer vector and j is the index of the winning recognition layer neuron.

10.8 THE GAIN CONTROL MECHANISM

Now, let us examine more closely the need for the gain control mechanism and the two-thirds rule.

One of the powerful abilities biological systems demonstrate is to often anticipate the next event. They have the ability to extract a trend in a series of events if one exists. For example, during a conversation we can often anticipate the next word that the other person is going to say.

Biological neural systems commonly have a hierarchical structure. Much of Grossberg's work has been focused on modeling the actual processes that occur within the brain. Thus Grossberg's ART network (shown in Figure 10.8) is only one in a hierarchy of networks in a much larger system. Expectation can arise in the ART model when the recognition layer is stimulated by a higher level in the hierarchy. This could result in a top-down signal that would arrive at the comparison layer before an input signal arrives from below.

The appearance of the early excitatory signal from the recognition layer could cause hallucinations if the comparison layer produces outputs in the absence of any sensory input pattern. The addition of the gain control and the two-thirds rule allows the system to avoid the occurrences of such hallucinations.

As shown in Figure 10.9 any signal coming from the recognition layer results in an inhibition of the gain. Now recall the two-thirds rule, that inhibition from G means an output from the comparison layer will elicit only if a sensory input pattern appears along with the anticipatory excitation from the above. If the comparison layer neurons do receive inputs from these two sources, they will send an output up to the recognition layer to begin the matching cycle.

Listing 10.2

```
/*****************************************************************************
 *
 * ADAPTIVE RESONANCE THEORY (ART) NETWORK
 *
 *****************************************************************************/

#include <stdio.h>
#include <stdlib.h>
#include <string.h>
#include <conio.h>
#include <math.h>


// DEFINES
#define MAXCNEURONS 75              // MAX COMPARISON LAYER NEURONS
#define MAXRNEURONS 30              // MAX RECOGNITION LAYER NEURONS
#define MAXPATTERNS 30     // MAX NUMBER OF PATTERNS IN A TRAINING SET
#define VERBOSE   1


class ARTNET
{
private:
 double Wb[MAXCNEURONS][MAXRNEURONS];  // Bottom up weight matrix
 int  Wt[MAXRNEURONS][MAXCNEURONS];    // Top down  weight matrix
 int  InData[MAXPATTERNS][MAXCNEURONS]; // Array of input vectors to be
```

```
                                        // presented to the network
int  NumPatterns;                       // Number of input patterns
double VigilThresh;                     // Vigilence threshold value
double L;                               // ART training const (see text)
int  M;                                 // # of neurons in C-layer
int  N;                                 // # of neurons in R-layer
int  XVect[MAXCNEURONS];                // Current in vect at C-layer.
int  CVect[MAXCNEURONS];                // Output vector from C-layer
int  BestNeuron;                        // Current best R-layer Neuron
int  Reset;                             // Active when vigilence has
                                        //    disabled someone
int  RVect[MAXCNEURONS];                // Output vector from R-layer
int  PVect[MAXCNEURONS];                // WeightedOutput vector from R-layer
int  Disabled[MAXRNEURONS];             // Resets way of disqualifying neurons
int  Trained[MAXRNEURONS];              // To identify allocated R-Neurons
void  ClearPvect();
void  ClearDisabled();
void  RecoPhase();                      // Recognition phase
void  CompPhase();                      // Comparison phase
void  SearchPhase();                    // Search Phase
void  RunCompLayer();                   // Calc comparison layer by 2/3 rule
void  RunRecoLayer();                   // Calc recognition layers R-vect
void  Rvect2Pvect(int);                 // Distribute winners result
int  Gain1();                           // Comp layer gain
int  Gain2();                           // Reco layer gain
double Vigilence();                     // Calc vigilence metric
void  InitWeights();                    // Initialize weights
void  Train();                          // Weight adjustment is done here
public:
 ARTNET(void);                          // Constructor/initializations
 int  LoadInVects(char *Fname);         // load all data vectors
 void  Run(int i);                      // Run net w/ ith pattern
 void  ShowWeights();                   // display top down and
                                        //  bottom up weights
 void  ShowInVect();                    // Display current input pattern
 void  ShowOutVect();                   // P-vector from Reco layer(see text)
};

//------------------------------------------------------------------------

// METHOD DEFINITIONS

ARTNET::ARTNET(){
int i;
L=2.0;
N=MAXRNEURONS;
for (i=0; i<N; i++) {                   //Set all neurons to untrained and enabled
  Trained[i]=0;
  Disabled[i]=0;
  } /* endfor */
}
```

```
int ARTNET::LoadInVects(char *Fname){
FILE *PFILE;
int  i,j,k;

PFILE = fopen(Fname,"r");  // batch
if (PFILE==NULL){
  printf("\nUnable to open file %s\n",Fname);
  exit(0);
  }
fscanf(PFILE,"%d",&NumPatterns);          //How many patterns
fscanf(PFILE,"%d",&M);                    //get width of input vector
fscanf(PFILE,"%lf",&VigilThresh);
for (i=0; i<NumPatterns; i++) {
  for (j=0; j<M; j++) {
    fscanf(PFILE,"%d",&k);                //Read all the pattern data and...

    InData[i][j]=k;                       // ...save it for later.
    } /* endfor */
  } /* endfor */
InitWeights();
return NumPatterns;
}

int ARTNET::Gain2(){
  int i;
for (i=0; i<M; i++) {
  if (XVect[i]==1)
    return 1;
  } /* endfor */
}


void ARTNET::Rvect2Pvect(int best){
  int i;
for (i=0; i<M; i++) {
  //PVect[i]= Wt[best][i]*RVect[i];
  PVect[i]= Wt[best][i];
  } /* endfor */
}

int ARTNET::Gain1(){
  int i,G;
G=Gain2();
for (i=0; i<M; i++) {
  if (RVect[i]==1)
    return 0;
  } /* endfor */
return G;
}

void ARTNET::RunCompLayer(){
  int i,x;
for (i=0; i<M; i++) {
  x=XVect[i]+Gain1()+PVect[i];
  if (x>=2) {
    CVect[i]=1;
    }
```

```
   else {
     CVect[i]=0;
     } /* endif */
  } /* endfor */
}

double  ARTNET::Vigilence(){
  int i;
  double S,K,D;
// count # of 1's in p-vect & x-vect
K=0.0;
D=0.0;
for (i=0; i<M; i++) {
  K+=CVect[i];
  D+=XVect[i];
  } /* endfor */
S=K/D;
return S;
}

void ARTNET::RunRecoLayer(){
  int i,j,k;
  double Net[MAXRNEURONS];
  int BestNeruon=-1;
  double NetMax=-1;
for (i=0; i<N; i++) {                          //Traverse all R-layer Neurons
  Net[i]=0;
  for (j=0; j<M; j++) {    // Do the product
     Net[i] +=Wb[i][j]*CVect[j];
     } /* endfor */
  if ((Net[i]>NetMax) && (Disabled[i]==0)) { //disabled neurons cant win!
     BestNeuron=i;
     NetMax=Net[i];
     }
  } /* endfor */
for (k=0; k<N; k++) {
  if (k==BestNeuron)
     RVect[k]=1;                               // Winner gets 1
   else
     RVect[k]=0;                               // lateral inhibition kills the rest
  } /* endfor */
}

void  ARTNET::RecoPhase(){
  int i;
//First force all R-layer outputs to zero
for (i=0; i<N; i++) {
  RVect[i]=0;
  } /* endfor */
for (i=0; i<M; i++) {
  PVect[i]=0;
  } /* endfor */
//Now Calculate C-layer outputs
RunCompLayer();                               //C-vector now has the result
RunRecoLayer();                               //Calc dot prod w/ bot up weight & C
Rvect2Pvect(BestNeuron);
}
```

```
void  ARTNET::CompPhase(){
  double S;
RunCompLayer();                          //Cvector<-dif between  x & p
S=Vigilence();
if (S<VigilThresh){
  Reset=1;
  RVect[BestNeuron]=0;
  Disabled[BestNeuron]=1;
  }
else
  Reset=0;
}

void  ARTNET::SearchPhase(){
  double S;
while (Reset) {
//  Rvect2Pvect(0);                       //Rvect 0 turns on gain 1
  ClearPvect();
  RunCompLayer();                        //Xvect -> Cvect
  RunRecoLayer();                        //Find a new winner with prev winners disabled
  Rvect2Pvect(BestNeuron);               //new pvect based on new winner
  S=Vigilence();                         //calc vigilence for the new guy
if (S<VigilThresh){                      //check if he did ok
  Reset=1;                               //     if not disable him too
  RVect[BestNeuron]=0;
  Disabled[BestNeuron]=1;
  }
else
  Reset=0;                               //Current Best neuron is a good
                                         //  winner...Train him
  } /* endwhile */
if (BestNeuron!=-1) {
  Train();
  }
else {
  //Failed to allocate a neuron for current pattern.
  printf("Out of neurons in F2\n");
  } /* endif */
ClearDisabled();
}

void ARTNET::ClearDisabled() {
  int i;
for (i=0; i<M; i++) {
  Disabled[i]=0;
  } /* endfor */
}

void ARTNET::ClearPvect() {
  int i;
for (i=0; i<M; i++) {
  PVect[i]=0;
  } /* endfor */
}
```

```
void ARTNET::Train(){
  int i,z=0;
for (i=0; i<M; i++) {
  z+=CVect[i];
  } /* endfor */
for (i=0; i<M; i++) {
  Wb[BestNeuron][i]=L*CVect[i]/(L-1+z);
  Wt[BestNeuron][i]=CVect[i];
  } /* endfor */
Trained[BestNeuron]=1;
}


void ARTNET::Run(int tp){
  int i,j;

ClearPvect();
for (i=0; i<M; i++) {
  XVect[i]=InData[tp][i];
  } /* endfor */
RecoPhase();
CompPhase();
SearchPhase();
}

void  ARTNET::InitWeights(){                    // Initialize weights
  int i,j;
  double b;
for (i=0; i<N; i++) {                           // from R-neuron i
  for (j=0; j<M; j++) {                         // to C-neuron j
     Wt[i][j]= 1;                               // All init'd to 1
     } /* endfor */
  } /* endfor */
b=L/(L-1+M);
for (i=0; i<N; i++) {                           // from C-neuron i
  for (j=0; j<M; j++) {                         // to R-neuron j
     Wb[i][j]= b;
     } /* endfor */
  } /* endfor */

}
void ARTNET::ShowWeights(){
  int i,j;
printf("\nTop Down weights:\n");
for (i=0; i<N; i++) {
  if(Trained[i]==1){
     for (j=0; j<M; j++) {
        printf("%d ",Wt[i][j]);
        } /* endfor */
     printf("\n");
     } /* endif */
  } /* endfor */
```

```
printf("\nBottom up weights:\n");
for (i=0; i<N; i++) {
  if(Trained[i]==1){
     for (j=0; j<M; j++) {
        printf("%f ",Wb[i][j]);
        } /* endfor */
     printf("\n");
     } /* endif */
  } /* endfor */
}

void ARTNET::ShowInVect(){
  int i;
printf("BEST NEURON:%d\nIN:  ",BestNeuron);
for (i=0; i<M; i++) {
  printf("%d ",XVect[i]);
  } /* endfor */
printf("\n");
}

void ARTNET::ShowOutVect(){
  int i;
printf("OUT: ");
for (i=0; i<M; i++) {
  printf("%d ",CVect[i]);
  } /* endfor */
printf("\n");
}

//-----------------------------------------------------------------
ARTNET    ART;

/*******************************************************************************
*  MAIN                                                                        *
*******************************************************************************/


int main(int argc, char *argv[])
{
  int TstSetSize;
  int i;
if (argc>1) {
  TstSetSize=ART.LoadInVects(argv[1]);
  for (i=0; i<TstSetSize; i++) {
     ART.Run(i);                         //Evaluate ith test pattern
     printf("\n");
     ART.ShowInVect();
     ART.ShowOutVect();
     if (VERBOSE==1) ART.ShowWeights();
  } /* endfor */
  }
else {
  printf("USAGE: ART PATTERN_FILE_NAME\n");
  exit(0);
  }
```

```
return 0;
}
```

10.8.1 Gain Control Example 1

The following example shows in detail the application of the ART1 algorithm in section 10.7 and the computations performed by the C++ code for ART1 given above (see listing 10.2).

For this example we have used a relatively low value for the vigilance threshold ($\rho = 0.3$). This will result in the formation of relatively fewer clusters than would occur at a higher vigilance threshold. The effect of higher vigilance will be addressed in our next example.

Figure 10.15 shows the five 3×5 input patterns that are presented to the ART1 network with 15 input neurons.

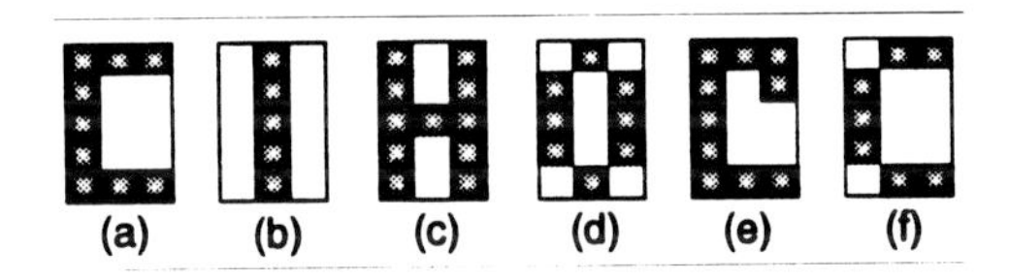

FIGURE 10.15 Example 1, input pattern.

At the first iteration pattern (a) in Figure 10.15 is presented to the ART1 network which is initialized, as discussed in section 10.6. The network enters the recognition phase and by applying the two-thirds rule (see equation 8) we get vector **C**, which is set equal to **X**. Note, that using equation 6, the Gain 1 is set to 1 at this stage (see Figure 10.11). Since no stored prototypes exist at this point, an unallocated neuron in the recognition layer wins (numbered 0 in this particular case). The results are shown below:

ITERATION 1

WINNING NEURON:0
INPUT PATTERN, X: 1 1 1 1 0 0 1 0 0 1 0 0 1 1 1
OUTPUT, C: 1 1 1 1 0 0 1 0 0 1 0 0 1 1 1

Top Down weights
1 1 1 1 0 0 1 0 0 1 0 0 1 1 1

Bottom up weights
0.200000 0.200000 0.200000 0.200000 0.000000
0.000000 0.200000 0.000000 0.000000 0.200000
0.000000 0.000000 0.200000 0.200000 0.200000

The first neuron in the recognition layer is assigned to this cluster and the prototype is stored by setting the top-down weights to the input pattern, using equation 12. Figure 10.16 shows the pictorial representation of the state of the network. Figure 10.18 shows the corresponding cluster formation.

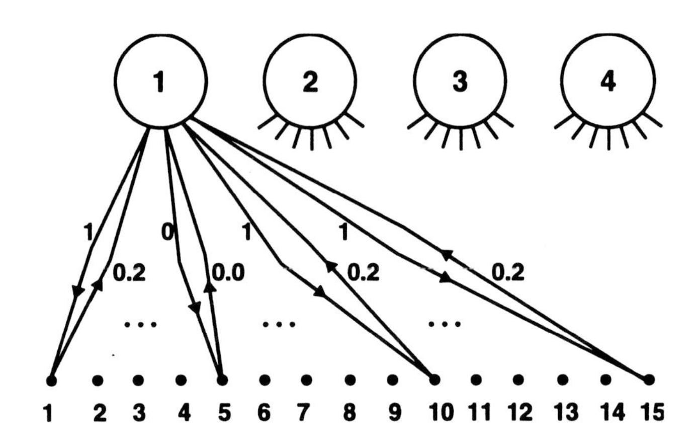

FIGURE 10.16 State of the example ART net after one iteration.

At the second iteration pattern (b) in Figure 10.15 is presented to the network. Again **C** is set to the same values as **X**.

ITERATION 2

WINNING NEURON:1
INPU NEURON, X: 0 1 0 0 1 0 0 1 0 0 1 0 0 1 0
OUTPUT, C: 0 1 0 0 1 0 0 1 0 0 1 0 0 1 0

Top Down weights
1 1 1 1 0 0 1 0 0 1 0 0 1 1 1
0 1 0 0 1 0 0 1 0 0 1 0 0 1 0

Bottom up weights
0.200000 0.200000 0.200000 0.200000 0.000000
0.000000 0.200000 0.000000 0.000000 0.200000
0.000000 0.000000 0.200000 0.200000 0.200000
0.000000 0.333333 0.000000 0.000000 0.333333
0.000000 0.000000 0.333333 0.000000 0.000000
0.333333 0.000000 0.000000 0.333333 0.000000

Note that the pattern presented to the network in iteration 2 (see Figure 10.15 [b]) was sufficiently different from the previously stored pattern (see Figure 10.15 [a]). Thus, a new cluster had to be created. As a consequence, the number of trained neurons now becomes two (rather that one) since recognition layer neurons have been allocated to a cluster center.

ITERATION 3
WINNING NEURON:0
INPUT PATTERN, X: 1 0 1 1 0 1 1 1 1 1 0 1 1 0 1
OUTPUT, C: 1 0 1 1 0 0 1 0 0 1 0 0 1 0 1

Top Down weights
1 0 1 1 0 0 1 0 0 1 0 0 1 0 1
0 1 0 0 1 0 0 1 0 0 1 0 0 1 0

Bottom up weights
0.250000 0.000000 0.250000 0.250000 0.000000
0.000000 0.250000 0.000000 0.000000 0.250000
0.000000 0.000000 0.250000 0.000000 0.250000
0.000000 0.333333 0.000000 0.000000 0.333333
0.000000 0.000000 0.333333 0.000000 0.000000
0.333333 0.000000 0.000000 0.333333 0.000000

The pattern presented in this iteration (Figure 10.15 [c]) was similar enough to the prototype for cluster 0. Thus the following occurred:

1. Neuron 0 in the recognition layer was the winner in the recognition phase.
2. The top-down weights of neuron 0 were similar enough to the input to pass vigilance.

As a result weights of neuron 0 (both top-down and bottom-up) were modified and no additional cluster center is formed; i.e., only the same two recognition layer neurons remain allocated.

ITERATION 4

WINNING NEURON:2
INPUT PATTERN, X: 0 1 0 1 0 1 1 0 1 1 0 1 0 1 0
OUTPUT, C: 0 1 0 1 0 1 1 0 1 1 0 1 0 1 0

Top Down weights
1 0 1 1 0 0 1 0 0 1 0 0 1 0 1
0 1 0 0 1 0 0 1 0 0 1 0 0 1 0
0 1 0 1 0 1 1 0 1 1 0 1 0 1 0
Bottom up weights
0.250000 0.000000 0.250000 0.250000 0.000000
0.000000 0.250000 0.000000 0.000000 0.250000
0.000000 0.000000 0.250000 0.000000 0.250000
0.000000 0.333333 0.000000 0.000000 0.333333

0.000000 0.000000 0.333333 0.000000 0.000000
0.333333 0.000000 0.000000 0.333333 0.000000
0.000000 0.222222 0.000000 0.222222 0.000000
0.222222 0.222222 0.000000 0.222222 0.222222
0.000000 0.222222 0.000000 0.222222 0.000000

The pattern in this iteration (Figure 10.15 [d]) requires a new cluster center resulting in the allocation of a new recognition layer neuron. The weights attached to this neuron (the third raw in both weight matrices) are accordingly updated to match that pattern.

ITERATION 5

WINNING NEURON:0
INPUT PATTERN, X: 1 1 1 1 0 1 1 0 0 1 0 0 1 1 1
OUTPUT, C: 1 0 1 1 0 0 1 0 0 1 0 0 1 0 1

Top Down weights
1 0 1 1 0 0 1 0 0 1 0 0 1 0 1
0 1 0 0 1 0 0 1 0 0 1 0 0 1 0
0 1 0 1 0 1 1 0 1 1 0 1 0 1 0

Bottom up weights
0.250000 0.000000 0.250000 0.250000 0.000000
0.000000 0.250000 0.000000 0.000000 0.250000
0.000000 0.000000 0.250000 0.000000 0.250000
0.000000 0.333333 0.000000 0.000000 0.333333
0.000000 0.000000 0.333333 0.000000 0.000000
0.333333 0.000000 0.000000 0.333333 0.000000
0.000000 0.222222 0.000000 0.222222 0.000000
0.222222 0.222222 0.000000 0.222222 0.222222
0.000000 0.222222 0.000000 0.222222 0.000000

At this point all five patterns are stored. Pattern (a) in Figure 10.15 is presented once again to make sure that it indeed gets classified in to cluster 0. Note that the resultant top-down and bottom-up weights still get updated and the stored prototype resembles the presented pattern more closely.

ITERATION 6

WINNING NEURON:0
INPUT PATTERN, X: 0 1 1 1 0 0 1 0 0 1 0 0 0 1 1
OUTPUT, C: 0 0 1 1 0 0 1 0 0 1 0 0 0 0 1

Top Down weights
0 0 1 1 0 0 1 0 0 1 0 0 0 0 1
0 1 0 0 1 0 0 1 0 0 1 0 0 1 0
0 1 0 1 0 1 1 0 1 1 0 1 0 1 0

Bottom up weights
0.000000 0.000000 0.333333 0.333333 0.000000
0.000000 0.333333 0.000000 0.000000 0.333333
0.000000 0.000000 0.000000 0.000000 0.333333
0.000000 0.333333 0.000000 0.000000 0.333333
0.000000 0.000000 0.333333 0.000000 0.000000
0.333333 0.000000 0.000000 0.333333 0.000000
0.000000 0.222222 0.000000 0.222222 0.000000
0.222222 0.222222 0.000000 0.222222 0.222222
0.000000 0.222222 0.000000 0.222222 0.000000

The resultant network is shown in Figure 10.17 with the top-down and bottom-up weights. Figure 10.18 shows the clusters formed with each prototype pattern vector.

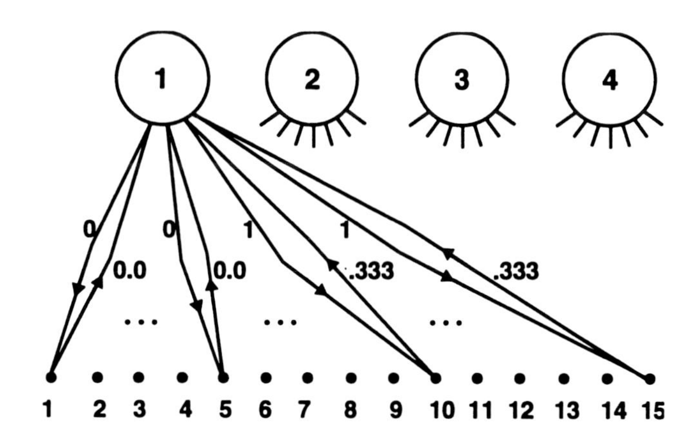

FIGURE 10.17 State of the example ART network after training with all patterns (vigilance, $\rho = 0.3$).

10.8.2 Gain Control Example 2

In this example we demonstrate the effect of raising the vigilance threshold. We repeat the same patterns used in example 1 (see Figure 10.15) but utilize a vigilance threshold of $\rho = 0.8$. We have observed that the number of clusters formed increases and results in a different distribution of patterns among the newly defined clusters.

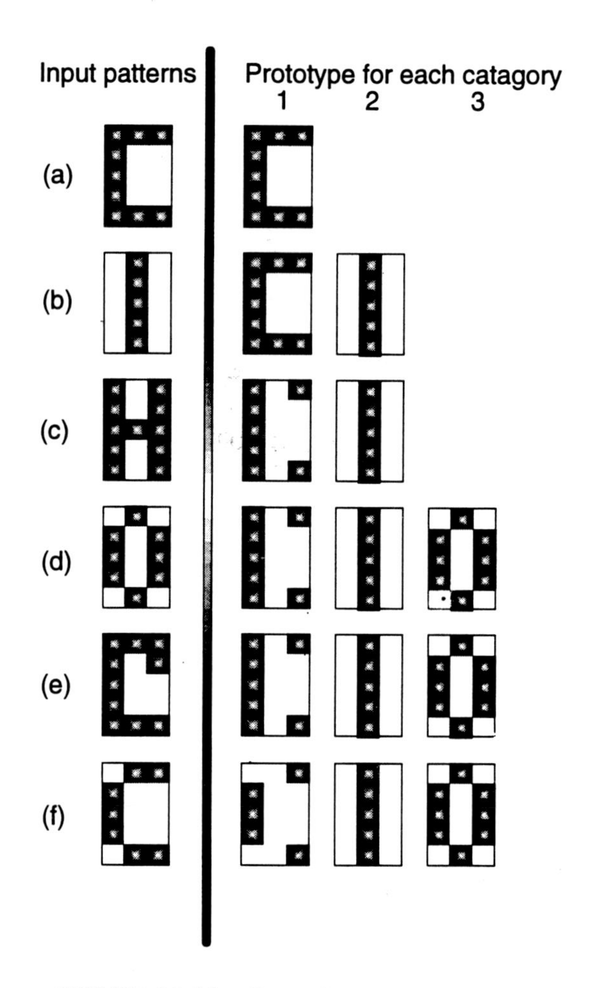

FIGURE 10.18 Cluster formation in example 1.

BEST NEURON:0
IN: 1 1 1 1 0 0 1 0 0 1 0 0 1 1 1
OUT: 1 1 1 1 0 0 1 0 0 1 0 0 1 1 1

Top Down weights:
1 1 1 1 0 0 1 0 0 1 0 0 1 1 1

Bottom up weights:
0.200000 0.200000 0.200000 0.200000 0.000000
0.000000 0.200000 0.000000 0.000000 0.200000
0.000000 0.000000 0.200000 0.200000 0.200000

BEST NEURON:1
IN: 0 1 0 0 1 0 0 1 0 0 1 0 0 1 0
OUT: 0 1 0 0 1 0 0 1 0 0 1 0 0 1 0

Top Down weights:
1 1 1 1 0 0 1 0 0 1 0 0 1 1 1
0 1 0 0 1 0 0 1 0 0 1 0 0 1 0

Bottom up weights:
0.200000 0.200000 0.200000 0.200000 0.000000
0.000000 0.200000 0.000000 0.000000 0.200000
0.000000 0.000000 0.200000 0.200000 0.200000
0.000000 0.333333 0.000000 0.000000 0.333333
0.000000 0.000000 0.333333 0.000000 0.000000
0.333333 0.000000 0.000000 0.333333 0.000000

BEST NEURON:2
IN: 1 0 1 1 0 1 1 1 1 1 0 1 1 0 1
OUT: 1 0 1 1 0 1 1 1 1 1 0 1 1 0 1

Top Down weights:
1 1 1 1 0 0 1 0 0 1 0 0 1 1 1
0 1 0 0 1 0 0 1 0 0 1 0 0 1 0
1 0 1 1 0 1 1 1 1 1 0 1 1 0 1

Bottom up weights:
0.200000 0.200000 0.200000 0.200000 0.000000
0.000000 0.200000 0.000000 0.000000 0.200000
0.000000 0.000000 0.200000 0.200000 0.200000
0.000000 0.333333 0.000000 0.000000 0.333333
0.000000 0.000000 0.333333 0.000000 0.000000
0.333333 0.000000 0.000000 0.333333 0.000000
0.166667 0.000000 0.166667 0.166667 0.000000
0.166667 0.166667 0.166667 0.166667 0.166667
0.000000 0.166667 0.166667 0.000000 0.166667

BEST NEURON:3
IN: 0 1 0 1 0 1 1 0 1 1 0 1 0 1 0
OUT: 0 1 0 1 0 1 1 0 1 1 0 1 0 1 0

Top Down weights:
1 1 1 1 0 0 1 0 0 1 0 0 1 1 1
0 1 0 0 1 0 0 1 0 0 1 0 0 1 0
1 0 1 1 0 1 1 1 1 1 0 1 1 0 1
0 1 0 1 0 1 1 0 1 1 0 1 0 1 0

```
Bottom up weights:
0.200000 0.200000 0.200000 0.200000 0.000000
0.000000 0.200000 0.000000 0.000000 0.200000
0.000000 0.000000 0.200000 0.200000 0.200000
0.000000 0.333333 0.000000 0.000000 0.333333
0.000000 0.000000 0.333333 0.000000 0.000000
0.333333 0.000000 0.000000 0.333333 0.000000
0.166667 0.000000 0.166667 0.166667 0.000000
0.166667 0.166667 0.166667 0.166667 0.166667
0.000000 0.166667 0.166667 0.000000 0.166667
0.000000 0.222222 0.000000 0.222222 0.000000
0.222222 0.222222 0.000000 0.222222 0.222222
0.000000 0.222222 0.000000 0.222222 0.000000

BEST NEURON:0
IN: 1 1 1 1 0 1 1 0 0 1 0 0 1 1 1
OUT: 1 1 1 1 0 0 1 0 0 1 0 0 1 1 1

Top Down weights:
1 1 1 1 0 0 1 0 0 1 0 0 1 1 1
0 1 0 0 1 0 0 1 0 0 1 0 0 1 0
1 0 1 1 0 1 1 1 1 1 0 1 1 0 1
0 1 0 1 0 1 1 0 1 1 0 1 0 1 0

Bottom up weights:
0.200000 0.200000 0.200000 0.200000 0.000000
0.000000 0.200000 0.000000 0.000000 0.200000
0.000000 0.000000 0.200000 0.200000 0.200000
0.000000 0.333333 0.000000 0.000000 0.333333
0.000000 0.000000 0.333333 0.000000 0.000000
0.333333 0.000000 0.000000 0.333333 0.000000
0.166667 0.000000 0.166667 0.166667 0.000000
0.166667 0.166667 0.166667 0.166667 0.166667
0.000000 0.166667 0.166667 0.000000 0.166667
0.000000 0.222222 0.000000 0.222222 0.000000
0.222222 0.222222 0.000000 0.222222 0.222222
0.000000 0.222222 0.000000 0.222222 0.000000

BEST NEURON:0
IN: 0 1 1 1 0 0 1 0 0 1 0 0 0 1 1
OUT: 0 1 1 1 0 0 1 0 0 1 0 0 0 1 1
```

Top Down weights:
0 1 1 1 0 0 1 0 0 1 0 0 0 1 1
0 1 0 0 1 0 0 1 0 0 1 0 0 1 0
1 0 1 1 0 1 1 1 1 1 0 1 1 0 1
0 1 0 1 0 1 1 0 1 1 0 1 0 1 0

Bottom up weights:
0.000000 0.250000 0.250000 0.250000 0.000000
0.000000 0.250000 0.000000 0.000000 0.250000
0.000000 0.000000 0.000000 0.250000 0.250000
0.000000 0.333333 0.000000 0.000000 0.333333
0.000000 0.000000 0.333333 0.000000 0.000000
0.333333 0.000000 0.000000 0.333333 0.000000
0.166667 0.000000 0.166667 0.166667 0.000000
0.166667 0.166667 0.166667 0.166667 0.166667
0.000000 0.166667 0.166667 0.000000 0.166667
0.000000 0.222222 0.000000 0.222222 0.000000
0.222222 0.222222 0.000000 0.222222 0.222222
0.000000 0.222222 0.000000 0.222222 0.000000

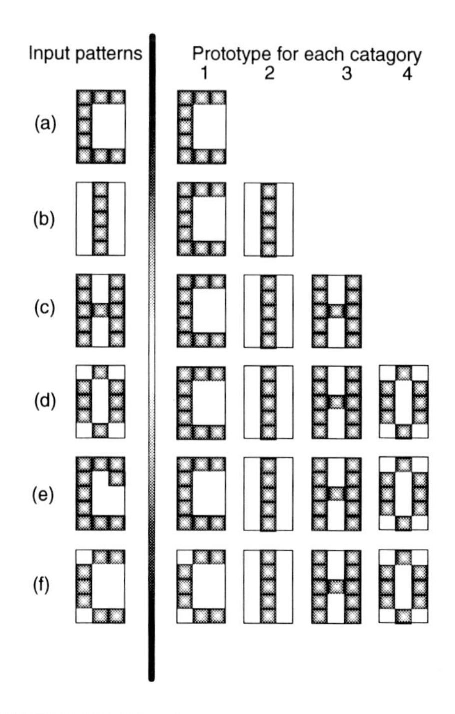

FIGURE 10.19 Cluster formation in example 2.

Observe that an additional cluster has been formed (cluster 3). Moreover, we may observe that cluster 3 (which used to belong to cluster 0 in the previous example) now also captures pattern 5 (also allocated to cluster 0 in example 1).

10.9 ART2 MODEL

The ART2 architecture builds on the ideas of ART1 with two layers and modifiable weights in both directions (i.e., both feed-forward and feedback weights). Both ART1 and ART2 contain an attentional subsystem and an orienting subsystem. Like ART1, the ART2 orienting subsystem consists of a comparison layer, F1, and a recognition layer, F2, and performs an equivalent function to that in ART1.

Several modifications were made to accommodate patterns with continuous-valued components. Figure 10.20 shows the ART2 architecture where the comparison layer of ART1 is replaced by a multilayer structure with several sets of neurons. Notice that the comparison or F1 layer has been split into several layers. Additionally the orienting subsystem has also been modified to accommodate real-valued data. The extra nodes are designed to include the following, as shown in Figure 10.20:

1. Allowance for noise suppression
2. Normalization, i.e., contrast to enhance significant parts of the pattern
3. Comparison of top-down and bottom-up signals needed for the reset mechanism
4. Dealing with real-valued data that may be arbitrarily close to one another.

Thus preprocessing in ART2 is much more complex than in ART1.

Although the ART2 network may seem complicated in comparison to its predecessor, the learning laws for ART2 are actually simpler. For ART2 the matching criterion, S (known as the similarity ratio), used for ART1 is no longer valid, since $\mathbf{X}$ and $\mathbf{C}$ are not binary. Instead ART2 utilizes a distance measure commonly used in the statistical pattern recognition techniques (also referred to as MINNET. Recall that we have seen such distance measures in the context of vector quantization, K-means clustering, and Kohonen networks). The cosine of the angle between the input vector and the prototype vectors is utilized as the quantity to be compared with the vigilance threshold.

For a more detailed discussion on ART2 along with the algorithm see Carpenter and Grossberg [1987b]. "C" language code segments illustrating the implementation of the ART2 network may be found in Freeman and Skapura [1992].

10.10 DISCUSSION

Some of the advantages ART has over competing pattern recognition techniques are as follows:

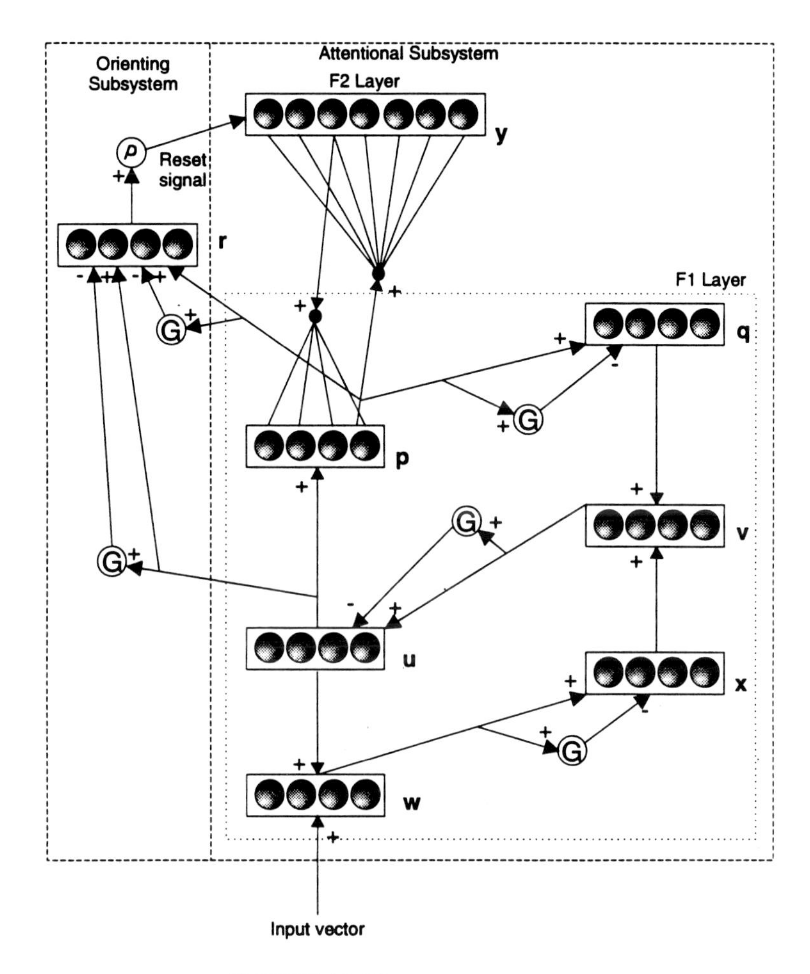

FIGURE 10.20 ART2 architecture.

1. Theorems have been proved [Carpenter and Grossberg, 1987a] to show that ART exhibits considerable stability and is not perturbed by an arbitrary barrage of inputs.
2. The network adapts to reflect the type of patterns most frequently observed in the environment, by updating the category prototypes adequately.
3. The ART architecture can easily integrate with other hierarchical theories of cognition. It allows for influences on the feature and category layers from the other subsystems external to the ART network, such as the attention, anticipation, or orientation systems.

The ART architecture is sensitive to the order in which the patterns are presented to the network. Kung [1993] has shown through an experiment that ART2 yields a different clustering on the same input pattern set when the patterns are presented in the reverse order.

The Table 10.1 shows the execution sequence of ART2 for a set of ten input patterns [from Kung, 1993]. The vigilance test fails for the third and the fifth patterns and new clusters are formed. Thus with a vigilance threshold of 1.5, three clusters are formed. Figure 10.21 shows the pattern space containing the actual patterns were presented in the reverse order (see Table 10.2), the stored prototypes have different values, the viligance test is failed only once (for the third pattern), and only two clusters are formed. Figure 10.22 shows the actual clusters formed corresponding to this sequence. Note that the vigilance threshold is the same in both cases.

TABLE 10.1
Execution Sequence for ART2

Order	Pattern	Winner	Test value	Decision	Cluster 1 centroid	Cluster 2 centroid	Cluster 3 centroid
1	(1.0, 0.1)	—	—	New cluster	(1.0, 0.1)		
2	(1.3, 0.8)	1	1.0	Pass vigilance test	(1.15, 0.45)		
3	(1.4, 1.8)	1	1.6	Fail → new cluster		(1.4, 1.8)	
4	(1.5, 0.5)	1	0.4	Pass vigilance test	(1.27, 0.47)		
5	(0.0, 1.4)	2	1.8	Fail → new cluster			(0.0, 1.4)
6	(0.6, 1.2)	3	0.8	Pass vigilance test			(0.3, 1.3)
7	(1.5, 1.9)	2	0.2	Pass vigilance test		(1.45, 0.85)	
8	(0.7, 0.4)	1	0.63	Pass vigilance test	(0.13, 0.45)		
9	(1.9, 1.4)	2	0.9	Pass vigilance test		(1.6, 1.7)	
10	(1.5, 1.3)	2	0.5	Pass vigilance test		(1.58, 1.6)	

TABLE 10.2
Patterns Presented in Reverse Order

Order	Pattern	Winner	Test value	Decision	Cluster 1 centroid	Cluster 2 centroid
1	(1.5, 0.3)	—	—	New cluster	(1.5, 1.3)	
2	(1.9, 1.4)	1	1.5	Pass vigilance test	(1.7, 1.35)	
3	(0.7, 0.4)	1	1.95	Fail → new cluster		(0.7, 0.48)
4	(1.5, 1.9)	1	0.75	Pass vigilance test	(1.63, 1.53)	
5	(0.6, 1.2)	2	0.9	Pass vigilance test		(0.65, 0.8)
6	(0.0, 1.4)	2	1.25	Pass vigilance test		(0.43, 1.0)
7	(1.5, 0.5)	1	1.17	Pass vigilance test	(1.6, 1.28)	
8	(1.4, 1.8)	1	0.72	Pass vigilance test	(1.56, 1.38)	
9	(1.3, 0.8)	1	0.84	Pass vigilance test	(1.52, 1.28)	
10	(1.0, 0.1)	2	1.47	Pass vigilance test		(0.58, 0.78)

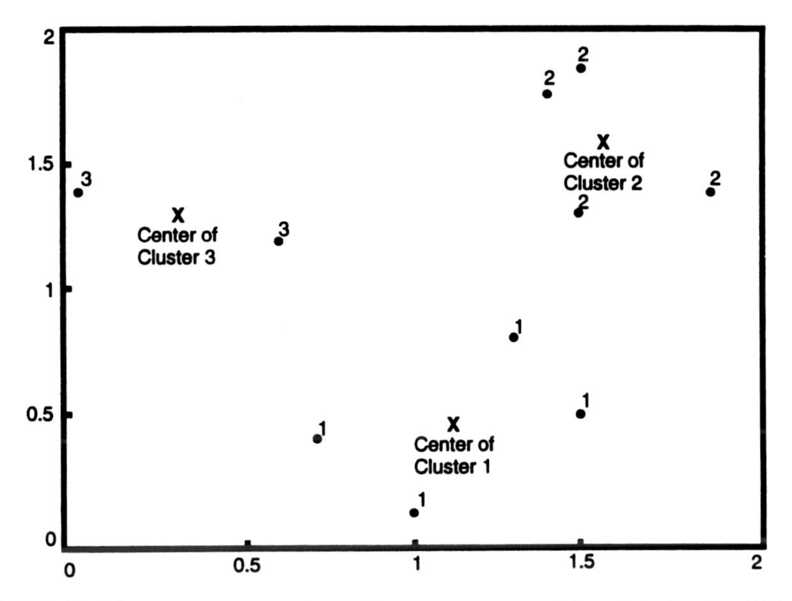

FIGURE 10.21 The results of ART with patterns presented in original order. (Adapted from Kung [1993].)

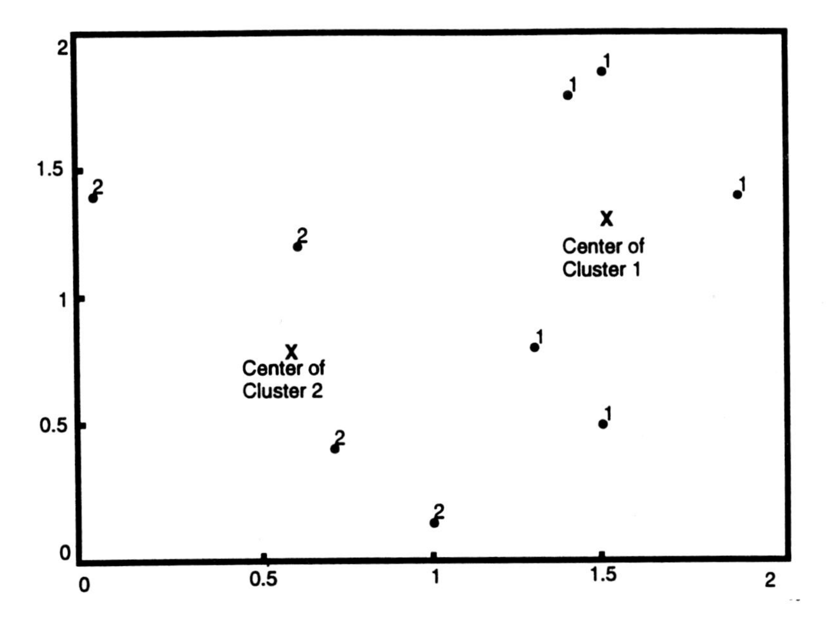

FIGURE 10.22 The results of ART with patterns presented in reverse order. (Adapted from Kung [1993].)

The ART architecture is suppose to maintain the plasticity required to learn new patterns, while preventing the modification of patterns that have been learned previously. However, for ART1 the code stability problem is not entirely solved. If the network receives a number of variations on a stored input pattern, it can gradually shift in a given direction in the pattern space. Each variation can match the previous stored category prototype closely enough to be placed into the same category. This would result in a large enough shift such that the network may not recognize the original pattern. Figure 10.23 shows how this can happen for a set of five characters with some noisy input patterns. Ryan and Winter [1987] have proposed a variation of ART designed to overcome this limitation.

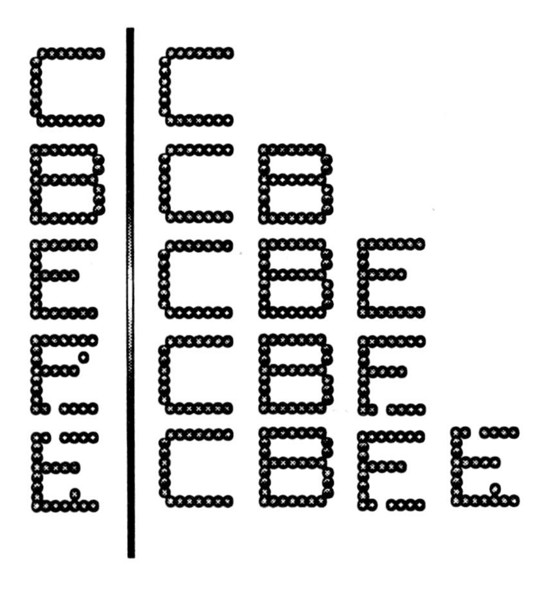

FIGURE 10.23 Evolution of an ART network trained on five noisy patterns.

Another problem can be caused by the fact that ART vigilance value is uniform across both feature and categories. For example, a vigilance setting that separates “O” from “Q” may impose intolerance to noise for other characters. Thus it may end up separating “a” and “a”. Weingard [1990] has proposed a variation of ART that allows for dynamic setting of the vigilance parameter.

Another feature of ART is that it always comes to a decision about where to categorize an input pattern. Unlike many other pattern recognition techniques it does not register the amount of ambiguity in the input, i.e., how close it is to the boundary between the categories. Levine [1989] and Levine and Penz [1990] have developed an improved model to address this issue and record the ambiguity.

The degree to which the ART system discriminates between different classes of input patterns can be varied by adequately choosing the value of the vigilance parameter. Thus the granularity with which the input patterns are classified is determined by the vigilance factor. A large value for the vigilance threshold will cause finer discrimination between classes for a given set of input patterns. That is, a larger vigilance threshold causes more clusters to be generated. On the other hand, a smaller value will allow for more noisy patterns to be classified within the same category. Figures 10.24 and 10.25 show the results of ART2 [Kung, 1993] on a set of input patterns for two different values of the vigilance threshold.

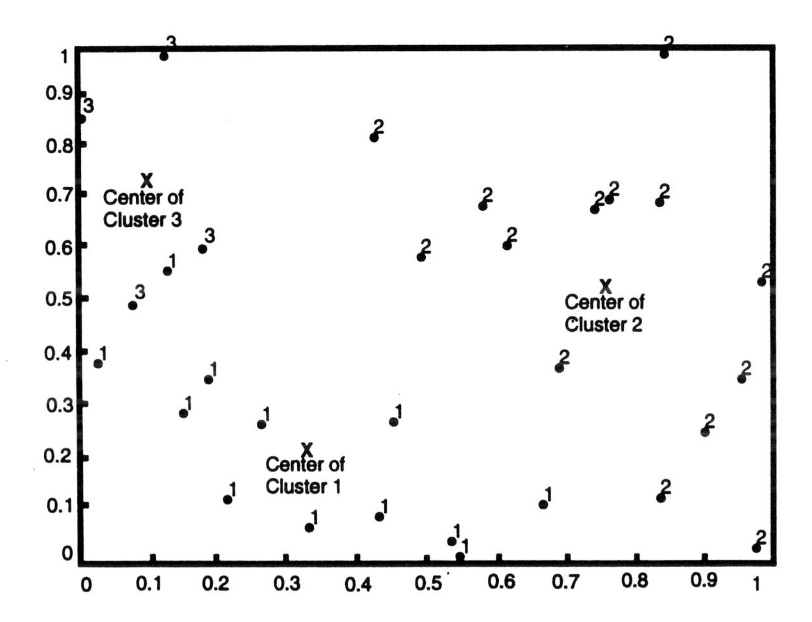

FIGURE 10.24 The results of ART with vigilance, ρ = 0.6. (Adapted from Kung [1993].)

10.11 APPLICATIONS

Carpenter and Grossberg [1987b] have applied the ART2 architecture to the problem of categorization of analog patterns which were drawn as graphs of function. Figure 10.26 shows the 50 input patterns which ART2 grouped into 34 clusters.

Gan and Lua [1992] have applied the ART2 architecture to the problem of character recognition for Chinese characters. ART2 was chosen because they used a real-valued feature set. The feature set consisted of 12 geometric features including intersection, turning points, and horizontal and vertical strokes. In this application the ART network was not used as the final classifier, but rather served to divide 3755 Chinese characters into 7 groups or classes in preparation for a final recognition stage. ART implementation issues include fast vs. slow training, threshold selection, and initial values for the bottom-up layer. A best-case classification accuracy of 97.23% for the training set and 90.20% for the test set is reported.

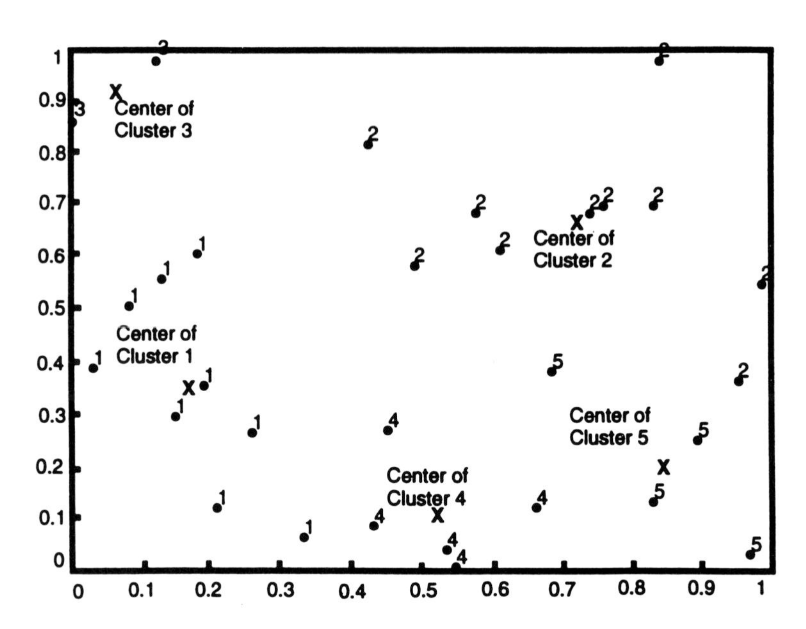

FIGURE 10.25 The results of ART with vigilance, $\rho = 0.4$. (Adapted from Kung [1993].)

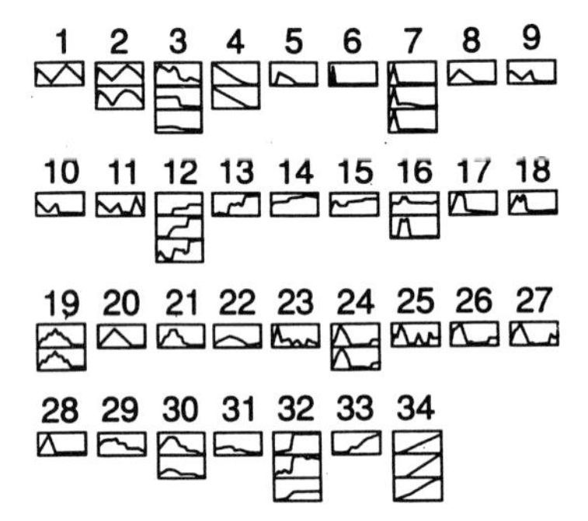

FIGURE 10.26 Example showing ART2 categorization of 50 analog input patterns into 34 clusters. (From Carpenter, G. A. and Grossberg, S., *Appl. Opt.* vol. 26, 1987. With permission.)

REFERENCES AND BIBLIOGRAPHY

Carpenter, G. A. and Grossberg, S., "A massively parallel architecture for a self-organizing neural pattern recognition machine," *Comput. Vision, Graphics Image Process.*, Vol. 37, pp. 54–115, 1987a.

Carpenter, G. A. and Grossberg, S., "ART 2: self-organization of stable category recognition codes for analog input patterns," *App. Optics*, vol. 26, pp. 4919–4930, 1987b.

Carpenter, G. A. and Grossberg, S., "ART 3: hierarchical search using chemical transmitters in self-organizing pattern recognition architecture," *Neural Networks*, Vol. 3, pp. 129–152, 1990.

Freeman, J. A. and Skapura, D. M., *Neural Networks: Algorithms, Applications and Programming Techniques*, Addison-Wesley, Reading, MA, 1992.

Gan, K. W. and Lua, K. T., "Chinese character classification using adaptive resonance network," *Pattern Recognition*, vol. 25, no. 8, 1992.

Grossberg, S., "Adaptive pattern classification and universal recoding:parallel development and coding of neural feature detectors," *Biol. Cybern.*, vol. 23, pp. 187–202, 1976.

Grossberg, S., "Competitive learning: from interactive activation to adaptive resonance," *Cognitive Sci.*, vol. 11, pp. 23–63, 1987.

Kung, S. Y., *Digital Neural Networks*, Prentice-Hall, Englewood Cliffs, NJ, 1993.

Levine, D. S., "Selective vigilance and ambiguity detection in adaptive resonance networks," In W. Webster (Ed.), *Simulation and AI*, Society for Computer Simulation, San Diego, CA, pp. 1–7, 1989.

Levine, D. S. and Penz, P. A., "ART 1.5 — a simplified adaptive resonance network for classifying low dimensional analog data," *Proc. of the Int. Joint Conf. Neural Networks*, vol. 2, pp. 639–642, 1990.

Pao, Y. H., *Adaptive Pattern Recognition and Neural Networks*, Addison-Wesley, Reading, MA, 1989.

Ryan, T. W. and Winter, C. L., "Variations on adaptive resonance," *Proc. IEEE First Int. Conf. Neural Networks*, vol. 2, pp. 767–776, 1987.

Tou, J. T. and Gonzalez, R. C., *Pattern Recognition Principles*, Addison-Wesley, Reading, MA, 1974.

Weingard, F. S., "Self-organizing analog fields (SOAF)," *Proc. Int. Joint Conf. Neural Networks*, vol. 2, pp. 34–38, 1990.

11 Neocognitron

11.1 INTRODUCTION

The networks we have discussed so far have been general networks that were adapted to address pattern recognition problems. In contrast to these, the neocognitron was specifically proposed to address the problem of handwritten character recognition. It is a hierarchical network with layers corresponding to simple, complex, and hypercomplex cells, with a very sparse and localized pattern of connectivity between layers. This model is also inspired by the biological model of the visual nervous system proposed by Hubel and Wiesel [1962]. This discussion of the neocognitron largely follows that found in Fukushima and Wake [1991]. Unlike most neural networks this implementation has a high tolerance to character distortion, position, and at least a moderate immunity to character size.

11.2 ARCHITECTURE

The neocognitron is organized into a hierarchy of levels. Each level consists of two layers. As shown in Figure 11.1, a level contains a layer of simple cells or S-cells and a layer of complex cells or C-cells.

Notice also that the layers are divided into planes. Each plane of S-cells is responsible for recognizing a specific feature. As a result, the S-cells within a plane share a common set of weights. The difference between S-cells within a plane is that each S-cell takes its input from a different location of the preceding layer. This is similar to the receptive field in a feature map as discussed in chapter 6 (see section 6.3). The first S-cell layer takes its input from the retina, a bit map of pixels, and extracts only relatively simple features. On subsequent levels the S-cells extract features with ever increasing complexity. As such the level of abstraction increases from level to level.

Figure 11.2 shows the interconnection scheme between layers of the neocognitron. Figure 11.2(a) shows S- and C-cells which provide excitatory input to the next layer. Note that each S-layer consists of S-cells as well as V-cells which provide inhibitory input to the S-cells in the same layer. Figure 11.2(b) shows the architecture of an actual system proposed by Fukushima and colleagues which was developed for recognizing alphanumeric characters (upper case and digits). Note that each S1-layer cell has three input connections so the receptive field is 3×3. For S2-S4 layer cells there are five connections resulting in a 5×5 receptive field. In the case of the C1-layer each cell has three connections; C2 cells have seven connections and so on.

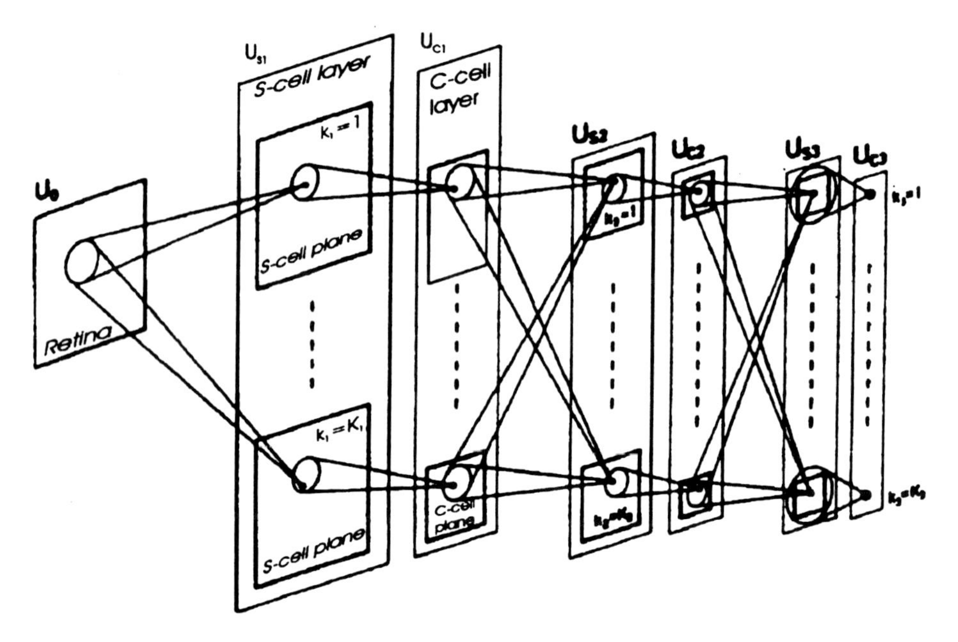

FIGURE 11.1 Hierarchical structure of the neocognitron network. The number of S-cells and C-cells in each layer are indicated at the bottom. (Adapted from Fukushima, K. and Wake, N., *IEEE Trans. Neural Networks,* vol. 2, no. 3, pp. 355–365, 1991.)

The C-cells receive input from a cluster of S-cells in a plane of the preceding S-cell layer. Since each S-cell in a plane detects the same feature at different locations, the C-cells become desensitized to the precise location of the feature on the input layer thus allowing the neocognitron system to identify characters without regard to their placement on the retina. It also provides some flexibility with regard to distorted characters.

The first feature map layer detects very simple features such as straight lines with different rotations, and these are only seen within the limited boundaries of the S-cells receptive field. Subsequent layers build these simple features into increasingly complex features as indicated in Figure 11.3.

The output of an S-cell is given by:

$$U_{S_l}(k_l, \mathbf{n}) = r_l \varphi \left[\frac{1 + \sum_{k_{l-1}=1}^{K_{l-1}} \sum_{\mathbf{v} \in A_l} a_l(k_{l-1}, \mathbf{v}, k_l) U_{C_{l-1}}(k_{l-1}, \mathbf{n} + \mathbf{v})}{1 + \frac{r_l}{1 + r_l} b_l(k_l) V_{C_l}(\mathbf{n})} - 1 \right] \quad (1)$$

Where ϕ is the linear threshold function:

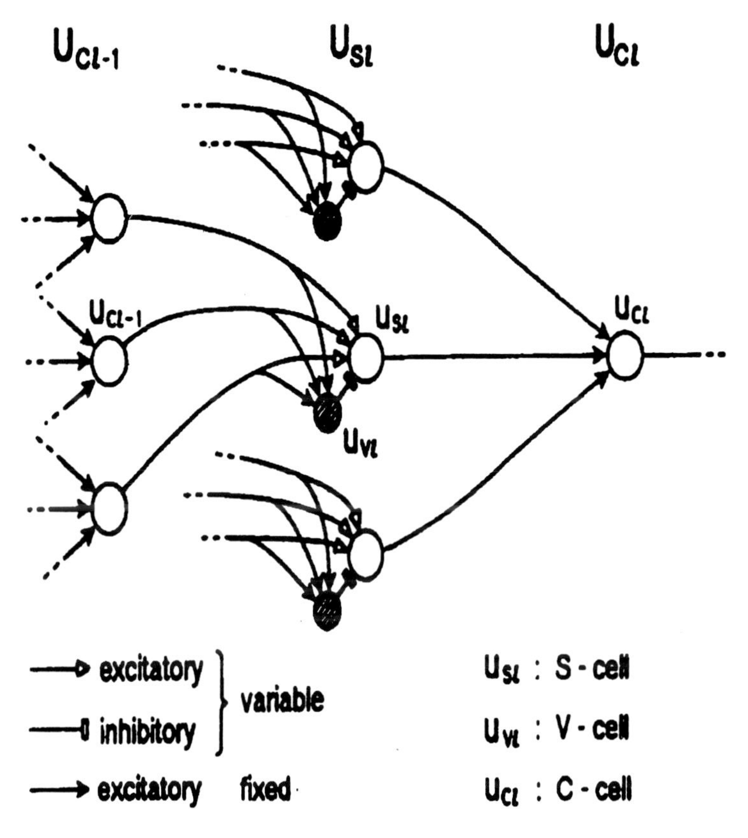

FIGURE 11.2(A) Connections between various types of cells in the neocognitron network. (Adapted from Fukushima, K. and Wake, N., *IEEE Trans. Neural Networks,* vol. 2, no. 3, pp. 355–365, 1991.)

$$\varphi(x) = \begin{cases} x & x \geq 0 \\ 0 & x < 0 \end{cases} \tag{2}$$

We use the following notation:

$U_{sl}(k_l,\mathbf{n})$ is the output of S-cell in layer S_l.
k_{l-1} is the size of each cell plane as indicated at the bottom of Figure 11.2(b).
$a_l(k_{l-1},\mathbf{v},k_l)$ is the strength of the variable excitatory connection from C-cell.
$U_{cl-1}(k_{l-1},\mathbf{n} + \mathbf{v})$ is the input from the C-cell of the previous level.
$b_l(k_l)$ is the strength of the variable inhibitory connection from the subsidiary V-cell.
$V_{cl}(\mathbf{n})$ is the input from the subsidiary V-cell.

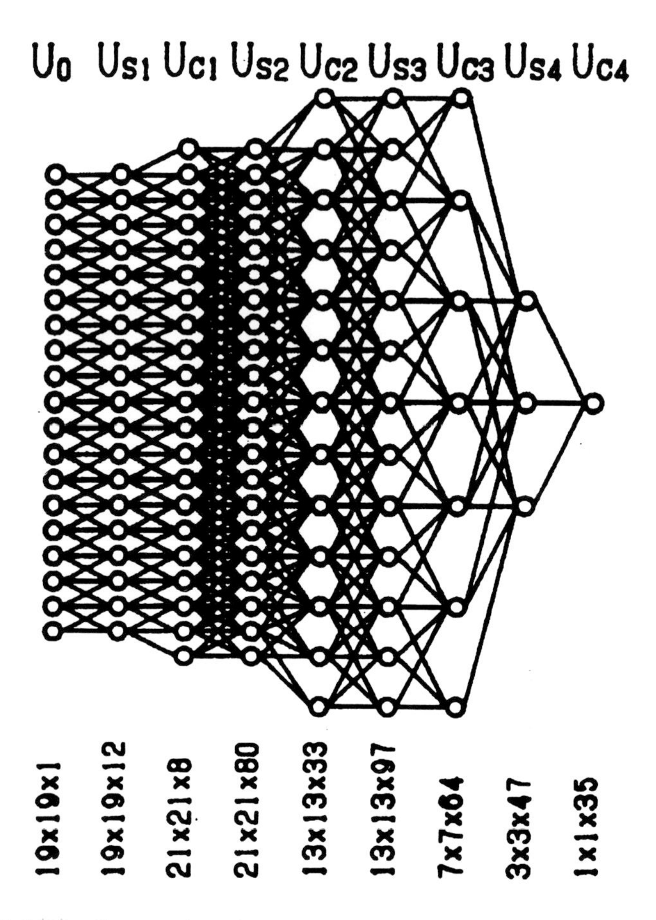

FIGURE 11.2(B) Cross section of interconnections between cells in each layer. Note that only one cell per plane is shown. (Adapted from Fukushima, K. and Wake, N., *IEEE Trans. Neural Networks,* vol. 2, no. 3, pp. 355–365, 1991.)

A_l determines the size of features to be extracted by S-cells. In Figure 11.2(b) $A_1 = 3 \times 3$ and $A_2 - A_4 = 5 \times 5$.

r_l controls selectivity of feature extraction. Larger r_l corresponds to a smaller tolerance of noise and deformation of feature.

In the above k_l refers to the kth plane on level 1, $\mathbf{n}$ is a two-dimensional vector that locates a neuron within the plane, and $\mathbf{v}$ is a vector and is the position relative to $\mathbf{n}$ of a cell in the preceding layer that is in the receptive field of $\mathbf{n}$. A_l represents the receptive field of $\mathbf{n}$ so that the inner sum over $\mathbf{v}$ incorporates all neurons in the given field. The outer sum over k_{l-1} incorporates all planes from the previous level. Thus the summation term in the numerator, sometimes called the excitatory term or e, is simply a sum of products. The output of those neurons with outputs incident on $\mathbf{n}$ are multiplied by their respective connection strength to $\mathbf{n}$.

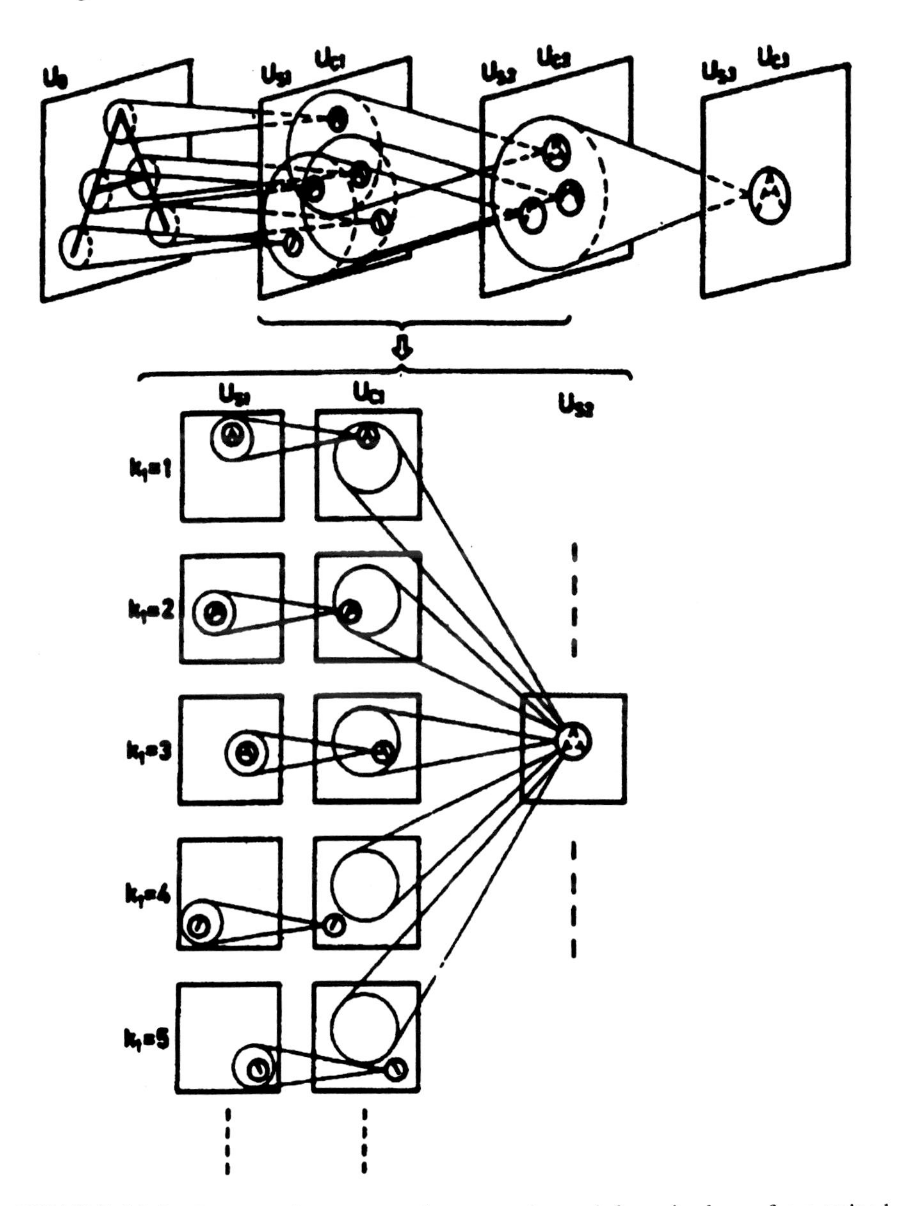

FIGURE 11.3 Response for an example pattern for each layer is shown for a trained neocognitron network. Interconnections between cells of different layers are also indicated. (Adapted from Fukushima, K., *Biol. Cybern.*, vol. 36, pp. 193–202, 1980.)

The term r_1 is a constant which operates cooperatively with an input to each neuron from a single inhibitory plane present in each S-layer. Output of each V_c-cell is distributed to the cells in all S-planes at the corresponding position within the plane. (Figure 11.4 shows the V_c-plane in relation to the first S-layer and the preceding layer, which in this case is the retina.) The larger the value of r_1 is, the

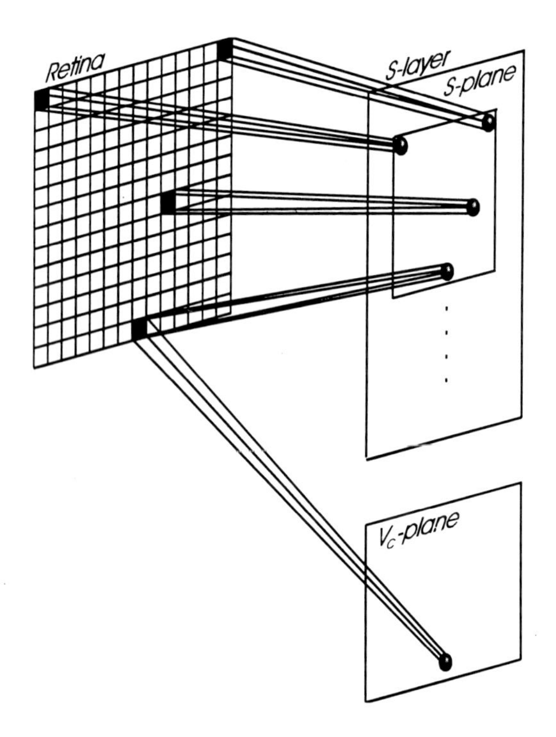

FIGURE 11.4 Interconnection scheme between retina and first S-cell layer.

greater the excitation must be relative to the inhibition to produce a nonzero output. That is, a rather good match is needed to fire the cell. However, since r_1 also multiplies $\phi(\ )$, large values of r_1 will also produce larger responses. Conversely, small values of r_1 allow the cell to fire with a less good match but also lead to a diminished response.

The output of the inhibitory V_c-cell at position **n** is

$$V_{C_l} = \sqrt{\sum_{k_{l-1}=1}^{K_{l-1}} \sum_{\mathbf{v} \in A_l} c_l(\mathbf{v}) U_{C_{l-1}}^2 \left(k_{l-1}, \mathbf{n} + \mathbf{v}\right)} \tag{3}$$

The weights $c_1(\mathbf{v})$ are the connection strengths to the cell at position **v** in the V_c-cell receptive field. These weights are not trained but are required to decrease monotonically as the magnitude of **v** increases. The following, where $r'(\mathbf{v})$ is the normalized distance from **v** to the center of the receptive field, is one option:

$$c_l = \frac{1}{C_l}(l)\alpha_l^{r'(\mathbf{v})} \tag{4}$$

where the normalization constant, C (*l*), is given by:

$$C(l) = \sum_{k_{l-1}=1}^{K_{l-1}} \sum_{v \in A_l} \alpha_l^{r'(v)} \tag{5}$$

The output of a C-cell is given by:

$$U_{C_l}(k_1, \mathbf{n}) = \psi\left[\frac{1 + \sum_{k_{l-1}=1}^{K_l} j_l(k_l, k_{l-1}) \sum_{v \in D_l} d_l(\mathbf{v}) U_{S_l}(k_l, \mathbf{n} + \mathbf{v})}{1 + V_{S_l}(\mathbf{n})} - 1\right] \tag{6}$$

where $\psi(\mathbf{x})$ is given by:

$$\psi(x) = \begin{cases} \dfrac{x}{\beta + x} & x \geq 0 \\ 0 & x < 0 \end{cases} \tag{7}$$

where β is a constant.

K_1 is the number of S-planes at level 1. D_1 is the receptive field of the C-cell. Thus it corresponds to the size of the feature. In Figure 11.1 $D_1 = 3 \times 3$, $D_2 = 5 \times 5$, $D_3 = 7 \times 7$, and $D_4 = 3 \times 3$. $d_1(\mathbf{v})$ denotes the strength of the fixed excitatory connections, and is a monotonically decreasing function of $|\mathbf{v}|$. $j_1(k_{1-1}, k_1)$ is 1 if the k_1th C-cell plane receives signals from the k_{1-1}th S-plane, or else it is 0.

Finally the output of S-layer V_s-cells is

$$V_{S_l} = \frac{1}{K_l} \sum_{k_{l-1}=1}^{K_{l-1}} \sum_{v \in D_l} d_l(\mathbf{v}) U_{S_l}(k_l, \mathbf{n} + \mathbf{v}) \tag{8}$$

The only weights that are trained are a_1 and b_1. The training of these weights can be supervised if the desired features have been predetermined. In this case each S-layer will be trained independently (for details see Fukushima and Wake, [1991]).

Alternately, a form of unsupervised training can be used as follows (for details see Fukushima, [1980]). The a_1 weights, those of the excitatory planes, are initialized to small positive random values. The b_1 weights are initialized to zero. When a training set vector is presented to the net, the S-cell whose response is the strongest within a given column is chosen as a representative for its plane. A representative is chosen for the inhibitory plane in a similar manner. Then the following equations are used to update the weights:

$$\Delta a_l\left(k_{l-1}, \mathbf{v}, \hat{k}_l\right) = q_l C_{l-1}(\mathbf{v}) U_{C_{l-1}}\left(k_{l-1}, \hat{\mathbf{n}} + \mathbf{v}\right) \tag{9}$$

$$\Delta b_l\left(\hat{k}_l\right) = q_l C_{l-1}\left(\hat{\mathbf{n}}\right)$$

q_1 is a positive constant determining the speed of reinforcement.

In these equations, $U_{C_{l-1}}(k_{l-1}, \mathbf{n} + \mathbf{v})$, the response of the C-cells, and, $V_{C_{l-1}}(\mathbf{n})$, the response of V-cells, are determined by presenting the training pattern to the input layer. The network is trained layer by layer in case of supervised learning, such that first connection strengths from the input units to the S1 units are trained and then frozen. Next the connection strengths from C1 units to the S2 units are trained, frozen, and so on till the output layer is reached.

It is reported [Fukushima and Wake, 1991] that the neocognitron system generalizes better when supervised training is done. However, it is also pointed out that the supervised training with its concomitant choices of parameters "requires hard labor" and that the problem increases dramatically with the number of training patterns.

11.3 EXAMPLE OF A SYSTEM WITH SAMPLE TRAINING PATTERNS

Fukushima and Wake [1991] have developed a network to recognize upperclass alphanumeric characters. Supervised learning was used to control the choice of local features to be extracted in intermediate stages of the hierarchical network. The local features can be determined by training patterns presented to the network in the following fashion.

Figure 11.5 shows the 12 training patterns for layer S1, one for each plane. As seen from Figure 11.3 each cell of this layer has a 3 × 3 receptive field, which is trained to extract a line component of a certain orientation. Layer S2 has 80 cell planes, each with a receptive field of 9 × 9 as seen from Figure 11.3 Figure 11.6 shows the 80 training patterns composed of typical examples of deformed patterns consisting of a part of an alphanumeric pattern. S3 layer extracts global features by combining local features extracted in the preceding layer. Figure 11.7 shows the 97 training patterns used to train these cell planes. Figure 11.8 shows the 47 training patterns used to train the 47 cell planes of layer S4. Figure 11.9 shows an example of the response of the network, which has been previously trained with training patterns shown in Figures 11.5–11.8.

FIGURE 11.5 Training templates for the S1 layer planes. (Adapted from Fukushima, K. and Wake, N., *IEEE Trans. Neural Networks,* vol. 2, no. 3, pp. 355–365, 1991.)

Computer simulations of the neocognitron system have been reported [Fukushima and Imagawa, 1993, Fukushima, 1992] but so far none have been reported which provide quantitative measures of recognition accuracy. Additionally, most were not done on appreciably large data sets.

The basic neocognitron model as described in this section does not function well when presented with a compound stimulus consisting of two or more patterns. To deal with compound images and to deal with the recognition of cursive handwriting [Fukushima and Imagawa, 1993] the model is extended to include "selective attention". When the model is presented with two or more patterns, the model attempts to focus selectively on one of them. In attempts to segment it from any other patterns present and then to recognize it. The model then switches its attention to another pattern. This modification was successful as long as the number of simultaneous patterns was small. To deal with situations in cursive writing in which the number of patterns was large the model was again extended. A search controller was added to limit the number of simultaneous patterns.

A considerably less elaborate recognition system based loosely on the neocognitron model is proposed by Lee and Choi [1992]. The size of the feature map is greatly increased to give features that are more complex and meaningful. The subsampling layer of U_c-cells is eliminated. Invariant properties are handled at the feature map level. To accomplish this front-end feature extraction is done. Once again, quantitative results are not available in the reference.

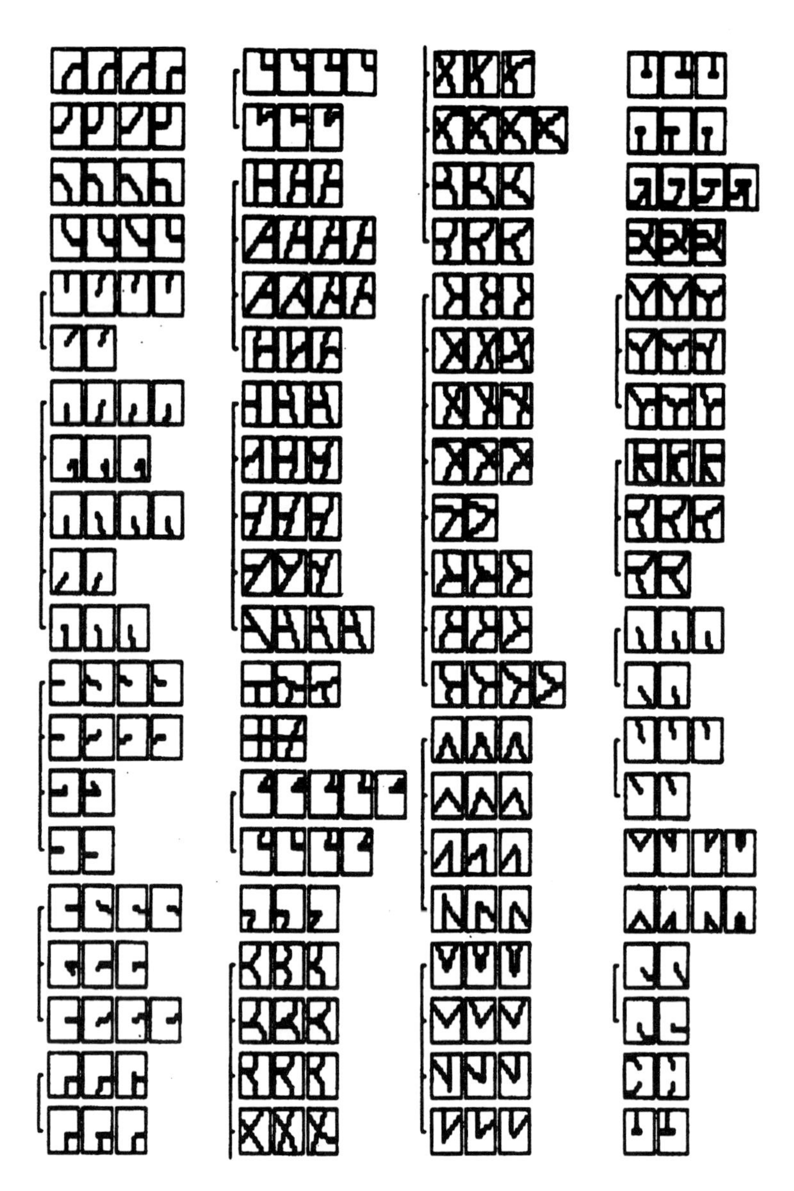

FIGURE 11.6 Eighty training patterns used to train each plane of the S2 layer. (Adapted from Fukushima, K. and Wake, N., *IEEE Trans. Neural Networks,* vol. 2, no. 3, pp. 355–365, 1991.) (Adapted from Fukushima and Wake [1991].

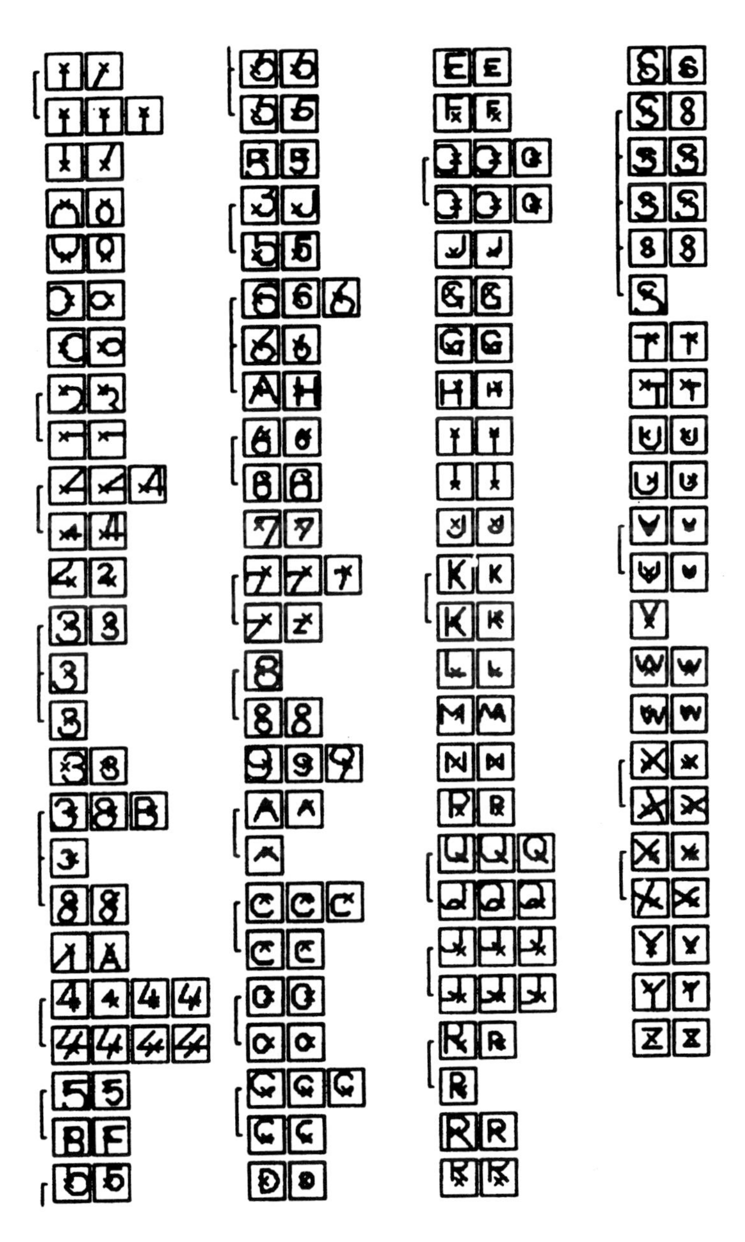

FIGURE 11.7 Ninety-seven training patterns used by Fukushima and Wake [1991] for each plane of S3 layer. The crosses mark centers of receptive fields of the seed cells. (Adapted from Fukushima, K. and Wake, N., *IEEE Trans. Neural Networks,* vol. 2, no. 3, pp. 355–365, 1991.)

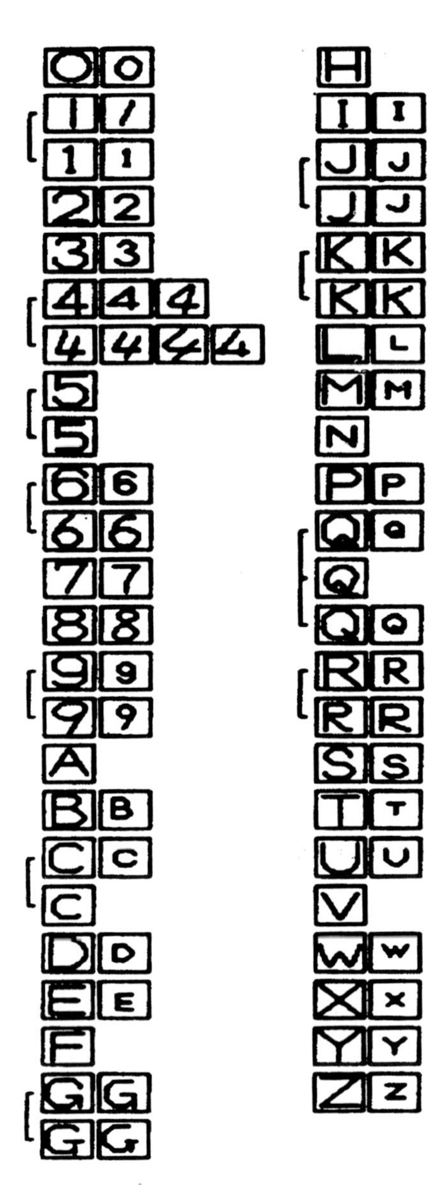

FIGURE 11.8 Forty-seven training patterns used to train cell planes of S4 layer. (Adapted from Fukushima, K. and Wake, N., *IEEE Trans. Neural Networks,* vol. 2, no. 3, pp. 355–365, 1991.)

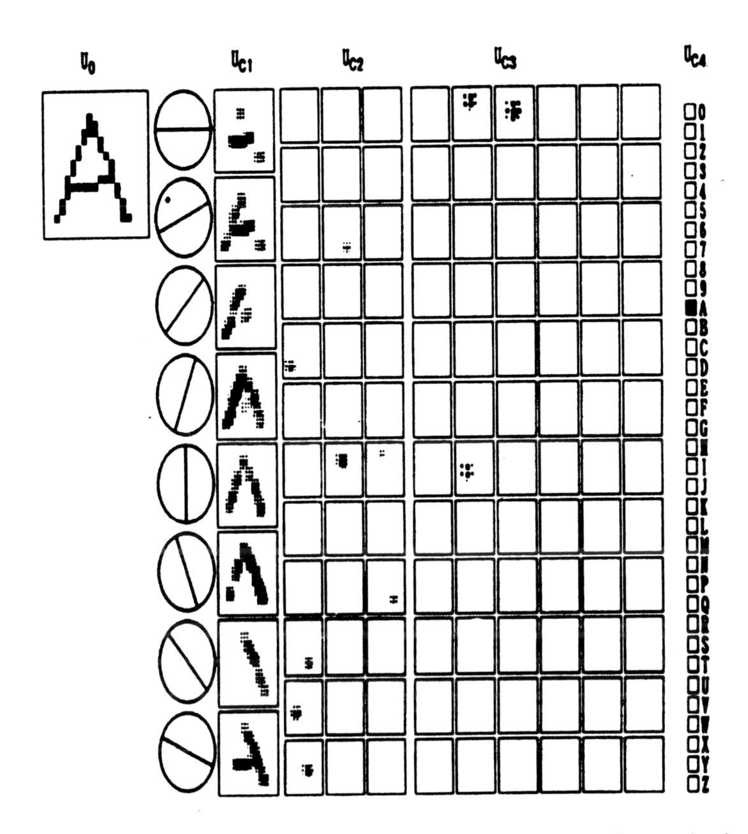

FIGURE 11.9 Responses of C-cells of different layers are shown for a trained network when the letter "A" is presented. (Adapted from Fukushima, K. and Wake, N., *IEEE Trans. Neural Networks,* vol. 2, no. 3, pp. 355–365, 1991.)

REFERENCES AND BIBLIOGRAPHY

Freeman, J. A. and Skapura, D. M., *Neural Networks: Algorithms, Applications and Programming Techniques,* Addison-Wesley, Reading, MA, 1992.

Fukushima, K., "Neocognition: a self-organizing neural network model for a mechanism of pattern recognition unaffected by shift in position," *Biol. Cybern.,* vol. 36, pp. 193–202, 1980.

Fukushima, K. and Wake, N., "Handwritten alphanumeric character recognition by neocognition," *IEEE Trans. Neural Networks,* vol. 2, no. 3, pp. 355–365, 1991.

Fukushima, K., "Character recognition with neural networks," *Neurocomputing,* vol. 4, pp. 221–233, 1992.

Fukushima, K. and Imagawa, T., "Recognition and segmentation of connected characters by selective attention," *Neural Networks,* vol. 6, pp. 33–41, 1993.

Hubel, D. H. and Wiesel, T. N., "Receptive fields, binocular interaction and functional architecture in cat's visual cortex," *J. Physiol.,* vol. 160, pp. 106–154, 1962.

Lee, S. and Choi, Y., "Robust recognition of handwritten numerals based on dual cooperative networks," *IEEE IJCNN,* vol. 3, pp. 760–768, 1992.

12 Systems with Multiple Classifiers

12.1 GENERAL

As we have seen there are many different kinds of pattern recognition classifiers, each having its own varying set of strengths and weaknesses. It is therefore reasonable to see whether results from diverse systems can be combined in some meaningful way to achieve an overall better result. Figure 12.1 illustrates the concept of multiple recognizers applied to a pattern recognition problem. Notice that from the figure that the individual classifiers are not necessarily homogeneous.

Although we have primarily dealt with neural network based systems, it is not necessarily true that all the systems we might wish to combine will be neural in character. Hybrid systems combining a variety of methodologies can provide significant benefits and do exist in commercial applications. Statistical, syntactic, and neural methodologies may all have a role to play within the various problem domains to which pattern recognition systems may be applied. Moreover, it has been observed that classifiers based on different classifiers and features are frequently complementary to one another [Duerr et al., 1980].

Clearly, a great deal of diversity exists within the recognition engines which we seek to utilize in an ensemble to achieve greater overall recognition accuracy. The recognition characteristics of a given recognizer may vary not only with the type or architecture of the recognition engine but also with a myriad of choices including:

1. The number and type of measurements performed to obtain the raw input vector; possibly varying for various reasons ranging from characteristics of the recognizer itself to historical (an existing system which happens to have some useful characteristics using a particular set of measurements)
2. The amount and type of preprocessing utilized to transform the input vector (e.g, skeletonizing, fuzzing, or edge detection)
3. The choice of input transformations or features utilized to map the input into the recognizer input space (e.g., the Fourier descriptors, the Gabor transformations, the Karhunen-Loeve transformation, etc.)
4. The size and selection of training data
5. The choice of training methods and parameters

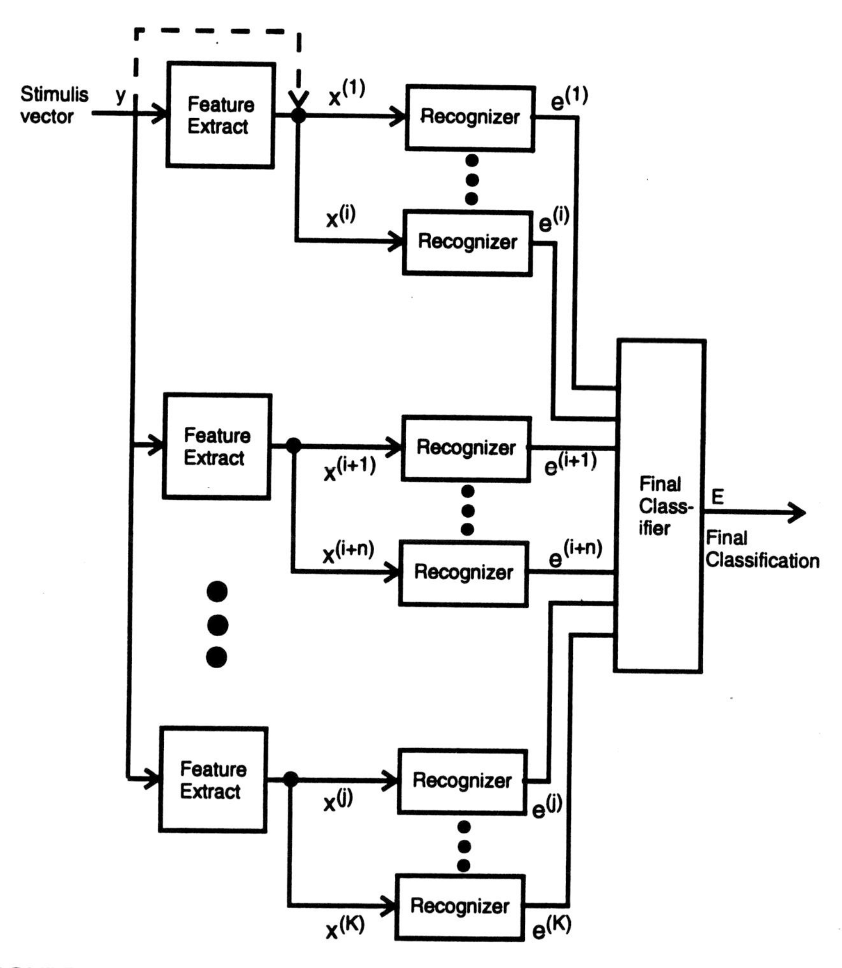

FIGURE 12.1 Application of multiple recognizers to a pattern recognition problem.

Clearly, the list of salient differences among various recognizer types could be much longer. The recognition accuracy of a system of recognizers will depend not only upon the characteristics of individual recognizers but also upon the manner in which such recognizers are combined.

To further illustrate the value in combining multiple recognition systems, in spite of the apparent difficulties, consider two recognition engines: recognizer 1 and recognizer 2. Each of these recognizers will be applied to the character recognition problem. Understanding that neither recognizer will be perfect, recognizer 1 may find ambiguity, say between *a* and *o*. The other recognizer, 2, with different qualities may do perfectly well with *a*'s and *o*'s but have difficulty with resolving *8* and *B*. Recognizer 1, of course, handles *8*'s and *B*'s perfectly well. If these recognition systems could be combined so that we would accept the results of each recognizer in the context in which it is most accurate, the overall recognition system would be improved.

As in the above example, recognizers within a specific problem domain can usually achieve different degrees of success. This leads to the question: how can the strengths of various systems be combined without incorporating their weaknesses? In the process of combining neural recognizers with others — neural or not — it will be necessary to give attention to the relationship between neuronal activation levels and the network certainty about its choices. In turn this leads us to consider certain aspects of training set design. Also covered in this chapter are the related ideas of reject categories, reliability measures for the recognizer, and the confusion matrix.

Other reasons, beyond the combination of competency of recognition engines, exist which motivate the use of several recognizers in combination. Where features differ in form (e.g., continuous valued vs. discrete), different recognizers may be more appropriate for handling different sets of features. Additionally, even where features are relatively homogeneous, the total feature vector may be of very high dimensionality. In such cases it may be useful to divide the high-dimension vector into several vectors of lower dimension. Several classifiers, each utilizing one of these lower dimensional vectors along with some method of combining the recognition results, would then comprise the overall recognition system. The arrangement is desirable in that high-dimension vectors are known to lead not only to computational complexity but also to implementation and accuracy problems. Recall the discussion of OCON networks in chapter 4 where a series of specialist networks were used to replace a single ACON network.

The combination of dissimilar recognition systems is far from a trivial problem and existing solutions may not be adequate to deal with given problem domains. With this in mind we will now discuss some available methods by which recognizers may be combined.

12.2 A FRAMEWORK FOR COMBINING MULTIPLE RECOGNIZERS

To proceed with the discussion of multiple recognizers we will put the problem into a more mathematical framework as discussed in Xu et al. [1992]. To accomplish this consider a decision space, P, with M mutually exclusive sets, C_i, $\forall i \in \Lambda = \{1,2, \ldots, M\}$. Each set, C_i, represents a class or category into which patterns will be grouped or classified. The decision space, P, may be written as:

$$P = C_1 \cup C_2 \cup \ldots \cup C_M \tag{1}$$

The decision space, P, therefore is the set of all possible patterns from all classes. Now a set of corresponding integer labels, Λ, is defined such that $\Lambda = \{1, 2, \ldots, M\}$. This set, Λ, therefore provides all possible integer labels for the defined classes. The decision space set, P, can then be extended to include $\Lambda \cup \{M + 1\}$, where $M + 1$ designate a reject class where by some established criteria the recognizer is believed to have failed to correctly identify the pattern. Throughout the following, K will be used to designate the total number of recognizers in an ensemble of recognizers. Each recognition engine in the system is designated, e, so that the recognizers in the system are $e_1, e_2, \ldots, e_k$. When an input pattern x, is evaluated by a recognizer e, the output of that recognizer will be an integer index, $j \in \Lambda \cup \{M + 1\}$ as a label indicating $\mathbf{x}$ belongs to class C_j (or potentially the reject class). This may be expressed as $e(\mathbf{x}) = j$. Notice that regardless of recognizer type of methodology, the classifier is regarded as a function box which receives a sample, x, and outputs the label of a classification, j. Most of the required notation has now been established.

Although the classification given by the index j is the final product of any given single recognizer, e, many existing classification systems can provide additional useful information in an intermediate level. A Bayes classifier, for example, will provide M *post-priori* probabilities. $P(i/\mathbf{x})$, $i = 1, \ldots, M$. That is, there is a probability of classification in C_i given that the vector $\mathbf{x}$ has occurred. In such a classifier the final label j is simply taken as the maximum of such probability. Once a classification has been made the probabilities may not be retained; however, such discarded information may be used beneficially in a multiple classifier system. The type of information available from different recognizers varies. The following is a taxonomy of the levels of information generally provided in identifying an input vector $\mathbf{x} \in P$:

1. The abstract level — The recognizer provides only its first or best choice of the label for **x**.
2. The rank level — The recognizer provides a ranked list of choices for the label for **x** (i.e., 1st, 2nd, 3rd, …, n^{th} choices). There is, however, no metric in the rank level which would indicate the degree to which each choice is favored.
3. The measurement level — The recognizer provides a metric for each label in representing the "degree to which **x** has that label".

It is perhaps tempting to think of certain neural recognizers, (e.g., multilayer perceptron nets) as operating on the measurement level. Unfortunately, the activation levels of output neurons from such networks do not in general represent anything like a reliable metric for the network certainty in its identification, $e(\mathbf{x}) = j$. The decision boundary, a surface in hyperspace, is guaranteed (through training) to separate classes. That is, the decision boundary is created such that an error measure — taken over a hopefully representative training set — will remain below some predetermined threshold. The difficulty with treating neuronal activation levels at the measurement level is that there is more than one way in which the separating hyperplanes may be drawn. This is illustrated in Figure 12.2 for a toy classification problem.

Regions A and B in Figure 12.2 represent the training set. Lines 1, 2, and 3 represent separating hyperplanes all of which will certainly meet training criteria. Each of the three indicated hyperplanes will separate the training data perfectly. Yet we will observe potentially very different activity levels outside of these regions

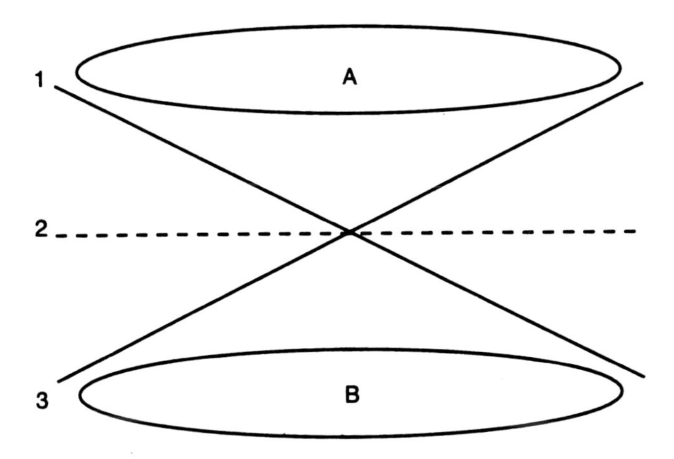

FIGURE 12.2 Three hyperplanes, each capable of separating the given classes.

depending upon the manner in which the error surface was formed. With a properly sized representative training set and with the application of modern training methods, the situation is less dire than this example might seem to indicate. The example is purposely extreme to make the point that the activity levels of a network do not per se represent a confidence level. Although some empirical observations may be utilized to order the classification results of neural classifiers, for now it is necessary to treat neural recognizers as belonging to the abstract level.

By way of contrast, Bayes' classifiers can provide a metric indicating the degree to which an input vector belongs to each of the available classes. For an individual Bayes classifier, e, the set of probabilities is available. This is the probability that $\mathbf{x}$ belongs to class C_i given that $\mathbf{x}$ has occurred.

$$P(\mathbf{x} \in C_i \mid \mathbf{x}), \quad i = 1, \ \ldots, \ M \tag{2}$$

Consider now how we may utilize an ensemble of such classifiers. (Note that this is not a solution to the general problem in that we are still dealing with a single type or recognizer). Each recognizer e_k within the ensemble contributes a postprobability as follows:

$$P_k(\mathbf{x} \in C_i \mid \mathbf{x}), \quad i = 1, \ \ldots, \ M, \quad k = 1, \ \ldots, \ K \tag{3}$$

The corresponding decision function for an individual recognizer, e_k, would can be expressed as:

$$e_k(\mathbf{x}) = j \ \text{ when } \ P_k(\mathbf{x} \in C_j) = \max_{i \in \Lambda} P_k(\mathbf{x} \in C_i \mid \mathbf{x}) \tag{4}$$

The decision functions for the individual recognizers are of relatively little importance to us when the objective is to create a decision function for the ensemble of recognizers, *E*. What we wish to do therefore is to combine the probabilities for the individual recognizers (see equation 3) in some fashion that will make sense over the ensemble. That is, for each class C_i, we wish to estimate the postprobabilities for combined recognizer, *E*, taken as a unit. One simple approach is to use the average of all recognizers as shown in equation 5.

$$P_E\left(\mathbf{x} \in C_i \mid \mathbf{x}\right) = \frac{1}{K}\sum_{k=1}^{K} P_k\left(\mathbf{x} \in C_i \mid \mathbf{x}\right), \quad i = 1, \ldots, M \tag{5}$$

Once P_E is known, the decision function, $E(\mathbf{x})$ for the overall classifier is straightforward:

$$E(\mathbf{x}) = j$$

When

$$P_E\left(\mathbf{x} \in C_j \mid \mathbf{x}\right) = \max_{i \in \Lambda} P_E = \left(\mathbf{x} \in C_i \mid \mathbf{x}\right) \tag{6}$$

The above decision function will always pick the classification for which the postprobability of the combined classifier is maximum. We may, however, wish to take into account the tradeoff between the substitution rate and the reject rate. A substitution error occurs when the wrong class is chosen. The rejection rate refers to the number of patterns which the recognizer declares itself to be unable to handle. Clearly, by using equation 6 no rejections will occur. We may therefore wish to extend equation 6 to take into account situations in which rejection is preferable to misclassification. The new decision function will then be as follows:

$$E(\mathbf{x}) = \begin{cases} j & P_E\left(\mathbf{x} \in C_j \mid \mathbf{x}\right) = \max_{i \in \Lambda} P_E\left(\mathbf{x} \in C_i \mid \mathbf{x}\right) \geq \alpha \\ M+1 & \text{Otherwise} \end{cases} \tag{7}$$

where $M + 1$ designates the reject category and is chosen in the range:

$$0 \leq \alpha \leq 1 \tag{8}$$

Low values of α lead to fewer rejections while high values of α increase the stringency of the test for category membership. The above discussion illustrates how measurement level (i.e., class 3) recognizers may be combined. Notice, however, that the recognizers being combined were homogeneous (all of one kind). The technique discussed could be extended to apply to other forms of recognizers where measurement level information is available. It may also be extended to heterogeneous recognizers if a means of normalizing the measurement level metric is available and appropriate. In the next section we will discuss the combination of heterogeneous classifiers at the abstract level.

12.3 VOTING SCHEMES

At the abstract level only the first or best choice is known from each recognizer, e_k. It is therefore natural to combine abstract level (i.e., type 1) classifiers by means of a voting method. We seek an overall decision, $E(\mathbf{x})$, for the combined recognizer given that the decision functions for the individual recognizers $e_k(\mathbf{x})$ may not agree. For this discussion, $e_k(\mathbf{x})$ is cast in an alternate form. $T_k(\mathbf{x})$, the binary characteristic function is defined as follows:

$$T_k(\mathbf{x} \in C_i) = \begin{cases} 1 & e_k(\mathbf{x}) = i, \ i \in \Lambda \\ 0 & \text{Otherwise} \end{cases} \tag{9}$$

The following C++ code fragment illustrates the calculation of the characteristic function given in equation 9. Here $e[K]$ is an array of the choices of K recognizers. We will also take $T[M + 1]$ as a global array which will contain the characteristic function.

```
void characteristic(){
int i,j,c;
for (j=0; j<M; j++)
  T[j]=0;
for (i=0; i<K;i++) {
  c=e[i]
  if ()
    T[c]=1;
  }
}
```

The most conservative form of voting would have it that all the individual recognizers must agree; otherwise the pattern is rejected. This is expressed

$$E(\mathbf{x}) = \begin{cases} j & \text{if } \exists j \in \Lambda, \bigcap_{k=1}^{K} T_k(\mathbf{x} \in C_j) > 0 \\ M+1 & \text{Otherwise} \end{cases} \tag{10}$$

where the symbol denotes the logical AND operator. Throughout this section we will also use the symbol to denote the logical OR operator.

The following C++ code fragment illustrates the voting scheme of equation 10 where $e[k]$ is an array of the choices of K recognizers.

```
int e[K];
int T[M+1];

int E(){
int c;
c=e[0];
for (i=0; i<K;i++) {
  if (e[i]!=c)
    return (M+1);
  }
return c;
}
```


Clearly, the requirement that all recognizers select the same class is a highly stringent condition and may lead us to reject pattern vectors which might otherwise have been successfully classified. However, such a method might be appropriate in situations where the result of a misclassification may be extreme.

If some of the K recognizers in the ensemble are allowed to declare ignorance, a slightly less extreme form of the above voting scheme would require that only all of those recognizers expressing a preference need to agree. That is, those choosing the reject category do not get a vote, but all recognizers declaring a classification must agree. This is expressed in equation 11 as follows:

$$E(\mathbf{x}) = \begin{cases} j & \text{if } \exists j \in \Lambda, \bigcap_{k=1}^{K} \left\{ T_k(\mathbf{x} \in C_j) \bigcup \left(1 - \bigcup_{q=1}^{M} T_k(\mathbf{x} \in C_q)\right) \right\} > 0 \\ M+1 & \text{Otherwise} \end{cases} \tag{11}$$

That is, $E(\mathbf{x})$ classifies $\mathbf{x}$ as belonging to C_j if and only if some classifiers indicate $\mathbf{x} \in C_j$ and no classifier supports $\mathbf{x} \in C_{ij}$.

The following C++ code fragment illustrates the voting scheme of equation 11.

```
int e[K];
int T[M+1];

int E(){
int c;
int i,j
```

```
i=0;flag=0;
while ((j<K) && (!flag)){
  c=e[j];
  if (c !=M+1)
     flag=1;
   else
     j++;
  }
for (i=j+1; i<K;i++) {
   if ((e[i]!=c)&& (e[i]!=M+1))
      return (M+1);
   }
return c;
}
```

A more common and less stringent form of voting will be more useful where substitution errors are not lethal. This is the majority voting rule which can be stated as follows:

$$E(\mathbf{x}) = \begin{cases} j & \text{if } T_E(x \in C_i) = \max_{i \in \Lambda} T_E(\mathbf{x} \in C_i) > \dfrac{K}{2} \\ M+1 & \text{Otherwise} \end{cases} \quad (12)$$

where

$$T_E(\mathbf{x} \in C_i) = \sum_{k=1}^{K} T_k(\mathbf{x} \in C_i), \quad i = 1, \ldots, M \quad (13)$$

The combined interpretation of equations 12 and 13 is that the pattern is classified as C_j if over half of the recognizers say $\mathbf{x} \in C_j$. Majority voting can be generalized to the case in which some fraction of the recognizers specified by $0 < \alpha < 1$ are required to agree:

$$E(\mathbf{x}) = \begin{cases} j & \text{if } T_E(\mathbf{x} \in C_j) = \max_{i \in \Lambda} T_E(\mathbf{x} \in C_i) \geq \alpha \bullet K \\ M+1 & \text{Otherwise} \end{cases} \quad (14)$$

Notice that the voting strategy defined by equation 14 requires only that T_E for the final selected label, j, exceeds threshold. No consideration is given to other labels. In cases where $\alpha < 0.5$, it is possible that several labels may in fact be maximal (i.e., there can be ties for first place). Certainly it is also possible, regardless of how α is chosen, that a second best choice would be comparable in value to the label receiving the maximum vote. In such cases the reliability of the given choice is suspect because there exists an alternative solution proposed by the system with comparable vigor. An approach to this problem involves a requirement that the best and runner-up choices be different by at least some threshold value. Let γ_1 and γ_2 represent the cumulative characteristic function as follows:

$$\gamma_1 = \max_{i \in \Lambda} T_E\left(\mathbf{x} \in C_i\right) \tag{15}$$

$$\gamma_2 = \max_{i \in \Lambda - \{j\}} T_E\left(\mathbf{x} \in C_i\right) \tag{16}$$

where j represents the label corresponding to γ_1.

Equation 16 provides the voting method taking into account contention among the recognizers:

$$E(\mathbf{x}) = \begin{cases} j & \text{if } T_E\left(\mathbf{x} \in C_j\right) = \gamma_1 \text{ and } \left(\gamma_1 - \gamma_2\right) > \beta \bullet K \\ M+1 & \text{Otherwise} \end{cases} \tag{17}$$

12.4 THE CONFUSION MATRIX

It is evident that the voting schemes described in section 12.3 do not take into consideration the characteristics of the individual recognizers. Each recognizer receives one vote in a highly democratic fashion. One way to accommodate differences among recognizers is by means of the confusion matrix. The errors for each classifier, e_k, can be described by a confusion matrix equation.

$$C\left(e_k\right) = \begin{pmatrix} \eta_{11}^{(k)} & \eta_{12}^{(k)} & \cdots & \eta_{1(M)}^{(k)} & \eta_{1(M+1)}^{(k)} \\ \eta_{21}^{(k)} & \eta_{22}^{(k)} & \cdots & \eta_{2(M)}^{(k)} & \eta_{2(M+1)}^{(k)} \\ \vdots & \vdots & \ddots & \vdots & \vdots \\ \eta_{M1}^{(k)} & \eta_{M2}^{(k)} & \cdots & \eta_{M(M)}^{(k)} & \eta_{M(M+1)}^{(k)} \end{pmatrix} \tag{18}$$

where the elements, η_{ij}, indicate the number of samples belonging to class C_i that were recognized as class C_j (i.e., $e_k(\mathbf{x} \in C_i) = j$). We may observe that the total number of samples in the confusion matrix will be

$$N^{(k)} = \sum_{i=1}^{M} \sum_{j=1}^{M+1} \eta_{ij}^{(k)} \tag{19}$$

The total number of samples in each class C_i will be

$$\eta_{i}^{(k)} = \sum_{j=1}^{M+1} \eta_{ij}^{(k)}, \quad i = 1, \ldots, M \tag{20}$$

and the total number of samples assigned by the recognizer to class C_j by e_k will be

$$\eta_{j}^{(k)} = \sum_{i=1}^{M} \eta_{ij}^{(k)}, \quad j = 1, \ldots, M+1 \tag{21}$$

The ideal recognizer would have only diagonal elements and would also have zeroes for all entries subscripted $M + 1$, the reject class. Equation 21 illustrates the confusion matrix for such an ideal recognizer.

$$C\left(e_k^{\text{ideal}}\right) = \begin{pmatrix} \eta_{11}^{(k)} & 0 & \cdots & 0 & 0 \\ 0 & \eta_{22}^{(k)} & \cdots & 0 & 0 \\ \vdots & \vdots & \ddots & \vdots & \vdots \\ 0 & 0 & \cdots & \eta_{M(M)}^{(k)} & 0 \end{pmatrix} \tag{22}$$

The confusion matrix would typically be obtained from a test set suite which is hoped to be representative of the pattern space. Confusion matrices are frequently used in the context of single recognizer systems to suggest where adjustments to the topology and/or training sets are indicated. For ensembles of recognizers used in a combined system the confusion matrix for the individual constituent recognizers can provide a measure of the trust to be placed in a given recognizer for classifications in particular categories. The confusion matrix has a great deal to tell us about the performance and properties of a recognizer. Let us briefly consider an example. Let us suppose we have five categories and the performance of a given recognizer over a test suite of 500 pattern vectors (100 for each class) results in the confusion matrix illustrated in Figure 12.3.

We will now consider the implications of the confusion matrix in Figure 12.3. A summary of the information which we may acquire is presented below in Table 12.1 where we will traverse the confusion matrix in a row-wise fashion as follows:

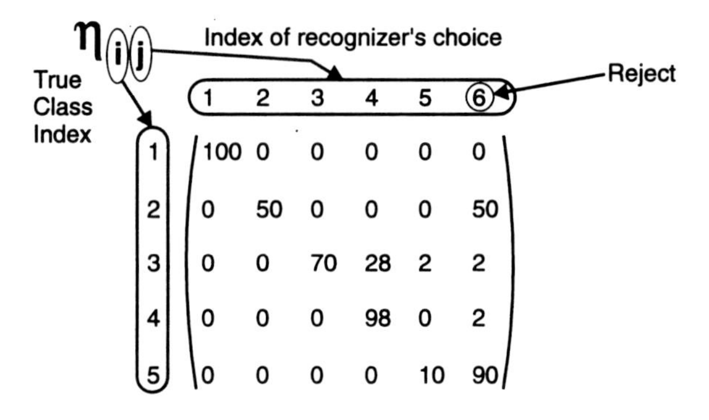

FIGURE 12.3 A sample confusion matrix.

Notice that from our inspection of the confusion matrix it is evident that a recognizer must not only be successful in correctly identifying a pattern vector of class C_j when it does in fact belong in C_j, but also it must avoid substitution errors. To this end it is crucial that when designing a training set both positive examples and counterexamples be included.

We may also create a confusion matrix for the ensemble to provide an insight into the effectiveness of the combined system. This would be represented as follows:

$$C(E) = \begin{pmatrix} \eta_{11} & \eta_{12} & \cdots & \eta_{1(M)} & \eta_{1(M+1)} \\ \eta_{21} & \eta_{22} & \cdots & \eta_{2(M)} & \eta_{2(M+1)} \\ \vdots & \vdots & \ddots & \vdots & \vdots \\ \eta_{M1} & \eta_{M2} & \cdots & \eta_{M(M)} & \eta_{M(M+1)} \end{pmatrix} \tag{23}$$

where

$$\eta_{ij} = \sum_{k=1}^{K} \eta_{ij}^{(k)} \tag{24}$$

Alternatively, the selection of a category, j, from among the set of candidate categories provided by each of the M recognizers can in itself be regarded as a

TABLE 12.1

$x \in C_i$	Analysis
1)	This recognizer always recognizes a vector $\mathbf{x} \in C_i$; there are no substitution errors or rejections; thus we are seeing ideal behavior for this case; since no other case exists wherein $x \in C_1$ and $e_k(\mathbf{x}) = 1$ (i.e., the column $\eta_{j1} = 0, \forall j \neq 1$), a classification of C_1 is always reliable ($e_k(\mathbf{x}) = 1 \rightarrow \mathbf{x} \in C_1$)
2)	Pattern vectors $\mathbf{x} \in C_2$ are recognized only 50% of the time; when $\mathbf{x} \in C_j, j \neq 2, e_k(\mathbf{x}) = 6$; that is, about half of the valid members of class C_2 are identified by this recognizer as members of the reject class; $\therefore$ this recognizer will frequently miss valid instance of class C_2; but notice that when C_2 is selected, $e_k(\mathbf{x}) = 2$, the result remains entirely reliable since $\eta_{2j} = 0, \forall j \neq 2$; further notice that because $e_k(\mathbf{x} \in C_2)$ is never confused with another class, patterns in $\mathbf{x} \in C_2$ do not degrade the accuracy or reliability when $\mathbf{x} \notin C_2$
3)	For patterns belonging to category 3, $\mathbf{x} \in C_3$; the recognizer correctly identifies the class 70% of the time; although on first inspection this seems better than the 50% recognition rate we saw when $\mathbf{x} \in C_2$, there is a difficulty; in this case when $e_k(\mathbf{x} \in C_3) \neq C_3$, the result is not a reject; rather it is a substitution error (most frequently involving class C_4); the recognition results for class 3 remain reliable (in the sense that when $e_k(\mathbf{x}) = 3$ we may believe $\mathbf{x} \in C_3$); however, we shall see that this behavior will compromise results when $\mathbf{x} \in C_4$
4)	Pattern vector such that $\mathbf{x} \in C_4$ are correctly classified 98% of the time; when not correctly classified, the pattern is rejected just as we would wish it to be; this represents very nearly ideal behavior; however, when this recognizer classifies a pattern, $e_k(\mathbf{x}) = 4$, we must regard the result with some considerable suspicion; due to the misclassifications (substitution errors) when $\mathbf{x} \in C_3$, there is a significant chance that when $e_k(\mathbf{x}) = 4$, in reality $\mathbf{x} \in C_3$
5)	The recognizer does a very poor job of recognizing pattern vectors belonging to class C_5; the best that can be said for its performance over vectors $\mathbf{x}$ C_5 is that it consistently rejects patterns it does not correctly classify and thus does not adversely affect results for other classes; even so, such a result clearly indicates that the performance of this recognizer on patterns in C_5 requires attention

classification problem. Relationships among subset of classifiers for certain patterns that are nonobvious may exist due to underlying characteristics of the individual recognizers. It is speculated that it is possible to express the confusion matrix information to a neural network (FFMLP) by the simple expedient of presenting to it the same test set suite as would have been used to create the confusion matrix along with a teaching set consisting of the correct classification. Input to the network would consist of the classifications obtained from each recognizer (perhaps augmented with additional measurement level information when available). It is reasonable to speculate that such a network could pick up this sort of nonobvious connection among recognizer traits. This network would then be used in place of the voting reviewed earlier. In effect, it would function very much like the back end in the OCON context as a postprocessing final classifier. Additionally, such a network could be made to incorporate a reject category with sensitivity established by means of the teaching inputs.

12.5 RELIABILITY

Another useful measure of a classifier that will be useful in evaluating recognizers both individually and in combination is reliability. The percentage of successful recognitions (the pattern was classified accurately), the percentage of substitution errors (the recognizer identified the wrong class), and the percentage of rejections (the recognizer by whatever means declared itself unable to classify the pattern) must sum up to 100%. Reliability is intended to serve as a metric of confidence that when the recognizer has indicated a classification (as opposed to a rejection), the classification is in fact correct. The reliability rate is defined by:

$$Reliability = \frac{Recognition}{100\% - Rejection} \tag{25}$$

As can be seen from the equation, the reliability measure both rewards correct recognition while penalizing bad guesses. Rejections are benign in that they do not reduce reliability. Notice therefore that reliability may be 1.0 (i.e., 100% reliable) even when the recognizer correctly classifies very few of the patterns presented to it as long as it does not misclassify any of them. The reliability is calculated based on a test set.

Reliability can be used in the context of single recognizers where misclassifications are particularly injurious. It may also be used to evaluate individual recognizers as candidates for potential inclusion in an ensemble. As we saw earlier in the discussion of confusion matrices, a recognizer with low reliability can impinge on the accuracy of the overall system. Frequently other measurements are reported from a context in which the rejection criteria is set to achieve a reliability of 100%.

12.6 SOME EMPIRICAL APPROACHES

Earlier, in section 12.2, it was observed that the activation levels in FFMLP networks are not suitable for analysis at the measurement level. Even so some empirical observations regarding activity levels may be sufficiently useful to serve as a heuristic which may enlighten our choices when applying a voting scheme for such networks. Let the neuron with the highest activation, the best neuron, be γ_1. Let the neuron with the second highest activation, the second best neuron, be γ_2. An empirical observation is that the certainty of a network appears to be as much a function of the difference $\gamma_1 - \gamma_2$ as it is of γ_1 alone. This was also observed in Yizhak and Chevalier [1991] where the three criteria used to establish a reject category were

1. The best neuron output exceeds a threshold.
2. The next best neuron output is below a threshold.
3. The difference between the best and next best neuron is below a threshold.

A continuous valued function, $f(\gamma_1, \gamma_2)$, incorporating these ideas could prove beneficial in weighing a network vote within a voting schema. It is to be emphasized that such a function would still not represent probability to be handled as type 1 would be in the above. Instead, the function may be utilized in a weighted voting scheme where the weight of a vote would be mediated by $f(\gamma_1,\gamma_2)$. A more conservative approach would be to simply utilize it in context of a threshold to establish a reject category in which the recognizer would not vote at all. An example of a function, $f(\gamma_1,\gamma_2)$, which takes into account both the strength of the maximum activation level and the relative closeness of the second best neuron is

$$f(\gamma_1,\gamma_2) = \gamma_1 \, e^{\frac{-\gamma_2^2}{2\pi\gamma_1^2(\gamma_1-\gamma_2)}} \qquad (26)$$

To illustrate, the family of curves generated by equation 26 is shown below in Figure 12.4. The voting scheme would then proceed by replacing $T_k(\mathbf{x})$ in equation 13 by $f(\gamma_1,\gamma_2)$ to be used in conjunction with Equation 12 or 14.

The combination of neural recognizers is not limited to voting schemes. Earlier in chapter 4 at least two such methods were discussed. We have seen MAXNET applied to the selection of results from ensembles of recognizers. Likewise front-end networks have been applied to select one of a number of specialist networks from within an ensemble [Bebbis et al., 1992]. Doubtlessly, given the appeal of combined recognizer systems, more shall be forthcoming as the field develops.

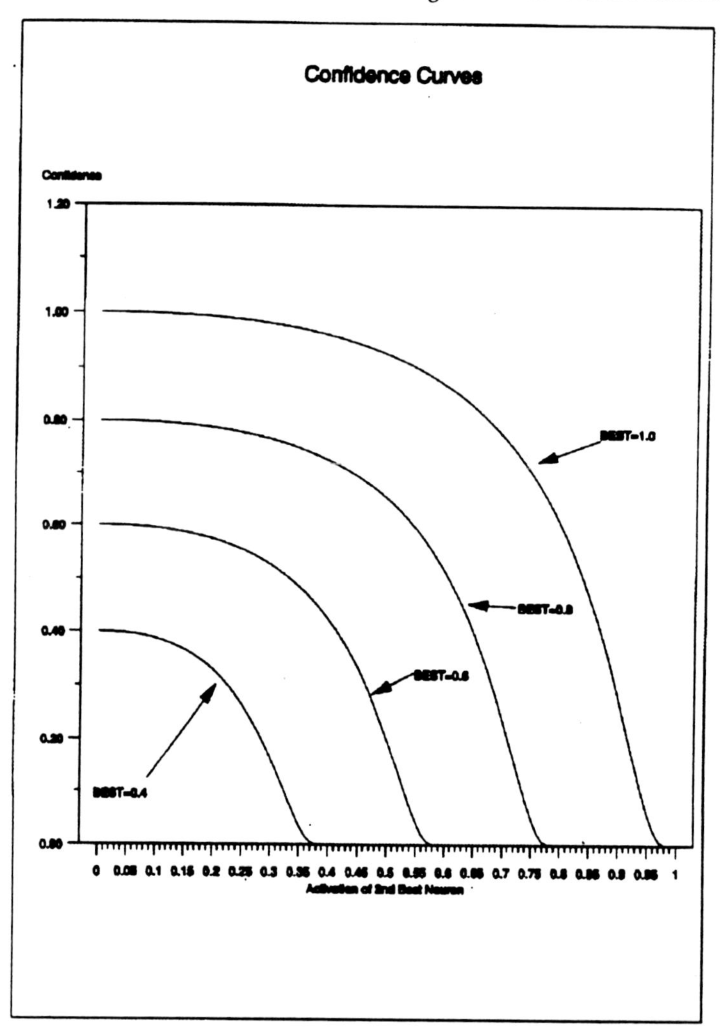

FIGURE 12.4 Confidence curves.

REFERENCES AND BIBLIOGRAPHY

Bebbis, G. N., Georgiopoulis, M., Papadourakis, G. M., and Heilman, G. L., "Increasing classification accuracy using multiple neural schemes," *Proc. SPIE Appl. Neural Networks III,* vol. 1709, pp. 221–231, 1992.

Duerr, B., Haettich, W., Tropf, H., and Winkler, G., "A combination of statistical and syntactical pattern recognition applied to classification of unconstrained handwritten numerals," *Pattern Recognition,* vol. 12, no. 3, pp. 189–199, 1980.

Perone, M. P. and Cooper, L. N., "When Networks Disagree: Ensemble Methods for Hybrid Neural Networks," U.S. Dept. of Commerce Report AF-S260 045, 1992.

Yizhak, I. and Chevalier, R. C., "Handwritten digit recognition by supervised Kohonen-like learning algorithm," *1991 IEEE Joint Conf. Neural Networks IJCNN,* pp. 2576–2581, 1991.

Xu, I., Krzyzak, A., and Suen, C. Y., "Methods of combining multiple classifiers and their applications to handwriting recognition," *IEEE Trans. Syst. Man, Cybern.,* vol. 22, no. 3, 1992.

Index

E

F

G